297

Topics in Current Chemistry

Topics in Current Chemistry
Recently Published and Forthcoming Volumes

Natural Products via Enzymatic Reactions
Volume Editor: Jörn Piel
Vol. 297, 2010

Nucleic Acid Transfection
Volume Editors: Wolfgang Bielke, Christoph Erbacher
Vol. 296, 2010

Carbohydrates in Sustainable Development II
Volume Editors: Amélia P. Rauter, Pierre Vogel, Yves Queneau
Vol. 295, 2010

Carbohydrates in Sustainable Development I
Volume Editors: Amélia P. Rauter, Pierre Vogel, Yves Queneau
Vol. 294, 2010

Functional Metal-Organic Frameworks: Gas Storage, Separation and Catalysis
Volume Editor: Martin Schröder
Vol. 293, 2010

C-H Activation
Volume Editors: Jin-Quan Yu, Zhangjie Shi
Vol. 292, 2010

Asymmetric Organocatalysis
Volume Editor: Benjamin List
Vol. 291, 2010

Ionic Liquids
Volume Editor: Barbara Kirchner
Vol. 290, 2010

Orbitals in Chemistry
Volume Editor: Satoshi Inagaki
Vol. 289, 2009

Glycoscience and Microbial Adhesion
Volume Editors: Thisbe K. Lindhorst, Stefan Oscarson
Vol. 288, 2009

Templates in Chemistry III
Volume Editors: Broekmann, P., Dötz, K.-H., Schalley, C.A.
Vol. 287, 2009

Tubulin-Binding Agents: Synthetic, Structural and Mechanistic Insights
Volume Editor: Carlomagno, T.
Vol. 286, 2009

STM and AFM Studies on (Bio)molecular Systems: Unravelling the Nanoworld
Volume Editor: Samorì, P.
Vol. 285, 2008

Amplification of Chirality
Volume Editor: Soai, K.
Vol. 284, 2008

Anthracycline Chemistry and Biology II
Mode of Action, Clinical Aspects and New Drugs
Volume Editor: Krohn, K.
Vol. 283, 2008

Anthracycline Chemistry and Biology I
Biological Occurence and Biosynthesis, Synthesis and Chemistry
Volume Editor: Krohn, K.
Vol. 282, 2008

Photochemistry and Photophysics of Coordination Compounds II
Volume Editors: Balzani, V., Campagna, S.
Vol. 281, 2007

Photochemistry and Photophysics of Coordination Compounds I
Volume Editors: Balzani, V., Campagna, S.
Vol. 280, 2007

Natural Products via Enzymatic Reactions

Volume Editor: Jörn Piel

With Contributions by

I. Abe · A. Bechthold · Z. Chen · V. Fedorenko · T. Fischer · T.A.M. Gulder · B.S. Moore · B. Ostash · J. Pietruszka · A.A. Roberts · M. Ruppert · K.S. Ryan · J. Stöckigt · X. Yan

Springer

Editor
Prof. Dr. Jörn Piel
Universität Bonn
Kekulé-Institut für Organische
Chemie und Biochemie
Gerhard-Domagk-Str. 1
53121 Bonn
Germany
joern.piel@uni-bonn.de

ISSN 0340-1022 e-ISSN 1436-5049
ISBN 978-3-642-26559-4 ISBN 978-3-642-16427-9 (eBook)
DOI 10.1007/978-3-642-16427-9
Springer Heidelberg Dordrecht London New York

Softcover reprint of the hardcover 1st edition 2010

Cover design: WMXDesign GmbH, Heidelberg, Germany

Printed on acid-free paper

Springer is part of Springer Science+Business Media (www.springer.com)

Volume Editor

Prof. Dr. Jörn Piel

Universität Bonn
Kekulé-Institut für Organische
Chemie und Biochemie
Gerhard-Domagk-Str. 1
53121 Bonn
Germany
joern.piel@uni-bonn.de

Editorial Board

Topics in Current Chemistry Also Available Electronically

Topics in Current Chemistry is included in Springer's eBook package *Chemistry and Materials Science*. If a library does not opt for the whole package the book series may be bought on a subscription basis. Also, all back volumes are available electronically.

For all customers with a print standing order we offer free access to the electronic volumes of the series published in the current year.

If you do not have access, you can still view the table of contents of each volume and the abstract of each article by going to the SpringerLink homepage, clicking on "Chemistry and Materials Science," under Subject Collection, then "Book Series," under Content Type and finally by selecting *Topics in Current Chemistry*.

You will find information about the

- Editorial Board
- Aims and Scope
- Instructions for Authors
- Sample Contribution

at springer.com using the search function by typing in *Topics in Current Chemistry*.

Color figures are published in full color in the electronic version on SpringerLink.

Aims and Scope

The series *Topics in Current Chemistry* presents critical reviews of the present and future trends in modern chemical research. The scope includes all areas of chemical science, including the interfaces with related disciplines such as biology, medicine, and materials science.

The objective of each thematic volume is to give the non-specialist reader, whether at the university or in industry, a comprehensive overview of an area where new insights of interest to a larger scientific audience are emerging.

Thus each review within the volume critically surveys one aspect of that topic and places it within the context of the volume as a whole. The most significant developments of the last 5–10 years are presented, using selected examples to illustrate the principles discussed. A description of the laboratory procedures involved is often useful to the reader. The coverage is not exhaustive in data, but rather conceptual, concentrating on the methodological thinking that will allow the non-specialist reader to understand the information presented.

Discussion of possible future research directions in the area is welcome.

Review articles for the individual volumes are invited by the volume editors.

In references *Topics in Current Chemistry* is abbreviated *Top Curr Chem* and is cited as a journal.

Impact Factor 2009: 4.291; Section "Chemistry, Multidisciplinary": Rank 20 of 138

Preface

Natural products are one of the major sources of today's drugs. In addition, they provide intriguing insights into the taxonomy, evolution, biochemistry and social behavior of organisms, and their remarkable structural and biogenetic diversity and pharmacological properties are a rich inspiration for synthetic chemists. Traditionally, natural products were prepared in the laboratory by organic synthetic strategies using small molecules as reagents. This situation has changed today, since enzymes are incorporated into many synthetic routes at the research and industrial scale. Enzymes possess several advantages over most classical reagents: they are nontoxic, can be used at mild conditions and often exhibit exquisite stereo- and regioselectivity. These properties make them ideally suited to aid the construction of complex molecular scaffolds of natural products and their analogues. To fully exploit their potential, research at the interface of biochemistry, molecular genetics and organic synthesis is necessary. Biosynthetic studies uncover enzymatic catalysts with remarkable properties, such as fluorination "reagents", Diels-Alderases and multimodular enzymes, where a single protein can catalyze dozens of individual reactions. Molecular biologists develop novel methods by which enzymes can be improved in terms of substrate range, stability and turnover. In the field of synthesis, the concept of "total biosynthesis" has emerged, in which natural products are generated in vitro using exclusively enzymes. Insights gained into enzymatic structure and mechanism also heavily inspire scientists in organic catalysis and supramolecular chemistry to develop increasingly complex and efficient enzyme-like synthetic tools.

This volume entitled "Natural Products via Enzymatic Reactions" presents the current status of selected areas of this broad discipline. Contributors who are leaders in their fields review enzymatic methodology and the range of applications for natural product synthesis. Individual natural product classes, such as alkaloids, polyketides or carbohydrates, will be individually discussed to provide deeper insights into biosynthetic pathways and their potential to generate known and novel compounds. Finally, synthetic strategies of chemists and nature are compared, and recent breakthroughs in total biosynthesis will be highlighted. The chapters reflect that these are exciting times for chemistry, when researchers from diverse methodological backgrounds merge their expertise to create new ways of effective and sustainable synthesis.

Bonn, September 2010 Jörn Piel
University of Bonn

Contents

Top Curr Chem (2010) 297: 1–43
DOI: 10.1007/128_2010_62

Published online: 23 March 2010

Key Building Blocks via Enzyme-Mediated Synthesis

Thomas Fischer and Jörg Pietruszka

Abstract Biocatalytic approaches to valuable building blocks in organic synthesis have emerged as an important tool in the last few years. While first applications were mainly based on hydrolases, other enzyme classes such as oxidoreductases or lyases moved into the focus of research. Nowadays, a vast number of biotransformations can be found in the chemical and pharmaceutical industries delivering fine chemicals or drugs. The mild reaction conditions, high stereo-, regio-, and chemoselectivities, and the often shortened reaction pathways lead to economical and ecological advantages of enzymatic conversions. Due to the enormous number of enzyme-mediated syntheses, the present chapter is not meant to be a complete review, but to deliver comprehensive insights into well established enzymatic systems and recent advances in the application of enzymes in natural product synthesis. Furthermore, it is focused on the most frequently used enzymes or enzyme classes not covered elsewhere in the present volume.

Keywords Asymmetric catalysis, Biocatalysis, Chiral building blocks, Enzymes, Natural products

Contents

T. Fischer and J. Pietruszka (✉)
Institut für Bioorganische Chemie der Heinrich-Heine-Universität Düsseldorf im Forschungszentrum Jülich, Stetternicher Forst Geb. 15.8, 52426 Jülich, Germany
e-mail: j.pietruszka@fz-juelich.de

Abbreviations

2,5-/3,6-DKCMO	2,5-/3,6-Diketocamphane 1,2-monooxygenase
Ac	Acetyl
ADH	Alcohol dehydrogenase
BAL	Benzaldehyde lyase
BVMO	Baeyer–Villiger monooxygenase
CAL-B	*Candida antarctica* lipase B
CHD	Cyclohexadienediol
CHMO	Cyclohexanone monooxygenase
CPMO	Cyclopentanone monooxygenase
CYP450	Cytochrome P450 monooxygenase
DBU	1,8-Diazabicyclo[5.4.0]undec-7-ene
DERA	2-Deoxy-D-ribose 5-phosphate aldolase
DHAP	Dihydroxyacetone phosphate
DKR	Dynamic kinetic resolution
DMAP	4-*N*,*N*-Dimethylaminopyridin
EDC	*N*-(3-Dimethylaminopropyl)-*N*′-ethylcarbodiimide
ee	Enantiomeric excess
Et	Ethyl
FDH	Formate dehydrogenase
GDH	Glucose dehydrogenase
GluDH	Glutamate dehydrogenase
HHDH	Halohydrin dehydrogenase
HNL	Hydroxynitrile lyase
KRED	Ketoreductase
Me	Methyl
NCR	NAD(P)H-dependent 2-cyclohexen-1-one reductase
n-Pe	*n*-Pentyl
OYE	Old yellow enzyme
PAL	*Pseudomonas aeruginosa* lipase
PCC	Pyridinium chlorochromate
PDC	Pyruvate decarboxylase

Ph	Phenyl
PheDH	Phenylalanine dehydrogenase
PMB	*p*-Methoxybenzyl
PMP	*p*-Methoxyphenyl
PPL	Porcine pancreatic lipase
Pr	Propyl
rac	Racemic
TBS	*tert*-Butyldimethylsilyl

1 Introduction

Nowadays, an increasing number of new pharmaceuticals have a natural lead structure or are directly derived from nature [1–4]. Their often remarkable physiological activities are frequently linked to structures bearing a number of stereogenic centers. Consequently, asymmetric synthesis is becoming even more important both in industry and in academic research [5, 6].

For providing chiral intermediates, different possibilities of asymmetric catalysis have been established. Various examples of transition metal-catalyzed or organocatalytic reactions can be found in the literature [7–9]. Biocatalysis – the application of isolated enzymes or whole cells in biotransformations – has a long historical background, being already known in ancient Egypt and the Far East for the preparation of food and alcoholic beverages [8, 10]. Plants, fruit, and extracts thereof were used for these purposes in the past, but they still find applications in recent research [11–13]. For example, coconut water was investigated as a new biocatalytical system for organic synthesis with some very promising results (Fig. 1) [14].

Major advances in the knowledge of biochemical pathways and the establishment of computer-based predictions of three-dimensional structures of proteins led to the development of new microbiological methods, such as rational protein design or directed evolution, giving scientists the possibility to provide tailor-made biocatalysts [15–19]. These methodological works on the disclosure of new efficient biocatalysts are not explicitly mentioned in this review unless they were applied in natural product synthesis. Biotransformations do not always compete with known chemical syntheses, but rather complement the portfolio of catalytic methods in organic chemistry.

Enzymatic processes are basically considered to be environmentally benign due to the mild reaction conditions needed and the high enantio-, regio-, and chemoselectivity of the enzymes. The ecological effect of using water as a solvent is quite often counterbalanced by low solubility of the substrates, making organic solvent resistant enzymes necessary [20], or by problems in downstream-processing, in most cases extracting the product with huge amounts of organic solvents [21]. Nevertheless, a biotransformation can sometimes substitute heavy metal catalysts

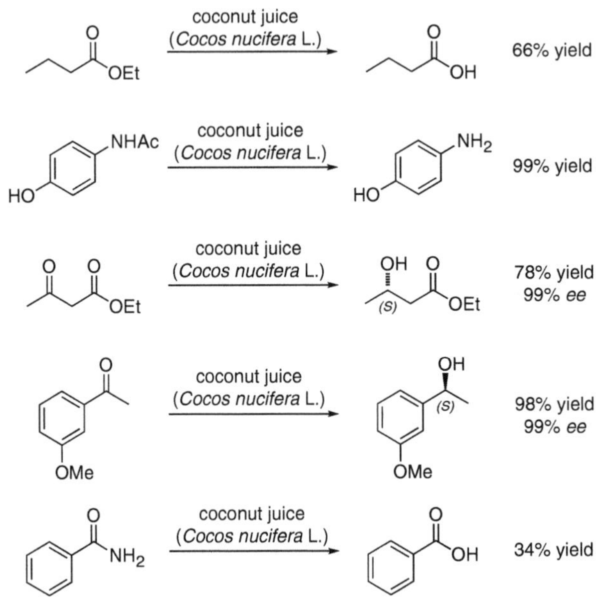

Fig. 1 Enzymatic reactions catalyzed by coconut juice (*Cocos nucifera*)

or raise immensely the efficiency of a process and thus deliver both ecological and economical advantages.

In general there are two principle possibilities using a biocatalyst in organic synthesis, namely as whole cells or as isolated enzymes – free or immobilized. The advantages and disadvantages of each can be intensively discussed, but the outcome of this consideration always depends on the whole system and the kind of application. There are numerous examples of both and thus there is no partitioning between whole cell biotransformation and isolated enzymes in this review.

Biocatalytic approaches to key building blocks in the synthesis of natural products and pharmaceuticals have been reviewed and summarized in various books in the last few years [22–48]. The variety of enzymes that were used is rising day by day, so this chapter is only meant to be instructive, not complete. The compilation is also restricted to the most frequently used enzyme classes, trying to give brief insights by selected examples of some older, but mainly recent, applications.

2 Lipases and Esterases in Organic Synthesis

Hydrolases are widely used enzymes in organic synthesis, with most applications concentrating on lipases and esterases. This chapter discloses the possibilities of asymmetric accesses to chiral building blocks for the synthesis of natural products using lipases and esterases. Other hydrolases, such as amidases or peptidases, are not dealt with explicitly.

Asymmetric synthesis with lipases and esterases can basically be performed by two different approaches – the desymmetrization of prochiral or meso compounds and the enzymatic kinetic resolution of racemic mixtures. The main bottleneck of kinetic resolutions, product yields of maximum 50%, can be overcome if an in situ racemization of the starting material is possible. In this case all starting material can theoretically be converted to the desired product [34].

2.1 Lipase-Catalyzed Desymmetrizations

Desymmetrization of prochiral or meso compounds is a powerful tool in organic synthesis, yielding key building blocks with high enantiopurity [49]. Therefore, chiral reagents or catalysts can be used to achieve the differentiation between enantiotopic groups. Some of the enantioselective enzymatic desymmetrizations using lipases that have been described in the literature represent interesting opportunities in natural product synthesis.

Enantiomerically pure cyclopropanes are a frequent motif in the structure of natural products. Their synthesis is often demanding and many approaches have been made [50, 51]. Porcine pancreatic lipase (PPL) was used for the stereoselective desymmetrization of a cyclopropane dibutanoate (Fig. 2). The asymmetric hydrolysis of the meso compound yielded the corresponding enantiopure alcohol almost quantitatively. The intermediate obtained was successfully applied in the total synthesis of dictyopterenes A and C, sexual pheromones of brown algae [52], and constanolactones (see below) [53].

Due to its structural complexity and interesting physiological activities as an antifungal agent and especially as an immunosuppressive drug, the macrolide rapamycin has been a target of many total syntheses [54–58]. In their approach, Ley and co-workers used a lipase-catalyzed desymmetrization of a meso diol (Fig. 3). In early studies, the selective acetylation of *meso*-2,4-dimethylpentane-1,5-diol was achieved by a PPL immobilized on celite with moderate yields and 92%

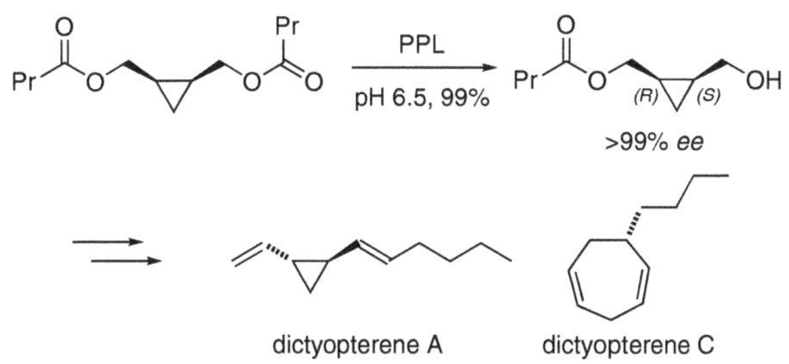

Fig. 2 Chemoenzymatic synthesis of dictyopterenes A and C

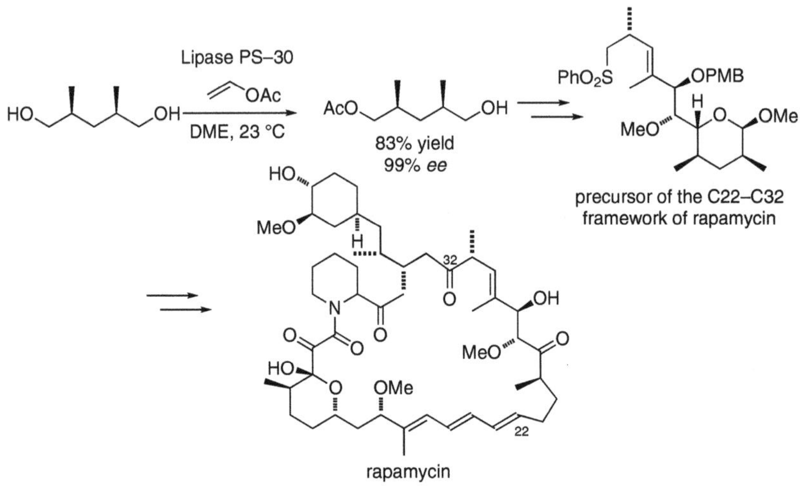

Fig. 3 Enantioselective acylation catalyzed by porcine pancreatic lipase (PPL)

enantiomeric excess (*ee*) [59]. Later, advances in the enzymatic desymmetrization led to the substitution of the catalyst with the also commercially available and inexpensive Lipase PS-30, providing the monoacetate in 83% yield and 99% *ee* [60, 61].

2.2 Kinetic Resolutions with Lipases and Esterases

Kinetic resolutions in general are regularly applied in organic synthesis. Since enzymes are highly attractive for asymmetric synthesis, various types of biocatalysts have been used in enzymatic (dynamic) kinetic resolutions, but the focus will remain on lipase- and esterase-mediated resolutions as the most common tools in early steps of natural product syntheses.

The enantiomeric excess of substrate and product shifts during the course of a kinetic resolution, making the determination of the efficiency of the reaction dependent on the reaction time. Hence, two kinetic resolutions can only be compared at the same extent of conversion [45, 62]. For a better comparison of two reactions, equations for the calculation of the enantioselectivity were established [63, 64]. In most cases, the enantiomeric excess of both substrate and product is determined for greater accuracy, but the determination of one and the conversion can also be used for the calculation of the E value (enantioselectivity):

$$E = \frac{ln[1 - c(1 + ee_p)]}{ln[1 - c(1 - ee_p)]}; \quad E = \frac{ln[(1 - c)(1 - ee_s)]}{ln[(1 - c)(1 + ee_s)]}; \quad E = \frac{ln\left[\frac{1 - ee_s}{1 + (ee_s/ee_p)}\right]}{ln\left[\frac{1 + ee_s}{1 + (ee_s/ee_p)}\right]} \quad (1)$$

E values above 100 are very inaccurate due to the logarithmic function, so that usually such high enantioselectivities are stated as $E \geq 100$. In these calculations the *E* value increases with increasing enantioselectivity. For the enzymatic kinetic resolutions, only *E* values greater than 20 indicate a potentially useful application in organic synthesis.

However, this theoretical approach does not cover all side effects of applied biocatalysis. The calculation of *E* values for enzyme mixtures or in systems with inherent inhibitory effects must be handled with care. Nevertheless, calculating *E* from (1) usually provides a very useful tool for estimating the efficiency of a resolution system.

An example for the application of enzymatic kinetic resolutions with high *E* values in natural product synthesis is the chemoenzymatic synthesis of the northern half of epothilones (also see Sect. 4.1). Various lipases and esterases could be found with outstanding enantioselectivity (up to >100); among these were lipase B from *Candida antarctica*, a lipase from *Burkholderia cepacia*, a lipase from *Pseudomonas* sp., and a lipase from *Streptomyces diastochromogenes*, all affording the desired (*S*)-configurated alcohol with >99% enantiomeric excess (Fig. 4) [65].

Allenes are versatile intermediates in organic synthesis and a variety of useful applications has been established [66]. A very interesting feature of allenes is their axial chirality along the cumulated diene system, so that optically active allenes were frequently used in asymmetric synthesis, in most cases being prepared from enantiomerically enriched precursors.

Since allenic systems are rather rare in nature, enzymatic approaches were hardly investigated due to a lack of known organisms metabolizing these structures. Recently, a mutant of a lipase from *Pseudomonas aeruginosa* (PAL) was generated by directed evolution, which was applied in the kinetic resolution of an allenic *p*-nitrophenolester [67]. Further investigations on the possibility of lipase-catalyzed kinetic resolutions of allenes with different lipases led to the application of PPL in the total synthesis of (–)-striatisporolide A, a fungal metabolite isolated from an

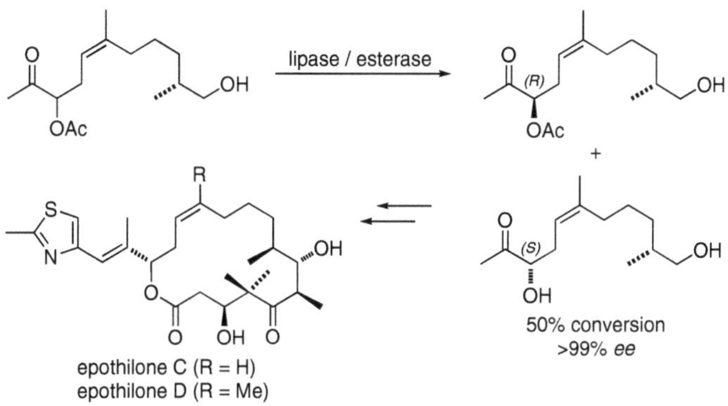

Fig. 4 Enzymatic kinetic resolution of epothilone precursors

Australian *Penicillium striatisporum* strain (Fig. 5) [68]. Before, neither a synthetic approach, nor the absolute configuration of this butenolide had been reported. Combining the enzymatic kinetic resolution with a palladium(II)-mediated racemization [69] could lead to a highly efficient dynamic kinetic resolution (DKR) of allenes.

The complex tetramic acid lactam cylindramide was isolated from the marine sponge *Halichondria cylindrata* and is known to possess a significant cytotoxic activity against human B16 melanoma cells [70]. A chemoenzymatic access to the characteristical substituted bicyclo[3.3.0]octane moiety was recently established, using lipase PS (Amano) for an enzymatic kinetic resolution (Fig. 6) [71]. The racemic diol was generated by a transannular Pd-catalyzed ring-closure of cycloocta-1,5-diene and removal of the acetyl groups. The resolution step provided the (1*S*,3a*R*,4*S*,6a*R*)-configured diol in 44% yield with 98% *ee*. Side products, the monoacetate (8%) and the (1*R*,3a*S*,4*R*,6a*S*)-configured diacetate (38%), could easily be separated by chromatography.

Psymberin (irciniastatin A), a cytotoxin of the pederin family, has been isolated from the marine sponges *Psammocinia* sp. and *Ircinia ramosa*. Its highly cytostatic

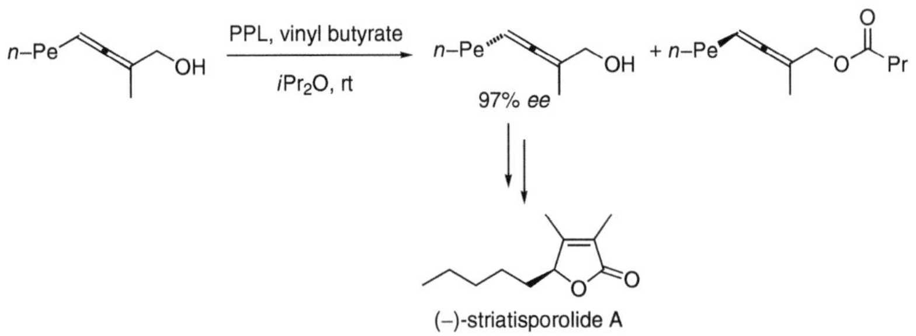

Fig. 5 Enzymatic kinetic resolution of primary allenic alcohols

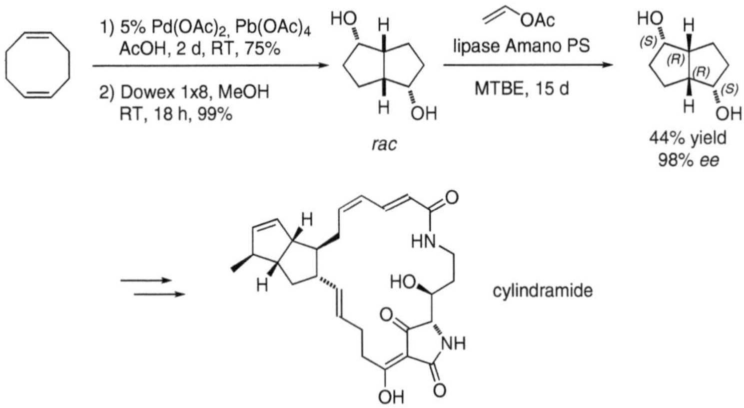

Fig. 6 Enzymatic kinetic resolution in the total synthesis of cyclindramide

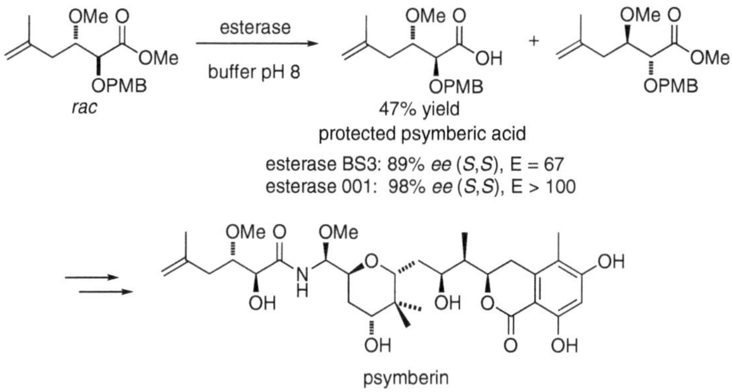

Fig. 7 Chemoenzymatic synthesis of PMB-protected psymberic acid

activity against various human cancer cell lines made it an attractive target for total syntheses [72–74]. Lately, a chemoenzymatic approach by an esterase-catalyzed kinetic resolution towards psymberic acid, one of the main structural elements of psymberin, was disclosed [75]. A number of lipases and esterases was screened affording esterase BS3 from *Bacillus subtilis* and commercially available esterase 001 (Codexis) as suitable biocatalysts for enantioselective hydrolysis. Especially esterase 001 showed superior performance, yielding the protected psymberic acid with 98% *ee* (Fig. 7).

The importance of enantiomerically pure cyclopropane derivatives has already been pointed out (see Sect. 2.1) and many examples of lipase- or esterase-catalyzed kinetic resolutions of cyclopropanes can be found in the literature, but, unfortunately, high *E* values are scarce [76–84]. For instance, several lipases were screened for their ability to perform a kinetic resolution of cyclopropylmethanols [85].

Among the enzymes tested, *Candida antarctica* lipase B (CAL-B) (commercially available as Novozyme 435 or Chirazyme L-2) was the most efficient. For the kinetic resolution of the corresponding acetate in tetrahydrofuran at 60°C, a comparably high *E* value was achieved ($E = 44$). Hence, this enzyme was applied in the total synthesis of constanolactones A and B, marine oxylipins isolated from the red algae *Constantinea simplex* [86]. An enantiomerically enriched cyclopropylmethanol was submitted to enzymatic kinetic resolution after acetylation, giving the enantiopure (>98% *ee*) in good yield (76%) over the two steps (Fig. 8). Application of enzymatic kinetic resolution to other cyclopropylmethanol derivatives delivered the series with opposite cyclopropane configuration, yielding building blocks for the total synthesis of solandelactones [87], halicholactone, and neohalicholactone [88].

The kinetic resolution of 5-hydroxymethyl-2-cyclohexenone towards the total synthesis of penienone, a fungal metabolite of *Penicillium* sp. no. 13 [89], has been performed by using Lipase PS from *B. cepacia* for the acetylation of the (*S*)-enantiomer, leaving the desired (*R*)-alcohol. As a main problem inherent with enzymatic kinetic resolutions, the reaction yielded both products in a 1:1 mixture

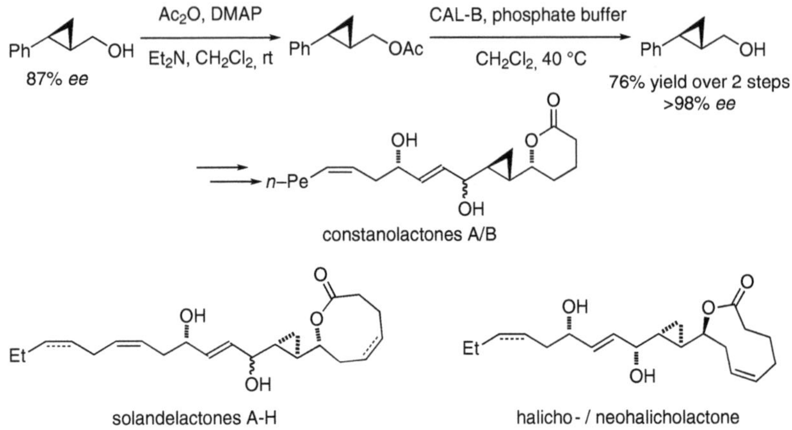

Fig. 8 Enzymatic kinetic resolution of cyclopropylmethanols

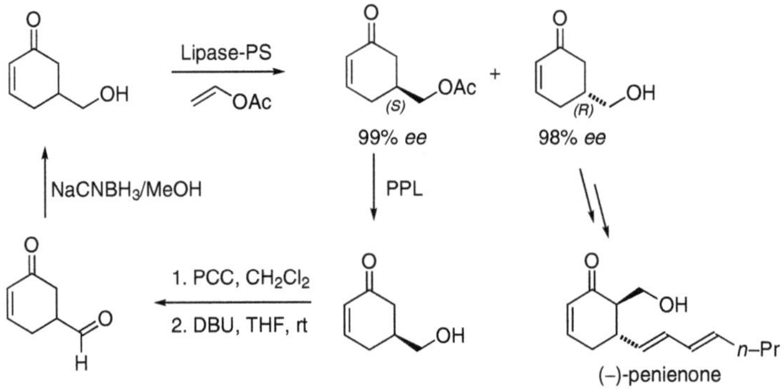

Fig. 9 Enzymatic kinetic resolution in the chemoenzymatic synthesis of (–)-penienone

almost quantitatively. For prohibiting the loss of half of the starting material, the acetate was racemized by hydrolysis with another lipase (PPL), followed by an oxidation–reduction sequence yielding the racemic starting material for the kinetic resolution (Fig. 9).

2.3 *Dynamic Kinetic Resolutions with Lipases and Esterases*

The enzymatic kinetic resolutions presented in Sect. 2.2 all clearly revealed the bottleneck of kinetic resolutions, a maximum yield of 50% for the desired enantiomer. In the total synthesis of penienone this problem was solved by choosing a rather inconvenient multistep racemization of the isolated remaining starting material.

In modern biotransformations, the insufficient yield is optimized by establishing an in situ racemization of the starting material, turning the system into a DKR [90–92]. Due to the high enantioselectivity of some enzymes, the number of applications of DKR is increasing; especially the utilization of the readily available lipases and esterases is most common. The racemization process can either be realized by biocatalysts or by chemocatalytic approaches. In some cases, the addition of base or acidification can shift an existing equilibrium towards racemization sufficiently. For catalyzed racemization, especially some ruthenium-complexes were proven to show high compatibility with enzymatic systems [93]. Other transition metals were also applied; for the DKR of allyl alcohols, an oxovanadium (V) catalyst was used in combination with a lipase to give high yields with excellent enantiomeric excesses [94].

The principle of kinetic resolutions (Fig. 10) can be explained by the different rates of conversion for the different enantiomers. In the special case of a DKR, the rate for the conversion of one enantiomer is many times higher ($k_R \gg k_S$). Without the possibility of racemization of the starting material ($k_{rac} = 0$), a kinetic resolution with a maximum yield of 50% is observed. As soon as k_{rac} exceeds the rate for the conversion of the unfavored enantiomer ($k_{rac} \geq k_S$), a DKR becomes efficient.

Bornscheuer and co-workers studied the lipase-catalyzed dynamic resolution of acyloins, using miscellaneous amine bases and ion-exchangers for racemization (Fig. 11) [95]. Unfortunately, acidic resins, e.g., Amberlyst 15, furnishing the best results for racemization, also deactivated the enzyme. As a solution, a two-compartment setup was established, separating ion exchanger and enzyme, the two vessels being connected by a pump loop. The enzymatic resolution of the acyloin and the racemization were thus carried out simultaneously furnishing a nice DKR with good yields and enantioselectivities. The enantiomerically enriched acyloins produced are important building blocks for natural products, e.g., epothilones [96].

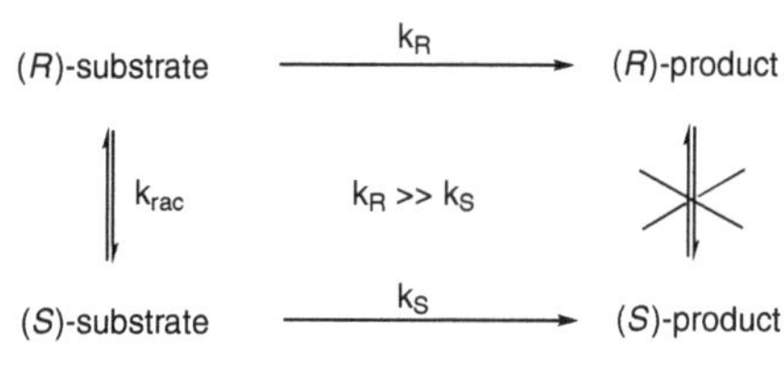

Fig. 10 Principle of dynamic kinetic resolutions

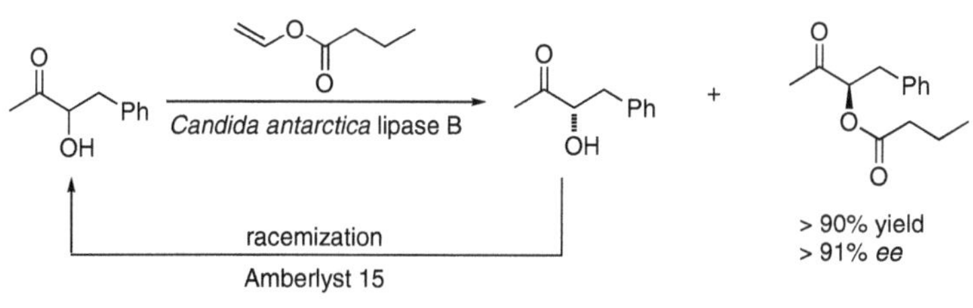

Fig. 11 Two-compartment enzymatic dynamic kinetic resolution of acyloins

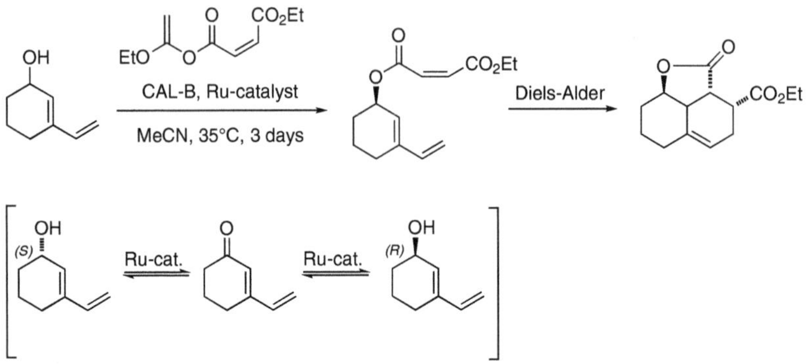

Fig. 12 Chemoenzymatic domino DKR and Diels–Alder reaction

Enzymatic DKRs have also been applied in domino one-pot processes [97]. The combination of a lipase-catalyzed resolution with an intramolecular Diels–Alder reaction led to interesting building blocks for the synthesis of natural products such as compactin [98, 99] or forskolin [100–102]. A ruthenium catalyst is employed for the racemization of the slow reacting enantiomer of the starting material. The DKR with lipase B from *C. antarctica* delivered high enantiomeric excesses which could mainly be contained through the Diels–Alder reaction (Fig. 12).

3 Enzymatic Reductions and Oxidations

In current research, oxidoreductases are second in the number of applications of enzymes in organic synthesis. The number of commercially available biocatalysts of this class has increased tremendously during the last few years and various screening kits for oxidation and reduction are sold. Many oxidoreductases are rather easy to handle, though, in contrast to hydrolases, they are dependent on cofactors [22].

3.1 Chiral Alcohols Through Enantioselective Reduction

The chemoenzymatic synthesis of chiral alcohols is a field of major interest within biocatalytic asymmetric conversions. A convenient access to secondary highly enantiomerically enriched alcohols is the usage of alcohol dehydrogenases (ADHs) (ketoreductases) for the stereoselective reduction of prochiral ketones. Here, as in many other cases in asymmetric catalysis, enzymes are not always only an alternative to chemical possibilities, but are rather complementary. Albeit biocatalysts might sometimes seem to be more environmentally friendly, asymmetric ketone reduction with Noyori's ruthenium catalysts [8, 103] or by chiral oxazaborolidins (Corey-Bakshi-Shibata) [104] are known to be very efficient for many applications and do not leave much room for improvement for these. Yet there are also examples in which chemical asymmetric reduction fails or shows low selectivity [105].

Ketoreductases (KREDs) are dependent on nicotinamide cofactors NADH or NADPH. Due to the reaction mechanism, these rather costly cofactors are needed in stoichiometric amounts, disclosing an economic problem that has to be dealt with when using these enzymes. Many different possibilities for cofactor recycling have been established with three major approaches finding application in research and industry (Fig. 13). Further regeneration systems, such as electrochemical methods, are not discussed within this review [22–24, 37, 106–108].

One possibility is the addition of a second substrate that itself is also a possible substrate for the KRED being used. Most isolated enzymes are quite robust to 2-propanol as a substrate, so that in some systems 2-propanol concentrations of more than 40% can be used, also bearing an improved solubility of the ketones for conversion. However, on a larger scale the acetone formed can show inhibitory effects. Removal of this by-product shifts the equilibrium of the reaction towards the desired products and prohibits a competitive inhibition [109, 110].

A further opportunity for cofactor regeneration is the addition of a second enzyme with a corresponding substrate to the reaction mixture. Often glucose dehydrogenase (GDH) is used for this purpose, converting glucose to gluconate by reducing the nicotinamide cofactor. Formate dehydrogenases (FDHs) also have a broad spectrum of applications, bringing the advantage of catalyzing the oxidation

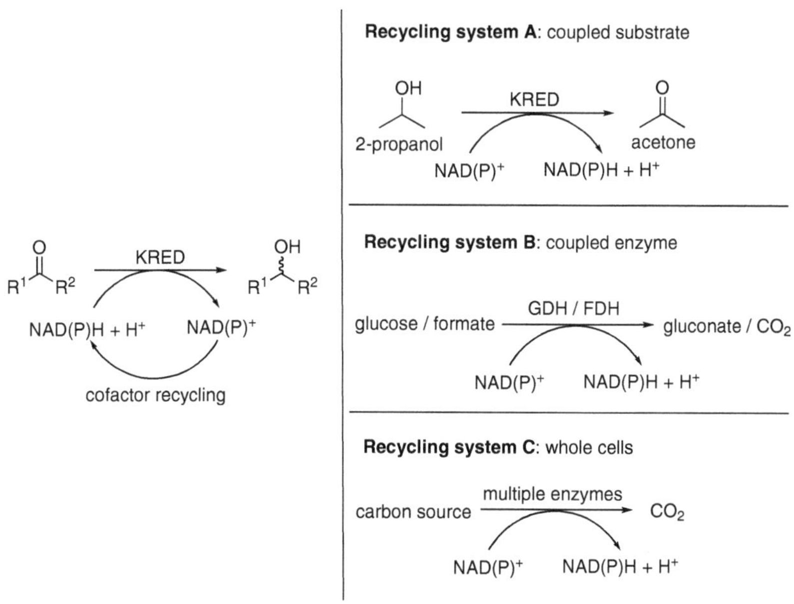

Fig. 13 Recycling of nicotinamide cofactors

of formate to carbon dioxide, which by leaving the reaction mixture draws the equilibrium to the desired side.

There is quite a lively discussion going on about the advantages and disadvantages of the usage of whole cells for biocatalysis vs isolated enzymes [23, 24, 43]. Intact cells contain all enzymes and the cofactors needed for the biotransformation, so that the addition of a carbon source should be sufficient. But whole cell biotransformations also cause some problems. One aspect is the transport of the substrate through the cell membrane, which might be a great hindrance. Partial lysis of the cells by adding organic solvents to the reaction medium sometimes allows transport of the substrate into the cell, but also provides a possibility of loss of the cofactor from the inner cell. Conclusively it is hardly possible to tell whether whole cell biotransformation or isolated enzymes are the method of choice, leading to both methods being almost equally applied [23, 24].

Chiral secondary alcohols are main targets of biotransformations since they are very common intermediates in various natural product syntheses. Often, kinetic or DKRs using lipases are used to obtain these key building blocks. The asymmetric reduction of prochiral ketones is more than an alternative to this approach. A huge number of applications, in which KREDs are used for reduction, can be found in the literature [22–24, 37]. For example, (*R*)-2-bromo-(4-nitrophenyl)ethanol, a precursor to the *β*-blocker (*R*)-nifenalol, which was obtained by kinetic resolution in the past, was conveniently synthesized in a direct approach from the corresponding ketone using whole cells of *Rhodotorula* sp. AS2.2241 (Fig. 14). This ketone and a number of derivatives thereof were converted with good yields and excellent enantioselectivities [111].

Whole cells were also used as biocatalysts in the biocatalytical synthesis of (*R*)-*o*-chloromandelates, highly valuable building blocks for biologically active substances, such as clopidogrel (Fig. 15) [112]. In this straightforward approach a versatile carbonyl reductase (SCR) was expressed in *Escherichia coli* and the cells fed with glucose for enabling the cofactor regeneration during the biotransformation. Methyl (*R*)-*o*-chloromandelate was isolated in enantiomerically pure form (>99% *ee*) in very high yield (89%).

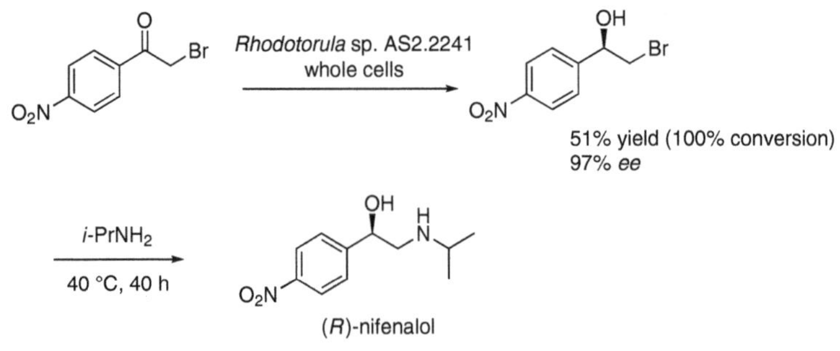

Fig. 14 Chemoenzymatic asymmetric synthesis of (*R*)-nifenalol with whole cells

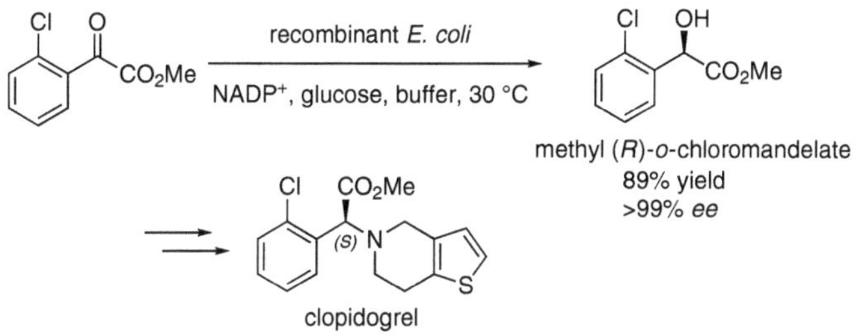

Fig. 15 Whole cell biocatalytic synthesis of methyl (*R*)-*o*-chloromandelate

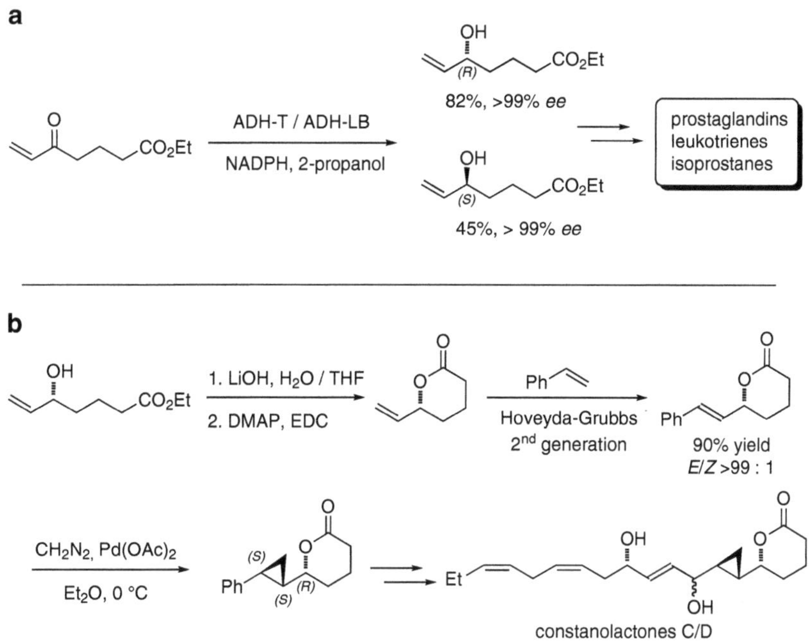

Fig. 16 (**a**) Enzymatic synthesis of building blocks for arachidonic acid metabolites. (**b**) Chemoenzymatic synthesis of constanolactones C/D

5-Hydroxyhept-6-enoates have been key intermediates in the synthesis of a variety of natural products, especially of the arachidonic acid metabolic pathway, including prostaglandins, leukotrienes, and isoprostanes [105, 113–118]. The usage of two enantiocomplementary enzymes allowed convenient access to both enantiomers via an ADH-catalyzed reduction of 5-oxo-hept-6-enoate. Alcohol dehydrogenase from *Lactobacillus brevis* (ADH-LB) furnished the (*S*)-enantiomer, *Thermoanaerobacter* sp. ADH (ADH-T) the (*R*)-enantiomer in excellent enantiomeric access respectively [119]. A cross-metathesis reaction followed by cyclopropanation led to the formal synthesis of constanolactones C and D (Fig. 16) [86, 120, 121].

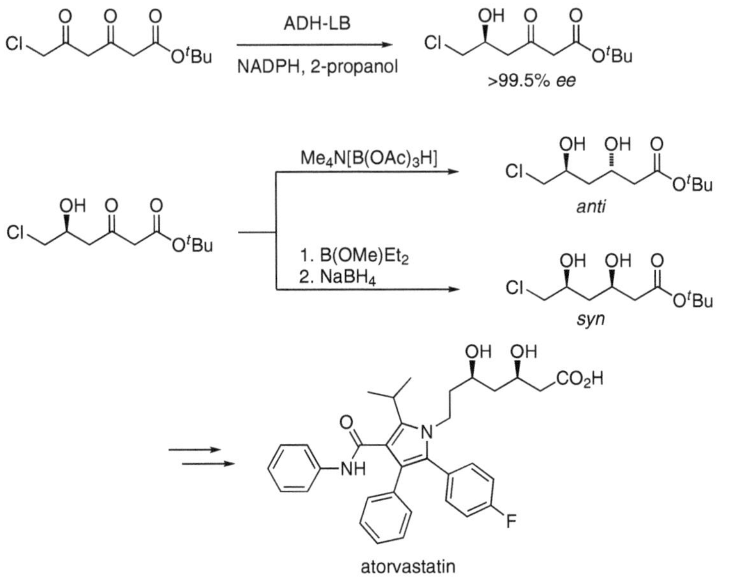

Fig. 17 Ketoreductase-catalyzed chemoenzymatic synthesis of statin side chains

Not only has the remarkable enantioselectivity of ADHs made them a very powerful tool in organic synthesis: Besides the excellent enantioselectivity of ADH from *L. brevis* already shown, the same enzyme's regioselectivity has been exploited in the chemoenzymatic synthesis of statin side chains [122]. A series of 3,5-dioxocarboxylates (β,δ-diketo esters) were successfully reduced selectively in the 5-position using ADH-LB as biocatalyst. The enantiomeric purity of the 5-hydroxy compound obtained was excellent [123–125]. The second ketone was then either reduced by Prasad's borohydride method [126] delivering the *syn*-dihydroxy building block, or according to Evans furnishing the *anti*-configured compound, respectively (Fig. 17) [127]. Furthermore, chiral building blocks for the synthesis of some δ-lactone containing natural products, such as goniothalamin, argentilactone, or callistatin A, could be accessed by this chemoenzymatic approach [125].

Similar building blocks have also been obtained by a variety of other enzymatic transformations. 4-Chloro-3-oxobutanoate esters have been reduced with outstanding enantioselectivity by ADHs from *Candida magnoliae* in combination with cofactor regeneration by a GDH (from *Bacillus megaterium*) or by enzymatic reduction using whole cells of *Geotrichum candidum* [128, 129]. This process was improved to a large scale application. The combination with a halohydrin dehydrogenase step led to versatile compounds for organic synthesis. Statistical analysis of protein sequence activity relationships (ProSAR) and recombination-based directed evolution led to an optimized enzyme catalyzing the epoxide formation in a first step and the addition of a nucleophile in a second step, accepting alternative nucleophiles, such as CN^- or N_3^- (Fig. 18) [130–132].

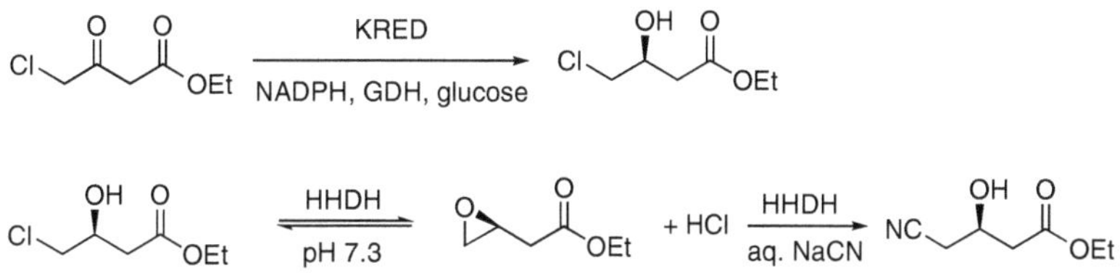

Fig. 18 Enzymatic synthesis of ethyl (*R*)-4-cyano-3-hydroxybutyrate

3.2 Reduction of C=C Bonds

The generation of stereogenic centers by asymmetric reduction of carbon–carbon double-bonds is a current topic in chemoenzymatic synthesis. Though enzymes of the old yellow enzyme (OYE) family were identified to perform alkene reduction and were characterized some years ago [133–135], applications of enoate reductases in natural product syntheses are still rare. Thus, potential applications are also shown in this chapter. With an increasing number of new enoate reductases, such as YqjM reductase from *B. subtilis*, more and more possible targets for biotransformations can be found.

Enzymes from the OYE family were reported to mediate efficiently the reduction of ketoisophorone. A subsequent regio- and stereoselective reduction of the thus produced levodione leads to actinol, both compounds being important building blocks for carotenoids. While early approaches used baker's yeast in the fermentative production of levodione [136], more recent approaches used enzymes from different organisms. Besides OYE analogs from *Candida macedoniensis*, *Saccharomyces carlsbergensis*, and *Saccharomyces cerevisiae*, a newly discovered enoate reductase from *Zymomonas mobilis* was applied successfully [137, 138]. Wada et al. combined the OYE-mediated step with an enzymatic formation of actinol catalyzed by a levodione reductase from *Corynebacterium aquaticum* (Fig. 19) [139].

Some reductases isolated from tobacco (*Nicotiana tabacum*) were found to exhibit excellent enantioselectivities on the reduction of a number of α,β-unsaturated compounds [140, 141]. For example, reductase p44 catalyzed the asymmetric reduction of *N*-phenyl-2-methylmaleimide, yielding the enantiopure (*R*)-succinimide. Reductase p90 mediated the enantioselective hydrogenation of a number of methyl or ethyl substituted cyclopentenones and cyclohexanones (Fig. 20).

3.3 Reductive Amination Towards Amino Acids

Unnatural amino acids in general, but especially α-amino acids being important intermediates for various biologically active compounds, have enjoyed great popularity as a growing field of interest during recent years [142]. Many enzymatic

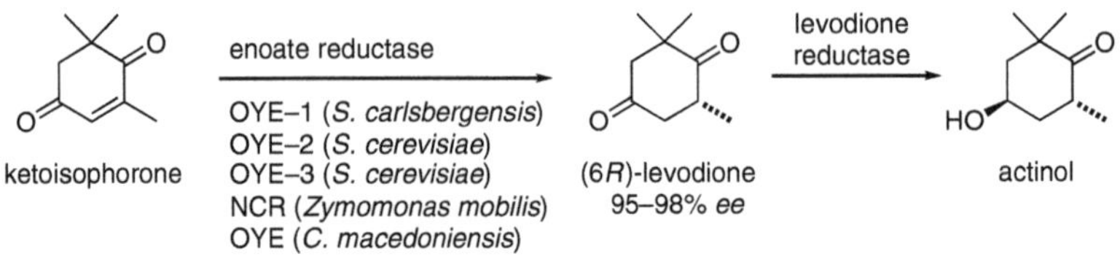

Fig. 19 Enzymatic synthesis of carotenoid building blocks

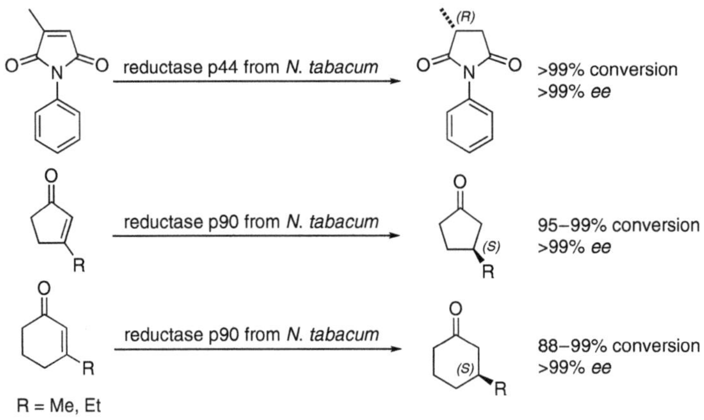

Fig. 20 Enantioselective reductions catalyzed by reductases from *N. tabacum*

approaches to these important building blocks, including resolutions with L-amino acid acylase or D-amino acid oxidase, the application of transaminases, and reductive amination by amino acid dehydrogenases, were studied [143]. For the production of enantiopure α-amino acids, reductive amination was most successfully applied.

The significance of chiral unnatural amino acids to drug and natural product synthesis is shown in the example of the antihypertensive drug omapatrilat (Vanlev®), which is composed of no less than three amino acid derived intermediates [144–147]. Diverse biocatalytical approaches to L-6-oxonorleucine were made (Fig. 21). Two different enzymes were applied in reductive amination reactions to produce derivatives of the desired intermediate.

With a protected aldehyde function, allysine ethylene acetal was synthesized via reductive amination catalyzed by phenylalanine dehydrogenase (PheDH). For regeneration of the nicotinamide cofactor, FDH was used as described before (see Sect. 3.1). Another similar approach was taken in the synthesis of L-6-hydroxynorleucine. In this case, reductive amination was catalyzed by glutamate dehydrogenase (GluDH) delivering the desired (*S*)-configured amino acid. The synthesis of the same compound has also been achieved by an interesting resolution process. Racemic 6-hydroxynorleucine was submitted to kinetic resolution using a D-amino

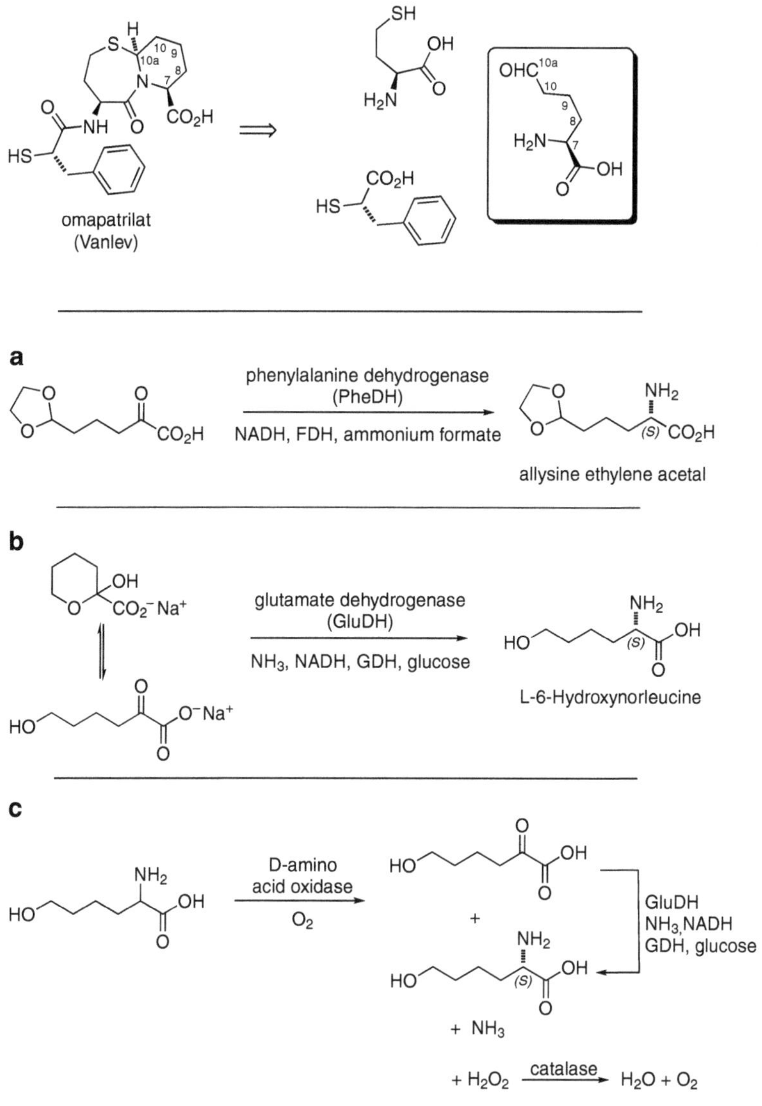

Fig. 21 Enzymatic synthesis of unnatural amino acids as building blocks for omapatrilat. (**a**–**c**) Enzymatic approaches to 6-oxonorleucine analogs

acid oxidase, generating an enantiomeric excess of >99% for the remaining enantiomer. The mixture was then converted completely to enantiopure L-hydroxynorleucine by reductive amination of 2-oxo-6-hydroxyhexanoic acid with GluDH. For the oxidation step, whole cells of *Trigonopsis variabilis*, bearing the oxidase and the catalase for the degradation of hydrogen peroxide to oxygen, were successfully used.

3.4 Cytochrome P450 Monooxygenases

Cytochrome P450 (CYP450) monooxygenases are very common oxidoreductases, frequently used in synthetic applications in research and industry. The versatility and potential as powerful industrial biocatalysts has been strengthened by important advances that have been made regarding mechanistic insights and optimization of the activity and high stereo- and regioselectivities of these hemoproteins recently [148–151].

Pravastatin, a 3-hydroxy-3-methyl glutaryl CoA reductase inhibitor applied as a therapeutic agent for hypercholesterolemia, can be synthesized by stereo- and regioselective hydroxylation of compactin by the soil microorganism *Streptomyces* sp. Y-110 (Fig. 22) [152]. The fermentative production of pravastatin has already been applied on an industrial scale by Sankyo Co. using different *Streptomyces* bacteria strains [153, 154].

Besides the stereo- and regioselective hydroxylations, CYP450 are also capable of performing regioselective oxidations to aldehydes or carboxylic acids. The sesquiterpene lactone artemisinin is highly effective against multidrug resistant *Plasmodium falciparum*, agent of malaria tropica. First, genes from the mevalonate pathway in *S. cerevisiae* were upregulated by means of metabolic engineering, increasing the production of farnesyl pyrophosphate, and the natural further metabolism to sterols was downregulated. Introduction of an amorphadiene synthase (ADS) from *Artemisia annua* led to the formation of amorpha-4,11-diene which could then be converted to artemisic acid, a precursor of artemisinin, via a three-step oxidation sequence catalyzed by a novel P450 from *A. annua* in *S. cerevisiae* (Fig. 23) [155].

3.5 Baeyer–Villiger Oxidations

Among the most popular oxidative biotransformations, Baeyer–Villiger monooxygenases (BVMOs) belong to the main fields of research. Nowadays, manifold enzymes catalyzing the Baeyer–Villiger oxidation are expressed in common recombinant organisms, such as *E. coli* or *S. cerevisiae*. The mechanism of the enzymatic

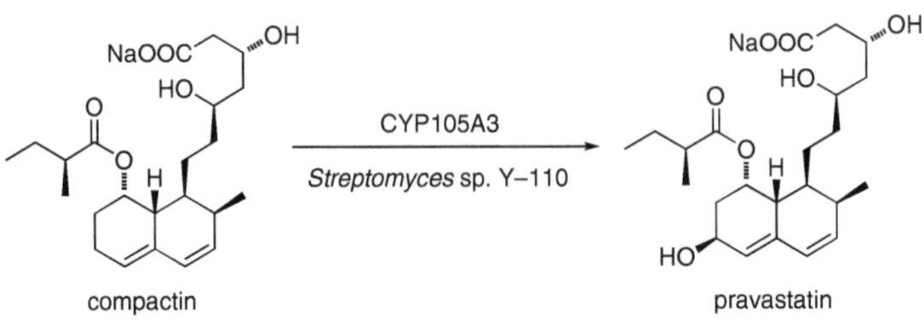

Fig. 22 Biocatalytic conversion of compactin to pravastatin

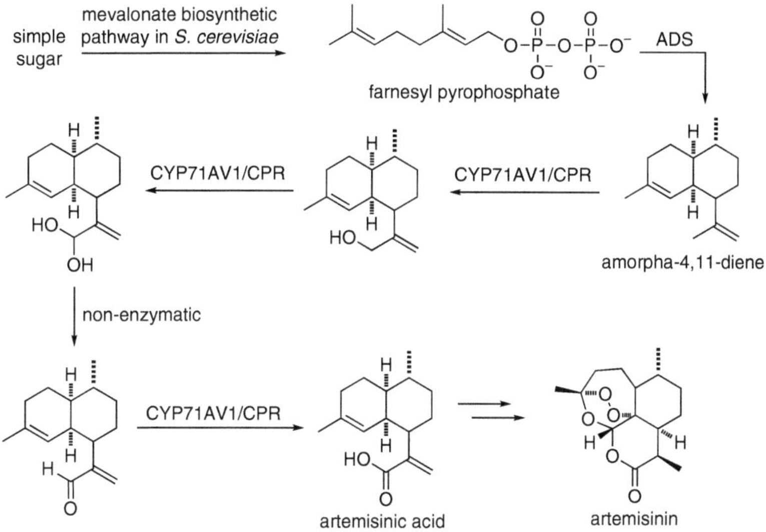

Fig. 23 Fermentative production of artemisinic acid

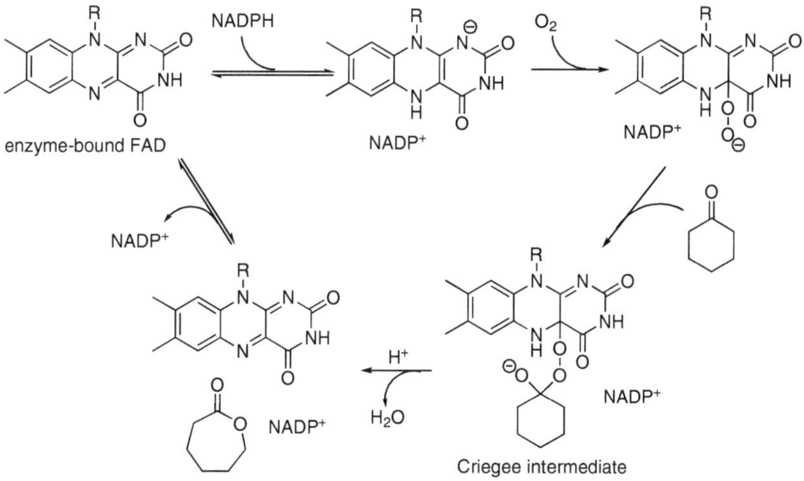

Fig. 24 Mechanism of flavin-dependent Baeyer–Villiger monooxygenases

Baeyer–Villiger oxidation involves NADPH and flavin (FAD) as cofactors and was originally proposed by Walsh et al. based on data obtained from cyclohexanone monooxygenase (CHMO) from *Acinetobacter calcoaceticus* (Fig. 24) [156]. In a first step, enzyme-bound flavin is reduced, followed by the addition of oxygen yielding a hydroperoxide anion. Reaction with the ketone substrate gives a Criegee intermediate, which is then converted into the product under dissociation of water. The cofactor FAD is recovered via oxidation with $NADP^+$.

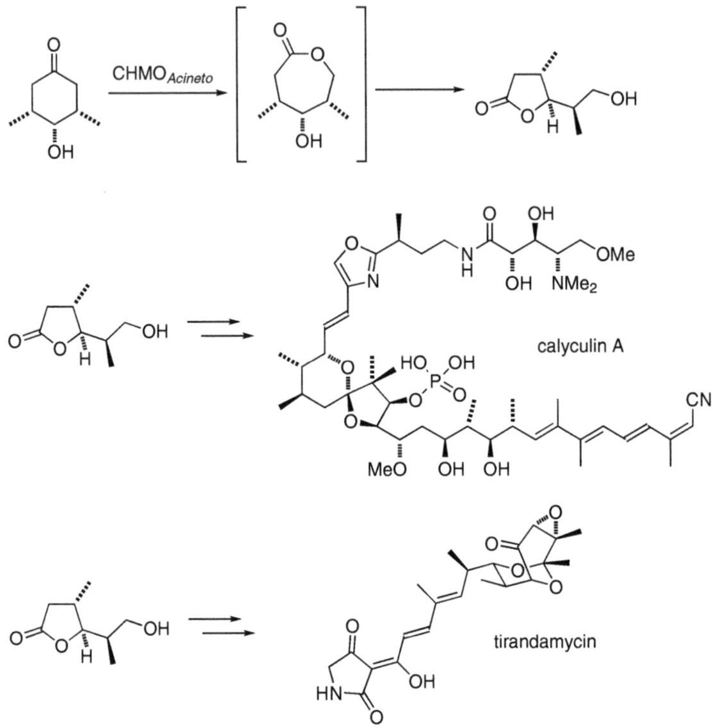

Fig. 25 BVMO-catalyzed synthesis of a precursor to calyculin A and tirandamycin

Enzymatic Baeyer–Villiger oxidations have been studied for a long time; some very useful applications in natural product synthesis date back more than two decades. For example, prochiral (3*R*,5*S*)-4-hydroxy-3,5-dimethylcyclohexanone was successfully oxidized using a CHMO from *Acinetobacter* sp. NCIMB 9871. In this case the seven-membered ring formed rearranges spontaneously into the thermodynamically more stable γ-lactone. The rearranged lactone has been used in the synthesis of natural products such as tirandamycin or calyculin A (Fig. 25) [157–159].

As most enzymes, BVMOs feature excellent enantio- and regioselectivities, making them very versatile tools for asymmetric synthesis. Bicyclic ketones such as functionalized norboranones have been resolved to the corresponding lactones with good regioselectivities and enantiomeric excesses of more than 95% [160]. Whole cells of *Pseudomonas putida* or isolated 2,5-/3,6-diketocamphane 1,2-monooxygenase (2,5-/3,6-DKCMO) were used for the biotransformation. Conversion into an acetal yielded a potential advanced intermediate to azadirachtin, a potent antifeedant and growth regulator (Fig. 26) [161–163].

The biocatalytical formation of γ-butyrolactones through Baeyer–Villiger oxidation is catalyzed by a number of different monooxygenases, yielding precursors to various natural products [29, 38]. Best enantioselectivities were obtained for the

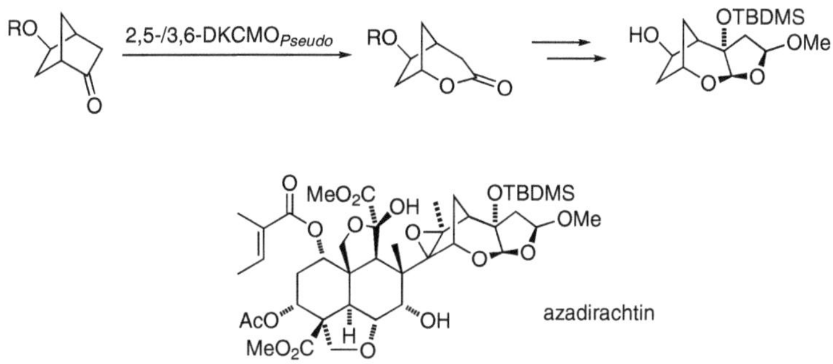

Fig. 26 Chemoenzymatic approach to azadirachtin

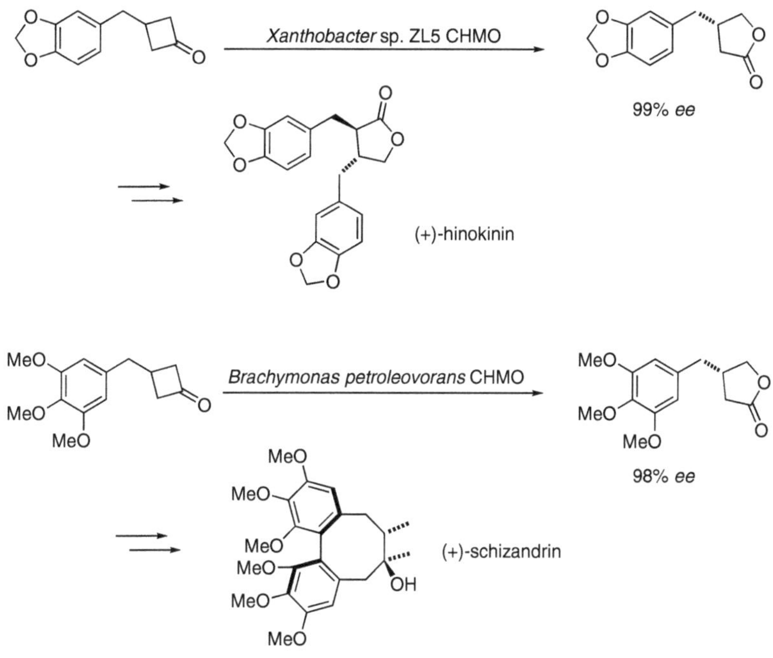

Fig. 27 Cyclohexanone monooxygenase-mediated synthesis of lignan building blocks

synthesis of precursors for lignans, which are known for their antileukemic, antiviral, antifungal, and antineoplastic activities [164–166]. For instance, building blocks for (+)-hinokinin and (+)-schizandrin have been generated with excellent enantioselectivities using CHMOs from *Xanthobacter* sp. ZL5 and *Brachymonas petroleovorans* respectively (Fig. 27).

More recently, the formal total syntheses of some bioactive natural products bearing a tetrahydrofurane moiety were achieved by enzymatic Baeyer–Villiger oxidation using cyclopentanone monooxygenases (CPMOs) from *Comamonas* sp.

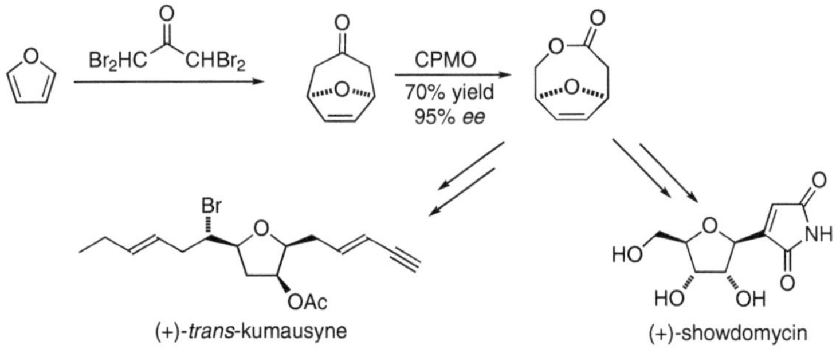

Fig. 28 Enzymatic Baeyer–Villiger oxidation in the synthesis of bioactive compounds

NCIMB 9872, overexpressed in *E. coli*, as a whole cell biotransformation [167]. The substrate, an oxabicycloketone, was successfully converted to the corresponding heterobicyclic lactone (Fig. 28) in 70% yield with an enantiomeric excess of 95%. This access to the chiral intermediate for several naturally occurring compounds, including (+)-*trans*-kumausyne and (+)-showdomycin, represents the first application of a whole cell biotransformation using recombinant *E. coli* with CPMO in natural product synthesis.

3.6 Cyclohexadienediols by Fermentation

There are only a few reactions known in which aromatic compounds suffer a permanent loss of aromaticity and at the same time the new stereogenic centers are formed selectively. The enzymatic oxidation is one of the few examples leading to exceptionally results: the reaction in eukaryotic systems is mainly catalyzed by cyctochrome-type monooxygenases, leading to epoxides. Furthermore, in prokaryotic systems, the processing of arenes by dioxygenases is well known [168], and especially *cis*-cyclohexadienediols (*cis*-CHDs) are generated by fermentation with recombinant *E. coli* strains. They have found broad application in several natural product and drug syntheses [28, 169–174].

For example, the fermentation of (2-bromoethyl)-benzene with recombinant *E. coli* furnished excellent yields of the corresponding *cis*-diol, enantiopure 3-(2-bromoethyl)-benzene-1,2-diol. The latter was used as a building block in the total synthesis of (+)-codeine (Fig. 29). Besides a Mitsunobu inversion of one of the stereogenic centers, two successive Heck cyclizations led to the enantiomer of the natural product [173]. Slight modifications of the reaction sequence, generating an epoxide intermediate, also furnished access to the naturally occurring enantiomer (−)-codeine [28].

Versatile epoxyquinol intermediates were conveniently accessed via metabolites derived from the enzymatic dihydroxylation of bromobenzene [172]. The chemoenzymatic approach led to shortened total syntheses of several epoxyquinols derived

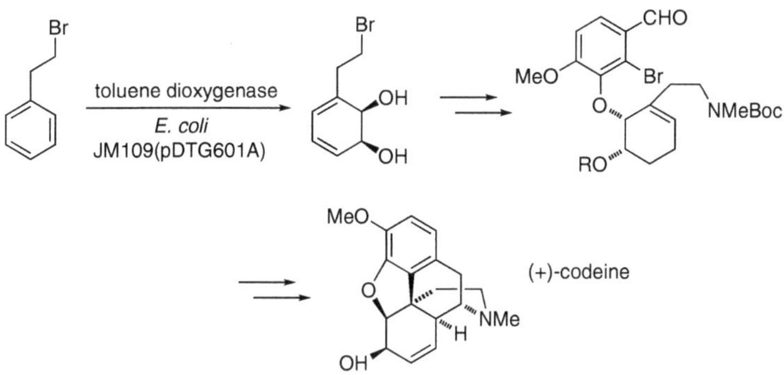

Fig. 29 Chemoenzymatic synthesis of (+)-codeine

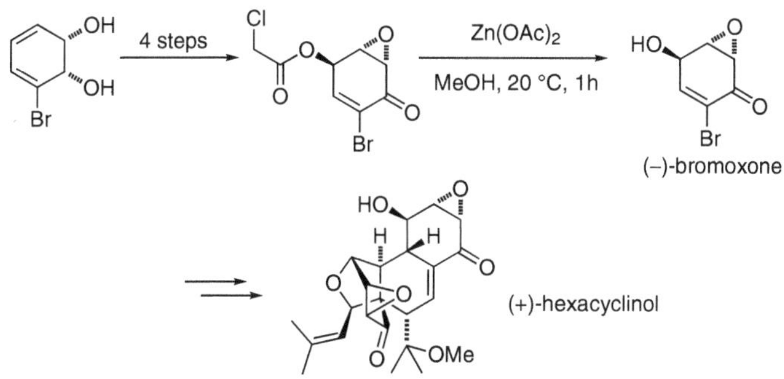

Fig. 30 Chemoenzymatic access to epoxyquinols

from bromoxone, such as (+)-hexacyclinol, a possible lead target for new antimalarial agents of fungal origin (Fig. 30).

Various approaches to the antiviral agent oseltamivir (Tamiflu®), which is one of the most effective drugs against avian influenza (H5N1) and the new influenza A (H1N1), have recently been published starting from biotechnologically provided *cis*-CHDs [175–177]. Hudlicky et al. started from ethyl benzoate, performing the oxidative biotransformation with a recombinant *E. coli* strain (Fig. 31).

The oxidation of benzenes to *trans*-cyclohexadienediols (*trans*-CHDs) has not been reported yet. Nevertheless, the fermentative production of *trans*-CHDs with recombinant *E. coli* strains has also been established and the value as a building block for natural product synthesis was demonstrated [178–180]. Starting from chorismate, *trans*-CHDs are metabolites derived from the shikimate–chorismate pathway in bacteria, plants, and fungi towards the biosynthesis of the iron chelator enterobactin (Fig. 32). Isochorismate synthase (EntC) and isochorismatase (EntB) catalyze the degradation to 2,3-*trans*-CHD or 3,4-*trans*-CHD respectively (Fig. 30). Using methods of metabolic pathway engineering, *E. coli* isochorismate synthase and isochorismatase were overexpressed in *E. coli* strains with a deficiency of *entA*, encoding 2,3-dihydroxybenzoate synthase (EntA).

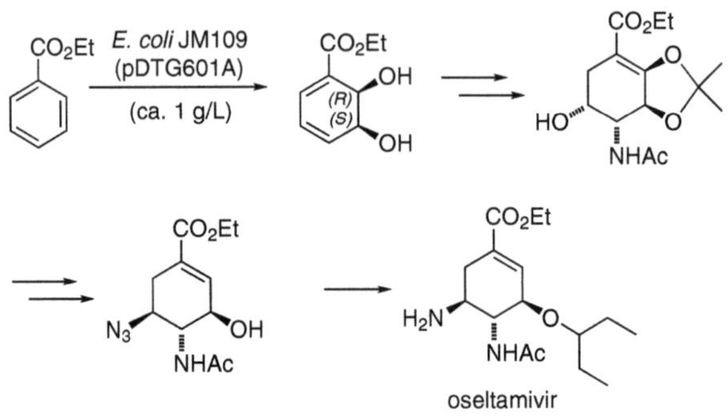

Fig. 31 Chemoenzymatic synthesis of oseltamivir

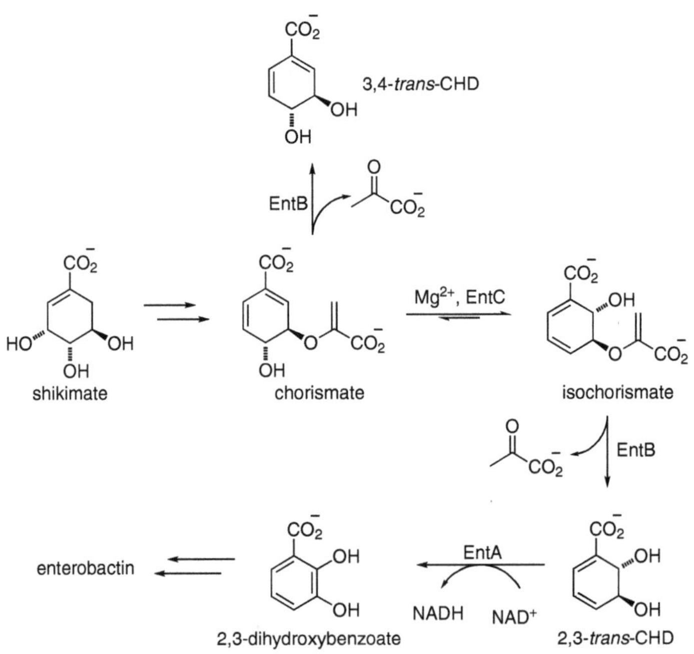

Fig. 32 Biosynthesis of *trans*-cyclohexadienediols

Microbially produced (2*S*,3*S*)-*trans*-dihydroxy-2,3-dihydrobenzoic acid was used in the synthesis of *ent*-streptol, *ent*-senepoxide, and *iso*-crotepoxide (Fig. 33). The short and efficient synthesis of these biologically active compounds included the esterification of the carboxylic acid and protection of the diol moiety, delivering control of the regio- and stereoselectivity of the following epoxidation or dihydroxylation steps [178, 180].

Recently, (2*S*,3*S*)-dihydroxy-2,3-dihydrobenzoic acid was isolated from the fermentation broth of a recombinant *E. coli* strain and the methyl ester was used in an interesting cyclopropanation sequence [181] yielding dicyclopropane building

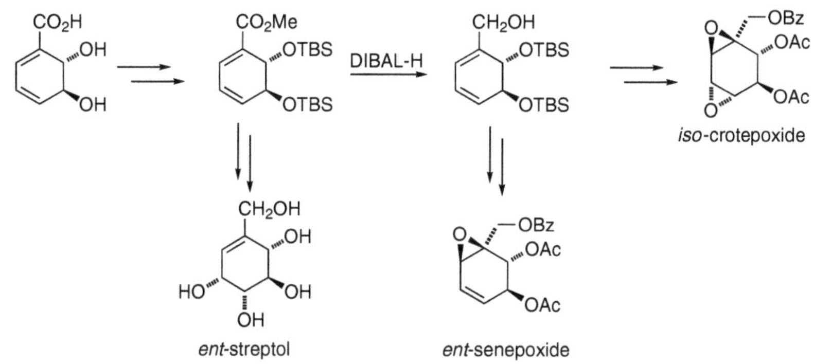

Fig. 33 Natural product syntheses starting from 2,3-*trans*-CHD

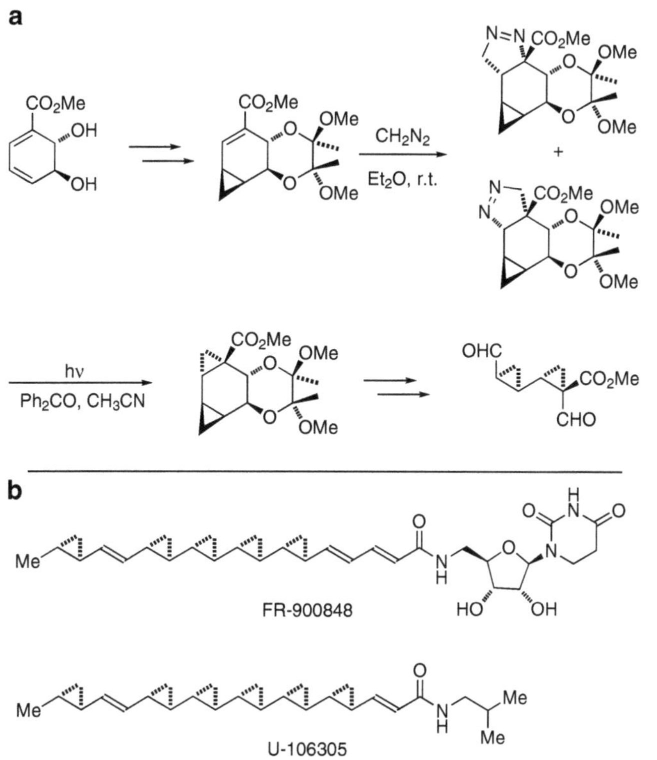

Fig. 34 (**a**) Synthesis of dicyclopropane building blocks via [2+3]-dipolar cycloaddition. (**b**) Oligocyclopropane-containing natural products

blocks that are potentially useful for the synthesis of oligocyclopropane natural products [51] such as the antifungal agent FR-900848 and the cholesteryl ester transfer protein inhibitor U-106305 (Fig. 34). For the formation of the second cyclopropane moiety, treatment with diazomethane led to a 1,3-dipolar cycloaddition to a five-membered heterocycle which furnished the desired second cyclopropane by photo induced nitrogen elimination.

4 C–C-Bond Formations

Enzymatic reactions forming new carbon–carbon bonds are a further important field of biotransformations in natural product synthesis. The construction of new, often complex carbon frameworks or their decomposition is performed by nature under catalysis of a set of enzymes. For organic chemists some of these enzymes, belonging to the enzyme class of lyases, such as aldolases, decarboxylases, hydroxynitrile lyases (HNLs), or benzaldehyde lyases (BALs), have been proven to represent versatile amendments to their synthetic toolbox.

4.1 Enzymatic Aldol Reactions

Aldol reactions enjoy great recognition as a useful tool for the synthesis of building blocks in natural product and drug synthesis [42, 182]. The stereochemistry of the stereogenic centers formed can be controlled by various means. Besides chiral auxiliaries, catalytic methods with chiral Lewis acids, organocatalysts, or catalytic antibodies were established for stereochemical control [183–187].

In nature, most aldolases are rooted in the sugar metabolic cycle and accept highly functionalized substrates for the aldol reaction. Nevertheless, the scope of enzymatic aldol reactions is limited, since aldolases strictly distinguish between the acceptor and the donor, yielding almost exclusively one product, and is furthermore restricted to only a few different possible natural donors. According to the donor molecules, aldolases are grouped in dihydroxyacetone phosphate-, phosphoenolpyruvate- or pyruvate-, acetaldehyde-, and glycine-dependent aldolases [41].

Fructose 1,6-biphosphate aldolase from rabbit muscle in nature reversibly catalyzes the addition of dihydroxyacetone phosphate (DHAP) to D-glyceraldehyde 3-phosphate. The tolerance of this DHAP-dependent enzyme towards various aldehyde acceptors made it a versatile tool in the synthesis of monosaccharides and sugar analogs [188], but also of alkaloids [189] and other natural products. For example, the enzyme-mediated aldol reaction of DHAP and an aldehyde is a key step in the total synthesis of the microbial elicitor (−)-syringolide 2 (Fig. 35a) [190].

A convenient chemoenzymatic access to sialic acid mimetics as important inhibitors of influenza sialidases has been established by Nelson et al. (Fig. 35b) [191, 192]. Application of a pyruvate-dependent sialic acid aldolase improved by directed evolution disclosed a new route to the core structure of important pharmaceuticals, such as zanamivir (Relenza®).

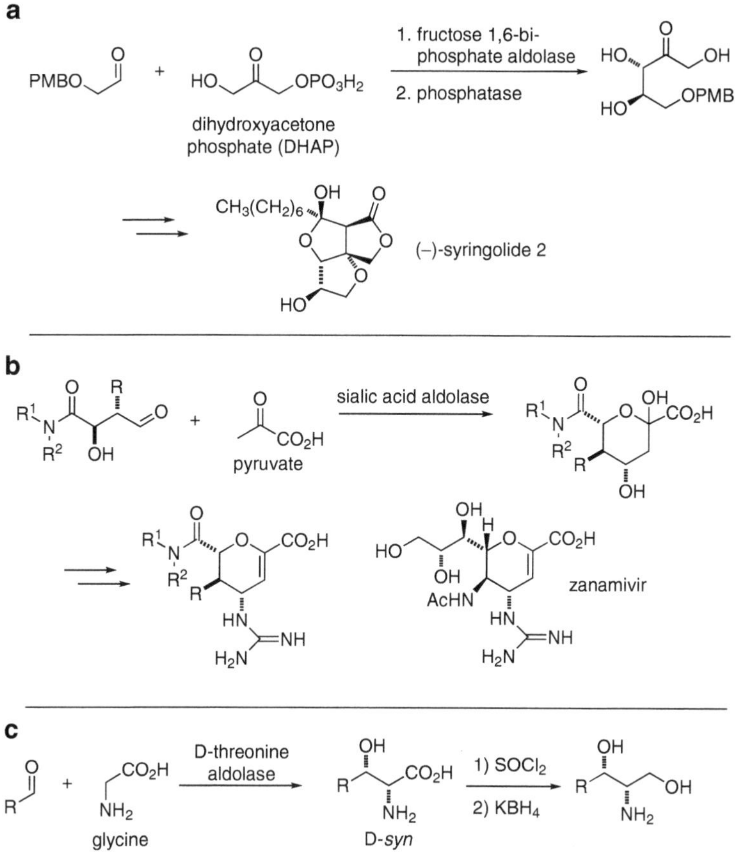

Fig. 35 Synthetic applications of (**a**) DHAP-, (**b**) pyruvate-, and (**c**) glycine-dependent aldolases

Glycine-dependent threonine aldolases have been used to synthesize a number of γ-halogenated and long-chain β-hydroxy-α-amino acids. For D-threonine aldolase *syn*-selectivity was observed exclusively. Further chemical conversion yielded the 2-amino-1,3-diols, potential precursors for the synthesis of short-chain sphingosine-derivatives (Fig. 35c) [193].

To date, 2-deoxy-D-ribose 5-phosphate aldolase (DERA) is the only acetaldehyde-dependent aldolase being applied in organic synthesis. Thus the stereoselectivity of DERA is significant, all known enzymes from different organisms showing the same preferences, limiting the field of application to syntheses in which specifically the DERA-catalyzed enantiomer is needed.

In comparison to other aldolases, DERA has a rather broad substrate range. DERA-catalyzed aldol reactions were used to get an access to key intermediates for epothilones (Fig. 36) [194]. According to retrosynthetic analysis, both fragments of the molecule could be obtained from aldol building blocks, and two out of seven stereocenters were established enzymatically. For the southern part of epothilone A,

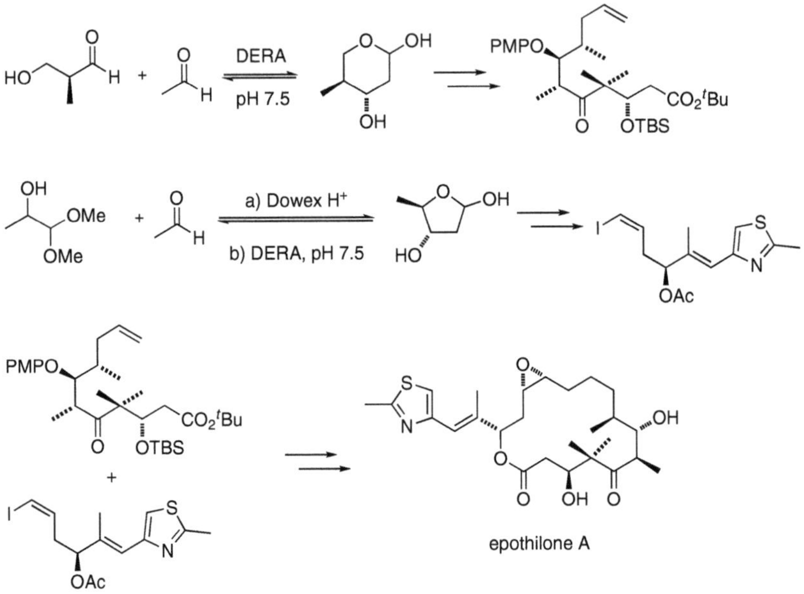

Fig. 36 DERA mediated access to key synthons in the synthesis of epothilone A

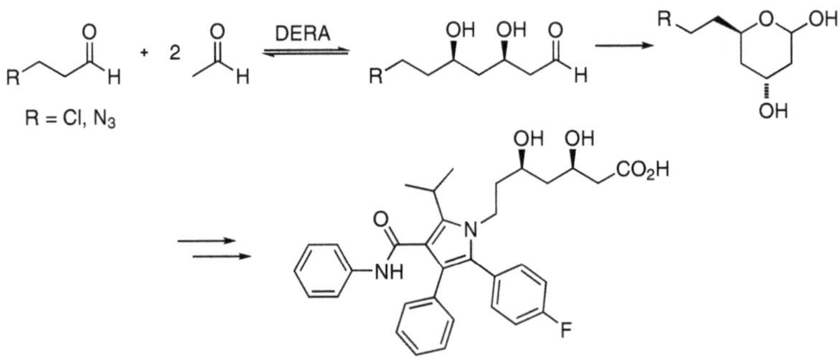

Fig. 37 Application of DERA in sequential one-pot reactions

(*S*)-3-hydroxy-2-methoxypropanal was successfully converted with acetaldehyde, with the primary open chain aldol product forming the lactol. The aldehyde needed as starting material for the northern part was generated in situ from the acetal: a DERA-catalyzed kinetic resolution led to conversion of the (*R*)-enantiomer in the aldol reaction, only.

Since the product of the aldol reaction still contains an aldehyde function, it is readily available for further reactions. Wong and co-workers very elegantly used a DERA mutant (Ser238Asp) for a sequential one-pot aldol condensation towards the synthesis of atorvastatin (Fig. 37) [195], a process also being applied by DSM [196]. When using wild-type DERA, the choice of residues in the acceptor aldehyde was limited to rather smaller negatively charged groups. The chosen mutation in the hydrophilic binding pocket preserved its nature, but went along with a 2.5-fold improvement in enzyme activity. Furthermore, the mutant presented a highly improved tolerance towards unnatural substrates. The performance in sequential aldol additions was also increased compared to the wild-type enzyme.

4.2 *Cyanohydrins by Hydroxynitrile Lyases*

Native HNLs from bitter almonds (*Prunus amygdalus*), cassava (*Manihot esculenta*), millet (*Sorghum bicolor*), and flax (*Linum usitatissimum*) were repeatedly used in the synthesis of chiral cyanohydrins [39, 41, 197]. Cyanohydrins are versatile building blocks in natural product synthesis, giving organic chemists the possibility of introducing all kinds of functional groups (Fig. 38) [198].

Effenberger and co-workers reported the biotransformation of pivaldehyde-derivatives to the corresponding cyanohydrins using wild-type *Pa*HNL [199]. In this context, a precursor to (*R*)-pantolactone could be obtained with a yield of 84% and 89% *ee*. Since (*R*)-pantolactone is the most important building block in the synthesis of vitamin B_5, further efforts were made for optimization. Recently, an acid stable almond HNL isoenzyme was generated by semirational design and applied in the synthesis described earlier, showing full conversion of hydroxypivalaldehyde to the corresponding cyanohydrin with an enantiomeric excess of 97% (Fig. 39) [200].

A large number of the most important modern insecticides are derived from cyanohydrins. Especially some esters formed from enantiopure cyanohydrins and chrysanthemum acid derivatives are known to be very potent [201]. Nowadays, the formation of 3-phenoxybenzaldehyde cyanohydrin is performed in a biocatalytic industrial process using *Me*HNL or *Hb*HNL isolated from rubber trees (*Hevea*

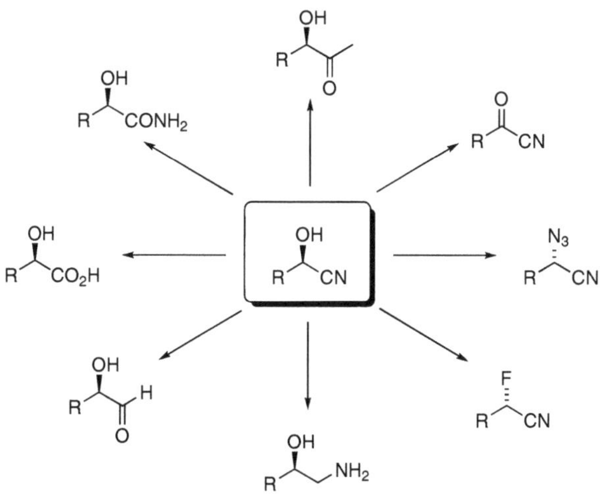

Fig. 38 Cyanohydrins as building blocks for organic synthesis

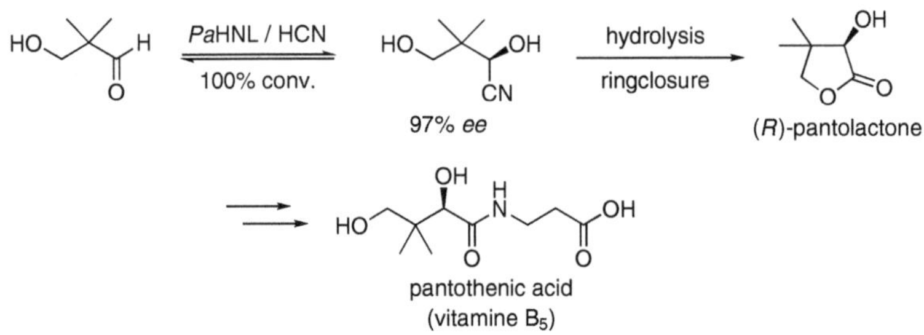

Fig. 39 Chemoenzymatic synthesis of (*R*)-pantolactone

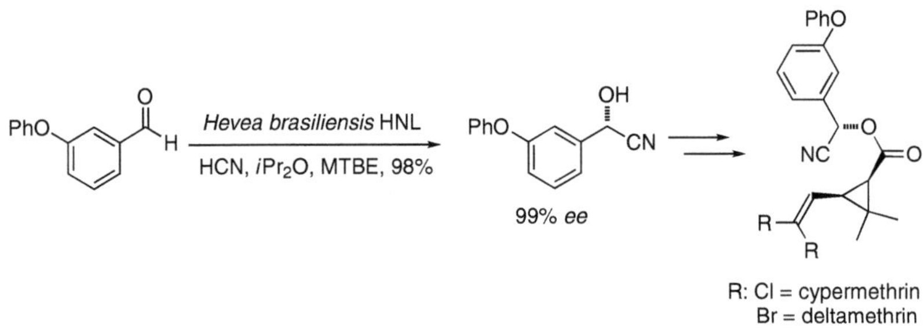

Fig. 40 Application of HNL in large-scale production of insecticides

brasiliensis) yielding precursors for cypermethrin or deltamethrin, two pyrethroids (Fig. 40) [21, 39, 41, 202].

Another application of HNLs can be found in the chemoenzymatic synthesis of the broadband antibiotic thiamphenicol and its fluorinated derivative florfenicol [203]. The conversion of 4-methylsulfanyl-benzaldehyde to the mandelonitrile was catalyzed by a novel enzyme from a Chinese almond (*Prunus communis* L. var. *dulcis* Borkh). A concentrated powder from the kernels was prepared and a mixture of the crude meal, aldehyde, and HCN was stirred in isopropyl ether at room temperature for 12 h, yielding the cyanohydrin with 99% *ee* after recrystallization. The building block formed was then successfully applied in the total synthesis of thiamphenicol and florfenicol (Fig. 41).

4.3 Biocatalytic Formation of Acyloins

Enantiopure, bifunctional acyloins (α-hydroxy ketones) are versatile intermediates in natural product synthesis (also see Sect. 2.3, Fig. 11). In nature, the formation of α-hydroxy ketones is efficiently catalyzed by thiamine diphosphate-dependent enzymes: transketolases, decarboxylases, and other lyases, such as BALs. A great portfolio of biotransformations, especially with benzaldehyde derivatives as starting materials, were realized [204].

A very short and efficient chemoenzymatic synthesis of (−)-ephedrine can be achieved by a decarboxylase-mediated reaction of benzaldehyde with pyruvate

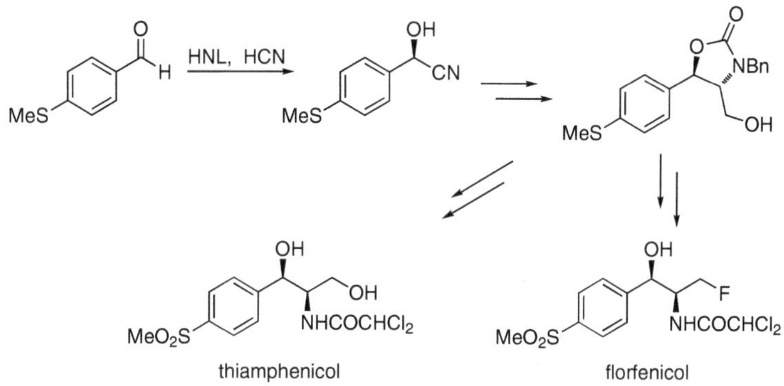

Fig. 41 Chemoenzymatic synthesis of broadband antibiotics using hydroxynitrile lyases

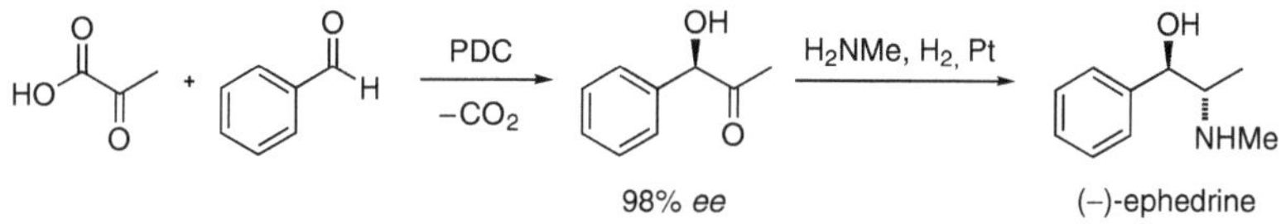

Fig. 42 Chemoenzymatic synthesis of (−)-ephedrine

[205–207]. The formation of the acyloin intermediate is catalyzed by pyruvate decarboxylase (PDC) with excellent enantioselectivity (Fig. 42).

2,3-Dioxygenated aryl propanones are very frequently used chiral building blocks for the synthesis of different biologically active compounds. They were synthons in the synthetic approach to 1,4-benzodioxane lignans, such as 5′-methoxyhydno-carpin or cytoxazone, or for the production of aryl isoserines, which are precursors to the side-chain of paclitaxel (Taxol®). A new access to these compounds was established by an enzymatic approach using BAL as the acyloin forming biocatalyst (Fig. 43) [208].

5 Conclusion

Nature with its complexity has provided a cornucopia of tools for improvements in twenty-first century natural science research. With new insights into metabolic pathways, chemists can adopt an increasing amount of evolutionary innovation and apply this knowledge to drug and natural product synthesis.

The usage of microorganisms or isolated enzymes thereof is gaining more and more in popularity, leading to an enormous variety of new enzymes being found and optimized for commercialization. Especially in asymmetric synthesis, enzymes can substitute or complement existing methods, often providing ecological advantages due to greater efficiency or less toxicity of the catalyst (enzyme vs transition metal). Thus, the number of applications of enzyme-mediated synthesis has risen enormously during the last few years and the trend does not seem to slow down.

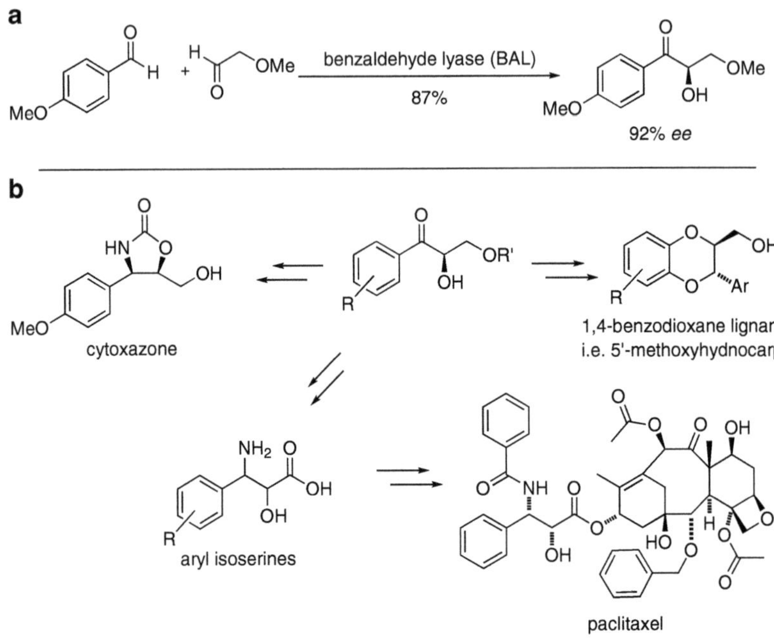

Fig. 43 (**a**) Benzaldehyde lyase-catalyzed acyloin formation. (**b**) Application of acyloins in natural product synthesis

Biocatalysis is a main part of white biotechnology and in combination with the usage of renewable primary products should lead to a more sustainable chemical and pharmaceutical industry.

Acknowledgments We gratefully acknowledge the Ministry of Innovation, Science, Research and Technology of the German federal state of North Rhine-Westphalia, the Deutsche Forschungsgemeinschaft, the Otto-Röhm-Gedächtnisstiftung, the Fonds der Chemischen Industrie, the Jürgen Manchot Stiftung, the Degussa Stiftung, the Heinrich-Heine-Universität Düsseldorf, the Forschungszentrum Jülich GmbH, and the Deutsche Bundesstiftung Umwelt (grant to T.F.) for their generous support of our projects. We also thank the BASF AG, the Bayer AG, the Cognis GmbH, the Evonik AG, the Wacker AG, the Umicore AG, the Chemetall GmbH, Codexis Inc., and evocatal for their kind donations and contributions to our research. Furthermore, we thank the analytical departments at the University of Stuttgart and the FZ Jülich as well as Monika Gehsing, Rainer Goldbaum, Verena Doum, Birgit Henßen, Erik Kranz, Christoph Lorenz, Vera Ophoven, Bea Paschold, Truc Pham, and Saskia Schuback for their ongoing support.

References

1. Cragg GM, Grothaus PG et al (2009) Impact of natural products on developing new anticancer agents. Chem Rev 109:3012–3043
2. Rollinger JM, Langer T et al (2006) Strategies for efficient lead structure discovery from natural products. Curr Med Chem 13:1491–1507

3. Gautam R, Jachak SM (2009) Recent developments in anti-inflammatory natural products. Med Res Rev 29:767–820
4. Miller SJ, Clardy J (2009) Beyond grind and find. Nat Chem 1:261–263
5. FDA (1992) FDA's policy statement for the development of new stereoisomeric drugs. Chirality 4:338–340
6. Carey JS, Laffan D et al (2006) Analysis of the reactions used for the preparation of drug candidate molecules. Org Biomol Chem 4:2337–2347
7. Kocovský P, Malkov AV (2006) Organocatalysis in organic synthesis. Tetrahedron 62:255
8. Noyori R (2002) Asymmetric catalysis: science and opportunities (Nobel lecture). Angew Chem Int Ed 41:2008–2022
9. Dalko PI, Moisan L (2001) Enantioselective organocatalysis. Angew Chem Int Ed 40:3726–3748
10. Ghanem A (2007) Trends in lipase-catalyzed asymmetric access to enantiomerically pure/enriched compounds. Tetrahedron 63:1721–1754
11. Bohman B, Unelius CR (2009) Synthesis of all four stereoisomers of 5-hydroxy-4-methyl-3-heptanone using plants and oyster mushrooms. Tetrahedron 65:8697–8701
12. Andrade LH, Utsunomiya RS et al (2006) Edible catalysts for clean chemical reactions: bioreduction of aromatic ketones and biooxidation of secondary alcohols using plants. J Mol Catal B Enzym 38:84–90
13. Cordell GA, Lemos TLG et al (2007) Vegetables as chemical reagents. J Nat Prod 70:478–492
14. Fonseca AM, Monte FJQ et al (2009) Coconut water (*Cocos nucifera* L.) – a new biocatalyst system for organic synthesis. J Mol Catal B Enzym 57:78–82
15. Bornscheuer UT, Pohl M (2001) Improved biocatalysts by directed evolution and rational protein design. Curr Opin Chem Biol 5:137–143
16. Jaeger K-E, Eggert T (2004) Enantioselective biocatalysis optimized by directed evolution. Curr Opin Biotechnol 15:305–313
17. Bornscheuer UT (2005) Trends and challenges in enzyme technology. Adv Biochem Eng Biotechnol 100:181–203
18. Reetz MT (2009) Directed evolution of enantioselective enzymes: an unconventional approach to asymmetric catalysis in organic chemistry. J Org Chem 74:5767–5778
19. Jäckel C, Kast P et al (2008) Protein design by directed evolution. Annu Rev Biophys 37:153–173
20. Carrea G, Riva S (2000) Properties and synthetic applications of enzymes in organic solvents. Angew Chem Int Ed 39:2226–2254
21. Hilterhaus L, Liese A (2007) Building blocks. Adv Biochem Eng Biotechnol 105:133–173
22. Matsuda T, Yamanaka R et al (2009) Recent progress in biocatalysis for asymmetric oxidation and reduction. Tetrahedron Asymmetry 20:513–557
23. Goldberg K, Schroer K et al (2007) Biocatalytic ketone reduction – a powerful tool for the production of chiral alcohols – part I: processes with isolated enzymes. Appl Microbiol Biotechnol 76:237–248
24. Goldberg K, Schroer K et al (2007) Biocatalytic ketone reduction – a powerful tool for the production of chiral alcohols – part II: whole-cell reductions. Appl Microbiol Biotechnol 76:249–255
25. Panke S, Wubbolts M (2005) Advances in biocatalytic synthesis of pharmaceutical intermediates. Curr Opin Chem Biol 9:188–194
26. Li Z, van Beilen JB et al (2002) Oxidative biotransformations using oxygenases. Curr Opin Chem Biol 6:136–144
27. Burton SG (2003) Oxidizing enzymes as biocatalysts. Trends Biotechnol 21:543–549
28. Hudlicky T, Reed JW (2009) Applications of biotransformations and biocatalysis to complexity generation in organic synthesis. Chem Soc Rev 38:3117–3132
29. Kayser MM (2009) 'Designer reagents' recombinant microorganisms: new and powerful tools for organic synthesis. Tetrahedron 65:947–974
30. Patel JM (2009) Biocatalytic synthesis of atorvastatin intermediates. J Mol Catal B Enzym 61:123–128

31. Patel RN (2006) Biocatalysis: synthesis of chiral intermediates for pharmaceuticals. Curr Org Chem 10:1289–1321
32. Patel RN (2008) Chemo-enzymatic synthesis of pharmaceutical intermediates. Expert Opin Drug Discov 3:187–245
33. Theil F (1995) Lipase-supported synthesis of biologically active compounds. Chem Rev 95:2203–2227
34. García-Urdiales E, Alfonso I et al (2005) Enantioselective enzymatic desymmetrizations in organic synthesis. Chem Rev 105:313–354
35. Chênevert R, Pelchat N et al (2006) Stereoselective enzymatic acylations (transesterifications). Curr Org Chem 10:1067–1094
36. Akita H (2009) Recent advances in the synthesis of biologically active natural products using biocatalyst. Heterocycles 78:1667–1713
37. Nakamura K, Matsuda T (2006) Biocatalytic reduction of carbonyl groups. Curr Org Chem 10:1217–1246
38. Mihovilovic MD (2006) Enzyme mediated Baeyer–Villiger oxidations. Curr Org Chem 10:1265–1287
39. Holt J, Hanefeld U (2009) Enantioselective enzyme-catalysed synthesis of cyanohydrins. Curr Org Synth 6:15–37
40. Molinari F (2006) Oxidations with isolated and cell-bound dehydrogenases and oxidases. Curr Org Chem 10:1247–1263
41. Sukumaran J, Hanefeld U (2005) Enantioselective C–C bond synthesis catalysed by enzymes. Chem Soc Rev 34:530–542
42. Dean Stephen M, Greenberg William A et al (2007) Recent advances in aldolase-catalyzed asymmetric synthesis. Adv Synth Catal 349:1308–1320
43. Faber K (2004) Biotransformation in organic chemistry – a textbook. Springer, Heidelberg
44. Liese A, Seelbach K, Wandrey C (eds) (2006) Industrial biotransformations. Wiley-VCH, Weinheim
45. Bornscheuer UT, Kazlauskas RJ (1999) Hydrolases in organic synthesis – regio- and stereoselective biotransformations. Wiley-VCH, Weinheim
46. Gotor V, Alfonso I, Garcia-Urdiales E (eds) (2008) Asymmetric organic synthesis with enzymes. Wiley-VCH, Weinheim
47. Schmid RD, Urlacher VB (eds) (2007) Modern biooxidation. Wiley-VCH, Weinheim
48. Garcia-Junceda E (ed) (2008) Multi-step enzyme catalysis. Wiley-VCH, Weinheim
49. Willis MC (1999) Enantioselective desymmetrisation. J Chem Soc Perkin Trans 1:1765–1784
50. Lebel H, Marcoux J-F et al (2003) Stereoselective cyclopropanation reactions. Chem Rev 103:977–1050
51. Pietruszka J (2003) Synthesis and properties of oligocyclopropyl-containing natural products and model compounds. Chem Rev 103:1051–1070
52. Grandjean D, Pale P et al (1991) Enzymatic hydrolysis of cyclopropanes. Total synthesis of optically pure dictyopterenes A and C′. Tetrahedron 47:1215–1230
53. Barloy-Da Silva C, Benkouider A et al (2000) Synthetic studies towards oxylipins: total synthesis of constanolactones A and B. Tetrahedron Lett 41:3077–3081
54. Maddess Matthew L, Tackett Miles N et al (2007) Total synthesis of rapamycin. Angew Chem Int Ed 46:591–597
55. Hayward CM, Yohannes D et al (2002) Total synthesis of rapamycin via a novel titanium-mediated aldol macrocyclization reaction. J Am Chem Soc 115:9345–9346
56. Romo D, Meyer SD et al (2002) Total synthesis of (−)-rapamycin using an Evans-Tishchenko fragment coupling. J Am Chem Soc 115:7906–7907
57. Nicolaou KC, Chakraborty TK et al (1993) Total synthesis of rapamycin. J Am Chem Soc 115:4419–4420
58. Smith AB, Condon SM et al (2002) Total synthesis of rapamycin and demethoxyrapamycin. J Am Chem Soc 117:5407–5408

59. Anderson JC, Ley SV et al (1994) Studies towards the total synthesis of rapamycin: a convergent and stereoselective synthesis of the C22–C32 carbon framework. Tetrahedron Lett 35:2087–2090
60. Ghosh AK, Xu X (2004) Assignment of absolute stereochemistry and total synthesis of (α)-spongidepsin. Org Lett 6:2055–2058
61. Ley SV, Tackett MN et al (2009) Total synthesis of rapamycin. Chem Eur J 15:2874–2914
62. Kagan HB, Fiaud JC (1988) Kinetic resolution. Top Stereochem 18:249–330
63. Chen C-S, Fujimoto Y et al (1982) Quantitative analyses of biochemical kinetic resolutions of enantiomers. J Am Chem Soc 104:7294–7299
64. Chen C-S, Wu S-H et al (1987) Quantitative analyses of biochemical kinetic resolution of enantiomers. 2. Enzyme-catalyzed esterifications in water–organic solvent biphasic systems. J Am Chem Soc 109:2812–2817
65. Scheid G, Ruijter E et al (2004) Synthesis and resolution of a key building block for epothilones: a comparison of asymmetric synthesis, chemical and enzymatic resolution. Tetrahedron Asymmetry 15:2861–2869
66. Ma S (2005) Some typical advances in the synthetic applications of allenes. Chem Rev 105:2829–2872
67. Carballeira JD, Krumlinde P et al (2007) Directed evolution and axial chirality: optimization of the enantioselectivity of *Pseudomonas aeruginosa* lipase towards the kinetic resolution of a racemic allene. Chem Commun 43:1913–1915
68. Deska J, Bäckvall J-E (2009) Enzymatic kinetic resolution of primary allenic alcohols. Application to the total synthesis and stereochemical assignment of striatisporolide A. Org Biomol Chem 7:3379–3381
69. Horvath A, Bäckvall J-E (2004) Mild and efficient palladium(II)-catalyzed racemization of allenes. Chem Commun 40:964–965
70. Kanazawa S, Fusetani N et al (1993) Cylindramide: cytotoxic tetramic acid lactam from the marine sponge *Halichondria cylindrata* Tanita & Hoshino. Tetrahedron Lett 34:1065–1068
71. Cramer N, Buchweitz M et al (2006) Total synthesis and NMR investigations of cyclindramide. Chem Eur J 12:2488–2503
72. Jiang X, García-Fortanet J et al (2005) Synthesis and complete stereochemical assignment of psymberin/irciniastatin A. J Am Chem Soc 127:11254–11255
73. Huang X, Shao N et al (2007) The total synthesis of psymberin. Org Lett 9:2597–2600
74. Smith AB, Jurica JA et al (2008) Total synthesis of (+)-psymberin (irciniastatin A): catalytic reagent control as the strategic cornerstone. Org Lett 10:5625–5628
75. Pietruszka J, Simon RC (2009) Chemoenzymatic synthesis of (protected) psymberic acid. Eur J Org Chem 2009:3628–3634
76. Nishizawa M, Shimizu M et al (1995) Stereoselective production of (+)-*trans*-chrysanthemic acid by a microbial esterase: cloning, nucleotide sequence, and overexpression of the esterase gene of *Arthrobacter globiformis* in *Escherichia coli*. Appl Environ Microbiol 61:3208–3215
77. Nishizawa M, Gomi H et al (1993) Purification and some properties of carboxylesterase from *Arthrobacter globiformis*; stereoselective hydrolysis of ethyl chrysanthemate. Biosci Biotech Biochem 57:594–598
78. Nanda S, Rao AB et al (1999) Enzyme catalysed kinetic resolution of racemic 2,2-dimethyl-3-(2,2-disubstituted vinyl) cyclopropane carboxylic acids anchored on polymer supports. Tetrahedron Lett 40:5905–5908
79. Yadav JS, Rao AB et al (1997) Enzymatic resolution of (+/−)-*cis*-3-(2,2-dichloro-3,3,3-trifluoropropyl)-2,2-dimethylcyclopropane carboxylate. Tetrahedron Asymmetry 8:2291–2294
80. Csuk R, Schabel MJ et al (1996) Synthesis of the enantiomer of the antidepressant tranylcypromine. Tetrahedron Asymmetry 7:3505–3512

81. Csuk R, Scholz Yv (1996) Enantiomerically pure cyclopropanoid nucleoside analogues: synthesis and analysis. Tetrahedron 52:6383–6396
82. Rosen TC, Haufe G (2002) Synthesis of enantiopure monofluorinated phenylcyclopropanes by lipase-catalyzed kinetic resolution. Tetrahedron Asymmetry 13:1397–1405
83. Tsuji T, Onishi T et al (1999) Lipase-catalyzed synthesis of a tri-substituted cyclopropyl chiral synthon: a practical method for preparation of chiral 1-alkoxycarbonyl-2-oxo-3-oxabicyclo[3.1.0]hexane. Tetrahedron Asymmetry 10:3819–2825
84. Beumer R, Bubert C et al (2000) The synthesis of diastereo- and enantiomerically pure β-aminocyclopropanecarboxylic acids. J Org Chem 65:8960–8969
85. Pietruszka J, Rieche ACM et al (2003) Kinetic enzymatic resolution of cyclopropane derivatives. Adv Synth Catal 345:1273–1286
86. Pietruszka J, Wilhelm T (2003) Total synthesis of marine oxylipins constanolactone A and B. Synlett 11:1698–1700
87. Pietruszka J, Rieche ACM (2008) Total synthesis of marine oxylipins solandelactones A–H. Adv Synth Catal 350:1407–1412
88. Bischop M, Doum V et al (2010) Total synthesis of halicholactone and neohalicholactone. Synthesis 42:527–537
89. Mahapatra T, Bhunya R et al (2009) Chemo-enzymatic asymmetric total synthesis of penienone. Tetrahedron Lett 50:5392–5394
90. Pellissier H (2003) Dynamic kinetic resolution. Tetrahedron 59:8291–8327
91. Pellissier H (2008) Recent developments in dynamic kinetic resolution. Tetrahedron 64:1563–1601
92. Kamaruddin AH, Uzir MH et al (2009) Chemoenzymatic and microbial dynamic kinetic resolutions. Chirality 21:449–467
93. Martín-Matute B, Edin M et al (2004) Highly compatible metal and enzyme catalysts for efficient dynamic kinetic resolution of alcohols at ambient temperature. Angew Chem Int Ed 43:6535–6539
94. Akai S, Tanimoto K et al (2006) A dynamic kinetic resolution of allyl alcohols by the combined use of lipases and [$VO(OSiPh_3)_3$]. Angew Chem Int Ed 45:2592–2595
95. Ödman P, Wessjohann LA et al (2005) Chemoenzymatic dynamic kinetic resolution of acyloins. J Org Chem 70:9551–9555
96. Kowalski RJ, Giannakakous P et al (1997) Activities of the microtubule-stabilizing agents epothilones A and B with purified tubulin and in cells resistant to paclitaxel (Taxol). J Biol Chem 272:2534–2541
97. Akai S, Tanimoto K et al (2004) Lipase-catalyzed domino dynamic kinetic resolution of racemic 3-vinylcyclohex-2-en-1-ols/Intramolecular Diels–Alder reaction: one-pot synthesis of optically active polysubstituted decalins. Angew Chem Int Ed 43:1407–1410
98. Brown AG, Smale TC et al (1976) Crystal and molecular structure of compactin, a new antifungal metabolite from *Penicillium brevicompactum*. J Chem Soc Perkin Trans 1:1165–1170
99. Robichaud J, Tremblay F (2006) Formal enantioselective synthesis of (+)-compactin. Org Lett 8:597–600
100. Calvo D, Port M et al (1996) Total synthesis of forskolin – part III studies related to an asymmetric synthesis. Tetrahedron Lett 37:1023–1024
101. Nagashima S, Kanematsu K (1990) A synthesis of an optically active forskolin intermediate via allenyl ether intramolecular cycloaddition strategy. Tetrahedron Asymmetry 1:743–749
102. Bhat SV, Bajqwa BS et al (1977) Structures and stereochemistry of new labdane diterpinoids from *coleus forskohlii briq*. Tetrahedron Lett 18:1669–1672
103. Matsumara K, Shohei H et al (1997) Asymmetric transfer hydrogenation of α,β-acetylenic ketones. J Am Chem Soc 119:8738–8739
104. Corey EJ, Helal CJ (1998) Reduction of carbonyl compounds with chiral oxazaborolidine catalysts: a new paradigm for enantioselective catalysis and a powerful new synthetic method. Angew Chem Int Ed 37:1986–2012

105. Corey EJ, Guzman-Perez A et al (1997) An enantioselective synthetic route to atractyligenin using the oxazaborolidine-catalyzed reduction of β-Siliyl- or β-stannyl-substituted α,β-enones as a key step. J Am Chem Soc 119:11769–11776
106. Daußmann T, Hennemann H-G et al (2006) Enzymatische technologien zur synthese chiraler alkohol-derivate. Chem Ing Tech 78:249–255
107. de Wildeman SMA, Sonke T et al (2007) Biocatalytic reductions: from lab curiosity to "first choice". Acc Chem Res 40:1260–1266
108. Ruinatscha R, Höllrigl V et al (2006) Productivity of selective electroenzymatic reduction and oxidation reactions: theoretical and practical considerations. Adv Synth Catal 348:2015–2026
109. Stillger T, Bönitz M et al (2002) Überwindung von thermodynamischen Limitierungen in substratgekoppelten Cofaktorregenerierungsverfahren. Chem Ing Tech 74:1035–1039
110. Goldberg K, Edegger K et al (2006) Overcoming the thermodynamic limitation in asymmetric hydrogen transfer reactions catalyzed by whole cells. Biotechnol Bioeng 95:192–198
111. Yang W, Xu J-H et al (2006) Asymmetric reduction of ketones by employing *Rhodotorula* sp. AS2.2241 and synthesis of the β-blocker (*R*)-nifenalol. Tetrahedron Asymmetry 17:1769–1774
112. Ema T, Ide S et al (2008) Highly efficient chemoenzymatic synthesis of methyl (*R*)-*o*-chloromandelate, a key intermediate for clopidogrel, *via* asymmetric reduction with recombinant *Escherichia coli*. Adv Synth Catal 350:2039–2044
113. Tanaka T, Okamura N et al (1986) Syntheses of (5E)-PGE2 and new 6-functionalized derivatives by the use of palladium-catalyzed decarboxylative allylic alkylation. Tetrahedron 42:6747–6758
114. Tanaka T, Hasato A et al (1987) Preparation of (5E)-prostaglandin E2 derivatives as potential drugs, JP 62195358 A2
115. Traverso G, Pirillo D et al (1982) Sull'epossidazione del metilestere dell'acido 5-idrossi-6-n. eptenoico e sui prodotti da essa derivati. Il Farmaco Ed Sc 37:192–198
116. Pirillo D, Gazzaniga A et al (1985) A route to the total synthesis of *iso*-LTB_4 system. Il Farmaco Ed Sc 40:249–252
117. Vig OP, Dhindsa AS et al (1972) Terpenoids. LXX. Synthesis of myrcenol. J Ind Chem Soc 49:163–166
118. Jacobo SH, Chang C-T et al (2006) Total synthesis of 8,12-*iso*-iPF3α-VI, an EPA-derived isoprostane: stereoselective introduction of the fifth asymmetric center. J Org Chem 71:1370–1379
119. Fischer T, Pietruszka J (2007) Efficient synthesis of either enantiomer of ethyl 5-hydroxyhept-6-enoate. Adv Synth Catal 349:1533–1536
120. Pietruszka J, Rieche ACM et al (2007) Synthesis of marine oxylipins constanolactones C and D. Synlett 18:2525–2528
121. Fischer T, Pietruszka J Heinrich-Heine-Universität Düsseldorf, unpublished results
122. Wolberg M, Filho M et al (2008) Chemoenzymatic synthesis of the chiral side-chain of statins: application of an alcohol dehydrogenase catalysed ketone reduction on a large scale. Bioprocess Biosyst Eng 31:183–191
123. Wolberg M, Hummel W et al (2001) Biocatalytic reduction of β,δ-diketo esters: a highly stereoselective approach to all four stereoisomers of a chlorinated β,δ-dihydroxy hexanoate. Chem Eur J 7:4562–4571
124. Wolberg M, Hummel W et al (2000) Highly regio- and enantioselective reduction of 3,5-dioxocarboxylates. Angew Chem 39:4306–4308
125. Müller M, Wolberg M et al (2005) Enzyme-catalyzed regio-and enantioselective ketone reductions. Adv Biochem Eng Biotechnol 92:261–287
126. Chen K-M, Hardtmann GE et al (1987) 1,3-Diastereoselective reduction of β-hydroxyketones utilizing alkoxydialkylboranes. Tetrahedron Lett 28:155–158
127. Evans DA, Chapman KT et al (2002) Directed reduction of β-hydroxy ketones employing tetramethylammonium triacetoxyborohydride. J Am Chem Soc 110:3560–3578

128. Wada M, Kawabata H et al (1999) Occurrence of multiple ethyl 4-chloro-3-oxobutanoate-reducing enzymes in *Candida magnoliae*. J Biosci Bioeng 87:144–148
129. Patel RN, McNamee CG et al (1992) Stereoselective reduction of β-keto esters by *Geotrichum candidum*. Enzyme Microb Technol 14:731–738
130. Spelberg JHL, Tang L et al (2004) Enzymatic dynamic kinetic resolution of epihalohydrins. Tetrahedron Asymmetry 15:1095–1102
131. Nakamura T, Nagasawa T et al (1991) A new catalytic function of halohydrin hydrogen-halide-lyase, synthesis of β-hydroxynitriles from epoxides and cyanide. Biochem Biophys Res Commun 180:124–130
132. Fox RJ, Davis SC et al (2007) Improving catalytic function by ProSAR-driven enzyme evolution. Nat Biotechnol 25:338–344
133. Brown BJ, Deng Z et al (1998) On the active site of old yellow enzyme. J Biol Chem 273:32753–32762
134. Fox KM, Karplus PA (1994) Old yellow enzyme at 2 Å resolution: overall structure, ligand binding, and comparison with related flavoproteins. Structure 2:1089–1105
135. Kohli RM, Massey V (1998) The oxidative half-reaction of old yellow enzyme. J Biol Chem 273:32763–32770
136. Leuenberger HGW, Boguth W et al (1976) Synthese von optisch aktiven, natürlichen Carotinoiden und strukturell verwandten Naturprodukten. I. Synthese der chiralen Schlüsselverbindung (4*R*,6*R*)-4-Hydroxy-2,2,6-trimethylcyclohexanon. Helv Chim Acta 59:1832–1849
137. Kataoka M, Kotaka A et al (2004) Cloning and overexpression of the old yellow enzyme gene of *Candida macedoniensis*, and its application to the production of a chiral compound. J Biotechnol 114:1–9
138. Hall M, Stueckler C et al (2008) Asymmetric bioreduction of activated C=C bonds using *Zymomonas mobilis* NCR enoate reductase and old yellow enzymes OYE 1–3 from yeasts. Eur J Org Chem 2008:1511–1516
139. Wada M, Yoshizumi A et al (2003) Production of a doubly chiral compound, (4*R*,6*R*)-4-hydroxy-2,2,6-trimethylcoclohexanone, by two-step enzymatic asymmetric reduction. Appl Environ Microbiol 69:933–937
140. Shimoda K, Kubota N et al (2004) Asymmetric reduction of α,β-unsaturated carbonyl compounds with reductases from *Nicotiana tabacum*. Tetrahedron Asymmetry 15:2443–2446
141. Hirata T, Takarada A et al (2005) Hydrogenation of the C–C double bond of maleimides with cultured plant cells. J Mol Catal B Enzym 32:131–134
142. Perdih A, Dolenc MS (2007) Recent advances in the synthesis of unnatural amino acids. Curr Org Chem 11:801–832
143. Hanson RL (2007) In: Gadamasetti K, Braish T (eds) Process chemistry in the pharmaceutical industry, vol 2. Challenges in an ever changing climate. CRC, Boca Raton
144. Robl JA, Sun C-Q et al (1997) Dual metalloprotease inhibitors: mercaptoacetyl-based fused heterocyclic dipeptide mimetics as inhibitors of angiotensin-converting enzyme and neutral endopeptidase. J Med Chem 40:1570–1577
145. Hanson RL, Howell JM et al (2000) Synthesis of allysine ethylene acetal using phenylalanine dehydrogenase from *Thermoactinomyces intermedius*. Enzyme Microb Technol 26:348–358
146. Hanson RL, Schwinden MD et al (1999) Enzymatic synthesis of L-6-hydroxynorleucine. Bioorg Med Chem 7:2247–2252
147. Patel RN (2001) Enzymatic synthesis of chiral intermediates for omapatrilat, an antihypertensive drug. Biomol Eng 17:167–182
148. Urlacher VB, Eiben S (2006) Cytochrome P450 monooxygenases: perspectives for synthetic application. Trends Biotechnol 24:324–330
149. Julsing MK, Cornelissen S et al (2008) Heme-iron oxygenases: powerful industrial biocatalysts? Curr Opin Chem Biol 12:177–186
150. Bernhardt R (2006) Cytochromes P450 as versatile biocatalysts. J Biotechnol 124:128–145

151. Tee KL, Schwaneberg U (2007) Directed evolution of oxygenases: screening systems, success stories and challenges. Comb Chem High Throughput Screen 10:197–217
152. Park J-W, Lee J-K et al (2003) Bioconversion of compactin into pravastatin by *Streptomyces* sp. Biotechnol Lett 25:1827–1831
153. Hosobuchi M, Kurosawa K et al (1993) Application of computer to monitoring and control of fermentation process: microbial conversion of ML-236B Na to pravastatin. Biotechnol Bioeng 42:815–820
154. Matsuoka T, Miyakoshi S et al (1989) Purification and characterization of cytochrome P_{450} from *Streptomyces carbophilus*. Eur J Biochem 184:707–713
155. Ro D-K, Paradise EM et al (2006) Production of the antimalarial drug precursor artemisinic acid in engineered yeast. Nature 440:940–943
156. Ryerson CC, Ballou DP et al (1982) Mechanistic studies on cyclohexanone oxygenase. Biochemistry 21:2644–2655
157. Taschner MJ, Aminbhavi AS (1989) α-Oxygenation of a trans-3,4-disubstituted γ-lactone. A comparative study. Tetrahedron Lett 30:1029–1032
158. Taschner MJ, Black DJ (1988) The enzymatic Baeyer–Villiger oxidation: enantioselective synthesis of lactones from mesomeric cyclohexanones. J Am Chem Soc 110:6892–6893
159. Yokokawa F, Hamada Y et al (1996) Total synthesis of calyculin A – construction of the C (9)–C(37) fragment. Chem Commun 32:871–872
160. Gagnon R, Grogan G et al (1995) Enzymatic Baeyer–Villiger oxidations of some bicyclo [2.2.1]heptan-2-ones using monooxygenases from *Pseudomonas putida* NCIMB 10007: enantioselective preparation of a precursor of azadirachtin. J Chem Soc Perkin Trans 1:1505–1511
161. Veitch GE, Beckmann E et al (2007) A relay route for the synthesis of azadirachtin. Angew Chem Int Ed 46:7633–7635
162. Veitch GE, Beckmann E et al (2007) Synthesis of azadirachtin: a long but successful journey. Angew Chem Int Ed 46:7629–7632
163. Veitch GE, Boyer A et al (2008) The azadirachtin story. Angew Chem Int Ed 47:9402–9429
164. Tanaka M, Mukaiyama C et al (2002) Synthesis of optically pure gomisi lignans: the total synthesis of (+)-schizandrin, (+)-gomisin A, and (+)-isoschizandrin in naturally occurring forms. J Org Chem 60:4339–4352
165. Honda T, Kimura N et al (1994) Chiral synthesis of lignan lactones, (−)-hinokinin, (−)-deoxypodorhizone, (−)-isohibalactone and (−)-savinin by means of enantioselective deprotonation strategy. J Chem Soc Perkin Trans 1:1043–1046
166. Alphand V, Mazzini C et al (1998) A new microorganism for highly stereospecific Baeyer–Villiger oxidation of prochiral cyclobutanones. J Mol Catal B Enzym 5:219–221
167. Mihovilovic MD, Bianchi DA et al (2006) Accessing tetrahydrofuran-based natural products by microbial Baeyer–Villiger biooxidation. Chem Commun 42:3214–3216
168. Hudlicky T, Reed JW (2009) Celebrating 20 years of SYNLETT – special account on the merits of biocatalysis and the impact of arene cis-dihydrodiols on enantioselective synthesis. Synlett 2009:685–703
169. Sullivan B, Hudlicky T (2008) Chemoenzymatic formal synthesis of (−)-balanol. Provision of optical data for an often-reported intermediate. Tetrahedron Lett 49:5211–5213
170. Fabris F, Collins J et al (2009) Investigation of steric and functionality limits in the enzymatic dihydroxylation of benzoate esters. Versatile intermediates for the synthesis of pseudo-sugars, amino cyclitols, and bicyclic ring systems. Org Biomol Chem 7:2619–2627
171. Banwell MG, Austin KAB et al (2007) Chemoenzymatic total syntheses of the linear triquinane-type natural products (+)-hirsutic acid and (−)-complicatic acid from toluene. Tetrahedron 63:6388–6403
172. Pinkerton DM, Banwell MG et al (2009) Chemoenzymatic access to versatile epoxyquinol synthons. Org Lett 11:4290–4293
173. Omori AT, Finn KJ et al (2007) Chemoenzymatic total synthesis of (+)-codeine by sequential intramolecular Heck cyclizations via C–B–D ring construction. Synlett 2007:2859–2862

174. Ley SV, Redgrave AJ (1990) Microbial oxidation in synthesis: concise preparation of (+)-conduritol F from benzene. Synlett 1990:393–394
175. Matveenko M, Willis AC et al (2008) A chemoenzymatic synthesis of the anti-influenza agent Tamiflu®. Tetrahedron Lett 49:7018–7020
176. Shie J-J, Fang J-M et al (2008) A concise and flexible synthesis of the potent anti-influenza agents tamiflu and tamiphosphor. Angew Chem Int Ed 47:5788–5791
177. Sullivan B, Carrera I et al (2009) Symmetry-based design for the chemoenzymatic synthesis of oseltamivir (tamiflu) from ethyl benzoate. Angew Chem Int Ed 48:4229–4231
178. Franke D, Lorbach V et al (2003) (*S*,*S*)-2,3-Dihydroxy-2,3-dihydrobenzoic acid: microbial access with engineered cells of *Escherichia coli* and application as starting material in natural-product synthesis. Chem Eur J 9:4188–4196
179. Franke D, Sprenger GA et al (2001) Synthesis of functionalized cyclohexadiene-*trans*-diols with recombinant cells of *Escherichia coli*. Angew Chem Int Ed 40:555–557
180. Lorbach V, Franke D et al (2002) Cyclohexadiene-*trans*-diols as versatile starting material in natural product synthesis: short and efficient synthesis of *iso*-crotepoxide and *ent*-senepoxide. Chem Commun 38:494–495
181. Hausmann T, Pietruszka J (2009) Enantiopure dicyclopropanes from trans-cyclohexadiene-diols. Synlett 21:3271–3274
182. Mukaiyama T (2004) Explorations into new reaction chemistry. Angew Chem Int Ed 43:5590–5614
183. Evans DA, Bartroli J et al (1981) Enantioselective aldol condensations. 2. Erythro-selective chiral aldol condensations via boron enolates. J Am Chem Soc 103:2127–2129
184. Guillena G, Nájera C et al (2007) Enantioselective direct aldol reaction: the blossoming of modern organocatalysis. Tetrahedron Asymmetry 18:2249–2293
185. Hanson CV, Nishiyama Y et al (2005) Catalytic antibodies and their applications. Curr Opin Biotechnol 16:631–636
186. Kiyooka S-i (1997) Development of a chiral lewis acid-promoted asymmetric aldol reaction using oxaborolidinone. Rev Heteroatom Chem 17:245–270
187. Machajewski TD, Wong C-H (2000) The catalytic asymmetric aldol reaction. Angew Chem Int Ed 39:1352–1375
188. Fessner W-D, Helaine V (2001) Biocatalytic synthesis of hydroxylated natural products using aldolases and related enzymes. Curr Opin Biotechnol 12:574–586
189. Romero A, Wong C-H (2000) Chemo-enzymatic total synthesis of 3-epiaustraline, australine, and 7-epialexine. J Org Chem 65:8264–8268
190. Chênevert R, Dasser M (2000) Chemoenzymatic synthesis of the microbial elicitor (−)-syringolide via a fructose 1,6-diphosphate aldolase-catalyzed condensation reaction. J Org Chem 65:4529–4531
191. Woodhall T, Williams G et al (2005) Synthesis of screening substrates for the directed evolution of sialic acid aldolase: towards tailored enzymes for the preparation of influenza A sialidase inhibitor analogues. Org Biomol Chem 3:1795–1800
192. Woodhall T, Williams G et al (2005) Creation of a tailored aldolase for the parallel synthesis of sialic acid mimetics. Angew Chem Int Ed 44:2109–2112
193. Steinreiber J, Fesko K et al (2007) Synthesis of γ-halogenated and long-chain β-hydroxy-α-amino acids and 2-amino-1,3-diols using threonine aldolases. Tetrahedron 63:8088–8093
194. Liu J, Wong C-H (2002) Aldolase-catalyzed asymmetric synthesis of novel pyranose synthons as a new entry to heterocycles and epothilones. Angew Chem Int Ed 41:1404–1407
195. Liu J, Hsu C-C et al (2004) Sequential aldol condensation catalyzed by DERA mutant Ser238Asp and a formal total synthesis of atorvastatin. Tetrahedron Lett 45:2439–2441
196. Jennewein S, Schürmann M et al (2006) Directed evolution of an industrial biocatalyst: 2-deoxy-D-ribose 5-phosphate aldolase. Biotechnol J 1:537–548
197. Effenberger F (1994) Synthesis and reactions of optically active cyanohydrins. Angew Chem Int Ed 33:1555–1564

198. Gregory RJH (1999) Cyanohydrins in nature and the laboratory: biology, preparations, and synthetic applications. Chem Rev 99:3649–3682
199. Effenberger F, Eichhorn J et al (1995) Enzyme catalyzed addition of hydrocyanic acid to substituted pivalaldehydes – a novel synthesis of (*R*)-pantolactone. Tetrahedron Asymmetry 6:271–282
200. Pscheidt B, Liu Z et al (2008) Efficient biocatalytic synthesis of (*R*)-pantolactone. Adv Synth Catal 350:1943–1948
201. Vashkov VI, Volkov YP et al (1968) Investigation of pyrethrins and related compounds. XV. Esters of *trans*-2,2-dichloro-3-arylcyclopropanecarboxylic acids as insecticides. Pharm Chem J 2:17–20
202. Breuer M, Ditrich K et al (2004) Industrial methods for the production of optically active intermediates. Angew Chem Int Ed 43:788–824
203. Lu W, Chen P et al (2008) New stereoselective synthesis of thiamphenicol and florfenicol from enantiomerically pure cyanohydrin: a chemo-enzymatic approach. Tetrahedron 64:7822–7827
204. Müller M, Gocke D et al (2009) Thiamin diphosphate in biological chemistry: exploitation of diverse thiamin diphosphate-dependent enzymes for asymmetric chemoenzymatic synthesis. FEBS J 276:2894–2904
205. Pohl M (1997) Protein design on pyruvate decarboxylase (PDC) by site-directed mutagenesis. Adv Biochem Eng Biotechnol 58:15–43
206. Rosche B, Breuer M et al (2004) Biphasic aqueous/organic biotransformation of acetaldehyde and benzaldehyde by *Zymomonas mobilis* pyruvate decarboxylase. Biotechnol Bioeng 86:788–794
207. Pohl M, Lingen B et al (2002) Thiamin-diphosphate-dependent enzymes: new aspects of asymmetric C–C bond formation. Chem Eur J 8:5288–5295
208. Demir AS, Sesenoglu O et al (2003) Benzaldehyde lyase-catalyzed enantioselective carboligation of aromatic aldehydes with mono- and dimethoxy acetaldehyde. Org Lett 5:2047–2050

Top Curr Chem (2010) 297: 45–66
DOI: 10.1007/128_2009_22

Published online: 21 January 2010

Engineered Biosynthesis of Plant Polyketides: Structure-Based and Precursor-Directed Approach

Ikuro Abe

Abstract Pentaketide chromone synthase (PCS) and octaketide synthase (OKS) are novel plant-specific type III polyketide synthases (PKSs) obtained from *Aloe arborescens*. Recombinant PCS expressed in *Escherichia coli* catalyzes iterative condensations of five molecules of malonyl-CoA to produce a pentaketide 5,7-dihydroxy-2-methylchromone, while recombinant OKS carries out sequential condensations of eight molecules of malonyl-CoA to yield octaketides SEK4 and SEK4b, the longest polyketides produced by the structurally simple type III PKS. The amino acid sequences of PCS and OKS are 91% identical, sharing 50–60% identity with those of other chalcone synthase (CHS) superfamily type III PKSs of plant origin. One of the most characteristic features is that the conserved active-site Thr197 of CHS (numbering in *Medicago sativa* CHS) is uniquely replaced with Met207 in PCS and with Gly207 in OKS, respectively. Site-directed mutagenesis and X-ray crystallographic analyses demonstrated that the chemically inert single residue lining the active-site cavity controls the polyketide chain length and the product specificity depending on the steric bulk of the side chain. On the basis of the crystal structures, an F80A/Y82A/M207G triple mutant of the pentaketide-producing PCS was constructed and shown to catalyze condensations of nine molecules of malonyl-CoA to produce an unnatural novel nonaketide naphthopyrone, whereas an N222G mutant of the octaketides-producing OKS yielded a decaketide benzophenone SEK15 from ten molecules of malonyl-CoA. On the other hand, the type III PKSs exhibited broad substrate specificities and catalytic potential. OKS accepted *p*-coumaroyl-CoA as a starter substrate to produce an unnatural novel C_{19} hexaketide stilbene and a C_{21} heptaketide chalcone. Remarkably, the C_{21} chalcone-forming activity was dramatically increased in the structure-guided OKS N222G mutant. In addition, OKS N222G mutant also yielded unnatural novel polyketides from phenylacetyl-CoA and benzoyl-CoA as a starter substrate. These results

I. Abe
Graduate School of Pharmaceutical Sciences, The University of Tokyo, 7-3-1 Hongo, Bunkyo-ku, Tokyo, 113-0033, Japan
e-mail: abei@mol.f.u-tokyo.ac.jp

suggested that the engineered biosynthesis of plant polyketides by combination of the structure-based and the precursor-directed approach would lead to further production of chemically and structurally divergent unnatural novel polyketides.

Keywords Chalcone synthase superfamily enzyme, Engineered biosynthesis, Precursor-directed biosynthesis, Structure-based engineering, Type III polyketide synthase

Contents

Abbreviations

2PS	2-Pyrone synthase
ALS	Aloesone synthase
BAS	Benzalacetone synthase
CHS	Chalcone synthase
CUS	Curcumin synthase
OKS	Octaketide synthase
PCS	Pentaketide synthase
PKS	Polyketide synthase
STS	Stilbene synthase
TAL	Triaceticacid lactone
THNS	Tetrahydroxynaphthalene synthase

1 Introduction

The chalcone synthase (CHS) (EC 2.3.1.74) superfamily of type III Polyketide synthases (PKSs) are pivotal enzymes in the biosynthesis of plant polyphenols. They are structurally and mechanistically different from the modular type I and the dissociated type II PKSs of bacterial origin; the simple homodimer of 40–45 kDa proteins performs a complete series of decarboxylation, condensation, cyclization, and aromatization reactions with a single active-site by utilizing CoA-linked substrates without involvement of the acyl carrier proteins [1–3]. For example, CHS is the ubiquitous plant-specific type III PKS that selects *p*-coumaroyl-CoA as a starter molecule and carries out three iterative decarboxylative condensations with the malonyl-CoA to produce a tetraketide naringenin chalcone, the biosynthetic precursor of flavonoids (Fig. 1) [4]. On the other hand, 2-pyrone synthase (2PS) selects acetyl-CoA as a starter, and catalyzes only two condensations with malonyl-CoA to produce a triketide triacetic acid lactone (TAL) (Fig. 1) [5, 6]. In addition, a growing number of functionally divergent plant type III PKSs have been cloned and characterized, including stilbene synthase (STS) [7, 8], acridone synthase (ACS) [9–11], benzalacetone synthase (BAS) [12–15], aloesone synthase (ALS) [16, 17], and curcumin synthase (CUS) [18–20] (Fig. 1). Moreover, several bacterial type III PKSs such as 1,3,6,8-tetrahydroxynaphthalene synthase (THNS or RppA) have been also reported [21, 22].

Recent crystallographic and site-directed mutagenesis studies have revealed that type III PKSs share a common three-dimensional overall fold with a conserved Cys-His-Asn catalytic triad [4, 6, 7, 23–29]. The polyketide assembly reaction is thus initiated by a starter molecule loading at the active-site Cys, which is followed by iterative decarboxylative Claisen-type condensations of malonyl-CoA and final cyclization and aromatization of the enzyme-bound reactive β-polyketo intermediate (Fig. 2). Most importantly, the enzyme utilizes a single SH group of a catalytic Cys, and each chain elongation involves cleavage of the once formed C–S thioester bond and the subsequent formation of a new C–C bond by insertion of an additional C_2 unit to generate a CoA-linked intermediate. As a result, the CHS-superfamily of type III PKSs produce a variety of polyphenol scaffolds including chalcone, stilbene, phloroglucinol, resorcinol, benzophenone, biphenyl, naphthalene, chromone, isocoumarin, acridone, quinolinone, pyrone, and curcumin, etc. The remarkable functional diversity of the type III PKSs derives from the differences in the selection of the starter molecules, the number of the malonyl-CoA condensations, and the mechanisms of cyclization reactions [1–3].

2 Aloe Polyketide Synthases

Aloe (*Aloe arborescens*) is a medicinal plant rich in aromatic polyketides such as pharmaceutically important aloenin (a hexaketide pyrone), aloesin (a heptaketide chromone), and barbaloin (an octaketide anthrone) (Fig. 4a). Pentaketide chromone synthase (PCS) and octaketide synthase (OKS) are novel plant-specific type III PKSs, which were obtained from the aloe plant by RT-PCR cloning using degenerate oligonucleotide primers based on the conserved sequences of known CHS enzymes [30–33]. The deduced amino acid sequences of PCS and OKS are 91% identical (368/403), and show 50–60% identity to those of other CHS superfamily type III PKSs of plant origin; OKS shares 60% identity (240/403) with CHS from

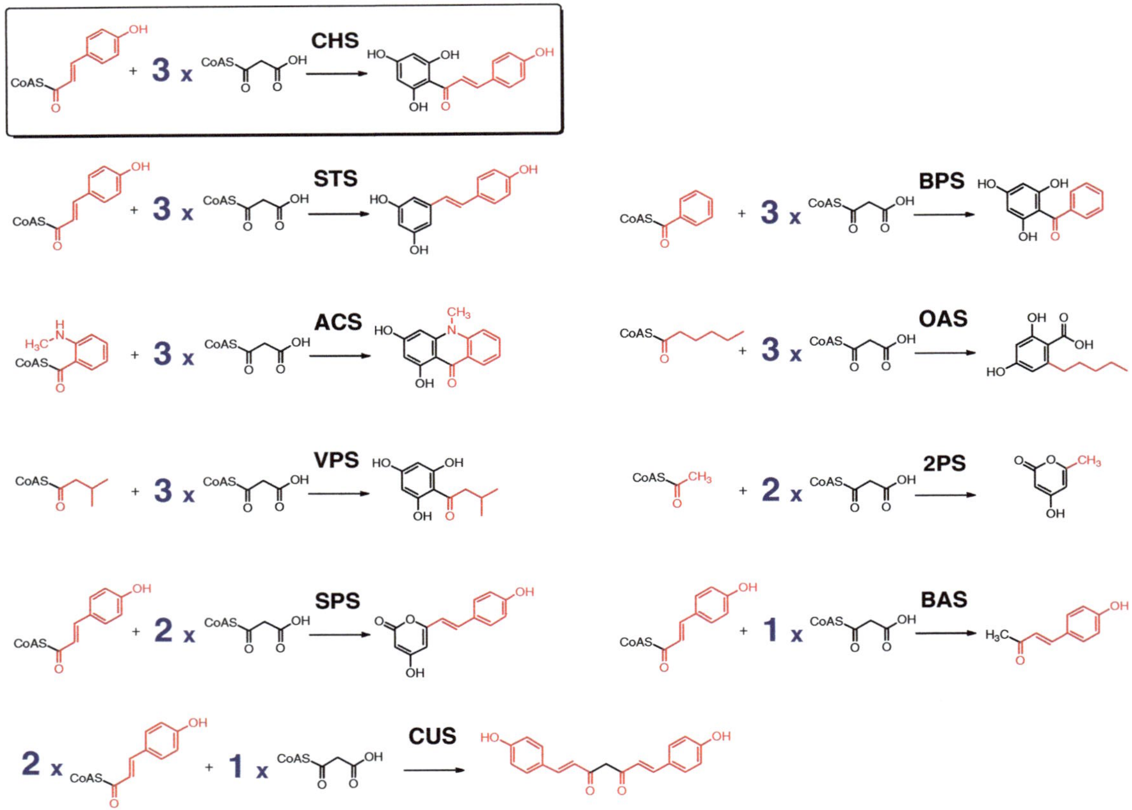

Fig. 1 Biosynthesis of plant polyphenols by CHS-superfamily type III PKSs. *CHS* chalcone synthase; *STS* stilbene synthase; *ACS* acridone synthase; *VPS* valerophenone synthase; *SPS* styrylpyrone synthase; *BPS* benzophenone synthase; *OAS* olivetolic acid synthase; *2PS* 2-pyrone synthase; *BAS* benzalacetone synthase; *CUS* curcumin synthase

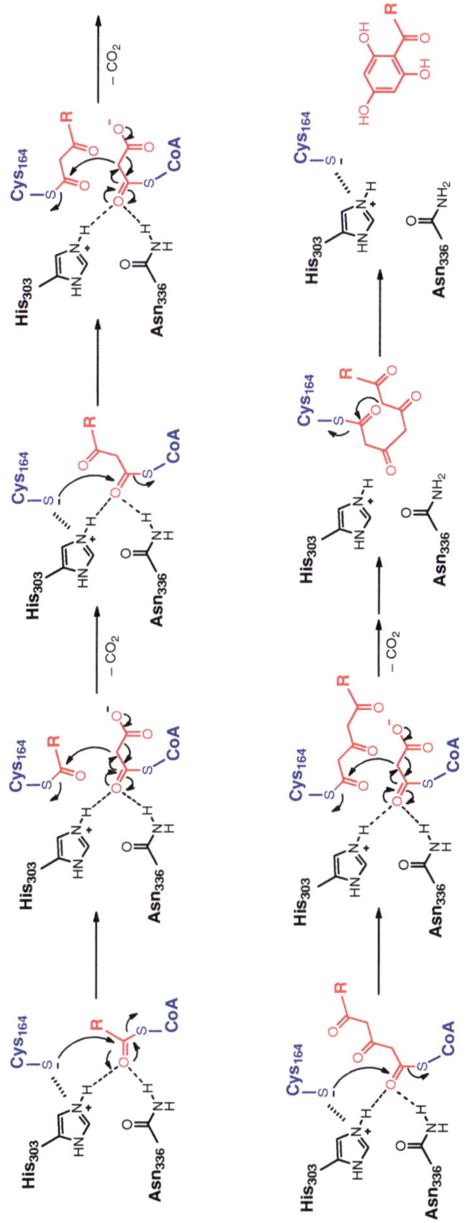

Fig. 2 Proposed mechanism of the enzyme reactions of type III PKSs

alfalfa (*Medicago sativa*) [4], and 54% identity (216/403) with a heptaketide-producing ALS from rhubarb (*Rheum palmatum*) [16, 17] (Fig. 3). In contrast, PCS and OKS show only low sequence similarity with type III PKSs of bacterial origin; OKS shows only 23% identity (93/403) with THNS from *Streptomyces griseus* [21, 22].

Recombinant PCS expressed in *Escherichia coli* does not produce naringenin chalcone from *p*-coumaroyl-CoA, but instead efficiently accepts malonyl-CoA as a sole substrate to carry out iterative condensations of five molecules of malonyl-CoA, producing a pentaketide 5,7-dihydroxy-2-methylchromone (Fig. 4b) [30]. The aromatic pentaketide is a biosynthetic precursor of the anti-asthmatic furochromones, kehellin and visnagin [30]. On the other hand, recombinant OKS catalyzes sequential condensations of eight molecules of malonyl-CoA to yield a 1:4 mixture of aromatic octaketides, SEK4 and SEK4b (Fig. 4c) [31]. The octaketides are known to be shunt products of the minimal type II PKS for actinorhodin (*act* from *Streptomyces coelicolor,* a heterodimeric complex of ketosynthase and chain length factor) [34, 35]. Since the aloe plant does not produce SEK4/SEK4b and their metabolites, but produces significant amount of anthrones and anthraquinones (octaketides), it is tempting to speculate that OKS is involved in the biosynthesis of anthrones/anthraquinones in the medicinal plant (Fig. 4c). However, maybe because of the absence of interactions with tailoring enzymes such as yet unidentified ketoreductase, OKS would have afforded the oktaketides as shunt products as in the case of the minimal type II PKS. The physiological function of OKS in the medicinal plant remains to be elucidated. The steady-state kinetics analysis revealed that the recombinant PCS showed a K_M value of 71.0 μM and k_{cat} value of 445×10^{-3} min^{-1} for malonyl-CoA, with a broad pH optimum within a range of 6.0–8.0, while OKS showed a K_M value of 95.0 μM and k_{cat} value of 94.0×10^{-3} min^{-1} with a pH optimum at 7.5. Interestingly, both PCS and OKS also accept acetyl-CoA, resulting from decarboxylation of malonyl-CoA, as a starter substrate for the enzyme reaction [30, 31].

Recently, three additional novel type III PKSs (PKS3, PKS4, and PKS5) were cloned and sequenced from the aloe plant [36]. These enzymes share 85–96% amino acid sequence identity with PCS and OKS. Recombinant PKS4 and PKS5 expressed in *E. coli* were functionally identical to OKS, catalyzing sequential condensations of eight molecules of malonyl-CoA to produce octaketides SEK4/SEK4b. As in the case of OKS, PKS4 and PKS5 are possibly involved in the biosynthesis of the octaketide barbaloin. On the other hand, PKS3 is a multifunctional enzyme that produces a heptaketide aloesone as a major product from seven molecules of malonyl-CoA. In addition, PKS3 also afforded a hexaketide 6-(2,4-dihydroxy-6-methylphenyl)-4-hydroxy-2-pyrone, heptaketide pyrones TW93a and 6-(2-(2,4-dihydroxy-6-methylphenyl)-2-oxoethyl)-4-hydroxy-2-pyrone, and octaketides SEK4/SEK4b (Fig. 4d). Notably, the heptaketide aloesone and the hexaketide pyrone are biosynthetic precursors of the anti-inflammatory *C*-glucoside aloesin and the anti-histamic *O*-glucoside aloenin, respectively. This is the first demonstration that a wild-type type III PKS obtained from *A. arborescens* catalyzed enzymatic formations of the precursors of aloesin and aloenin, the

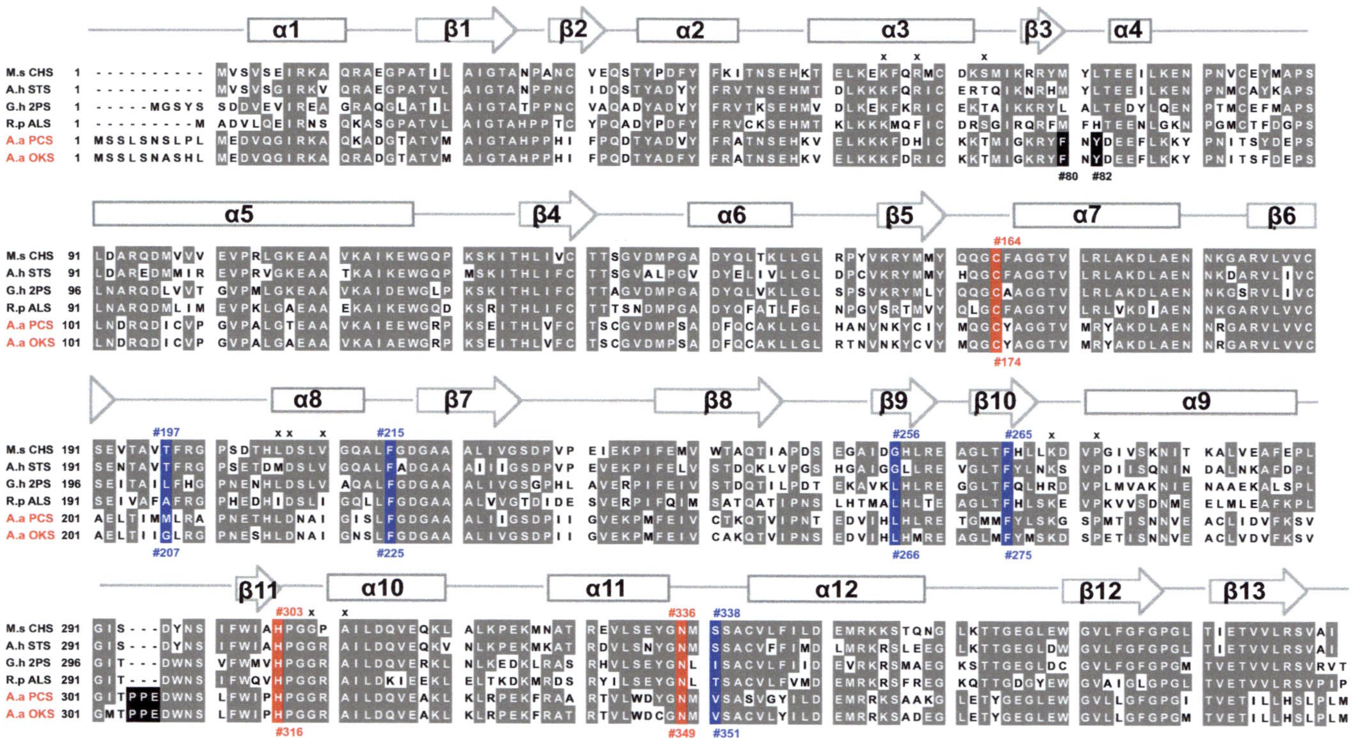

Fig. 3 Comparison of the amino acid sequences of *A. arborescens* PCS and OKS with other CHS superfamily type III PKSs. *M.s CHS M. sativa* CHS; *A.h STS Arachis hypogaea* stilbene synthase; *G.h 2PS G. hybrida* 2PS; *R.p ALS R. palmatum* ALS; *A.a PCS A. arborescens* PCS; *A.a OKS A. arborescens* OKS. The critical active-site residue 197, the catalytic triad (Cys164, His303, and Asn336), and the residues lining the active-site (Phe215, Gly256, F265, Ser338) are marked with # (numbering in *M. sativa* CHS), and residues for the CoA binding with +. Phe80 and Tyr82 of *A. arborescens* PCS are also marked. α-Helices (*rectangles*) and β-strands (*arrows*) of CHS are diagrammed

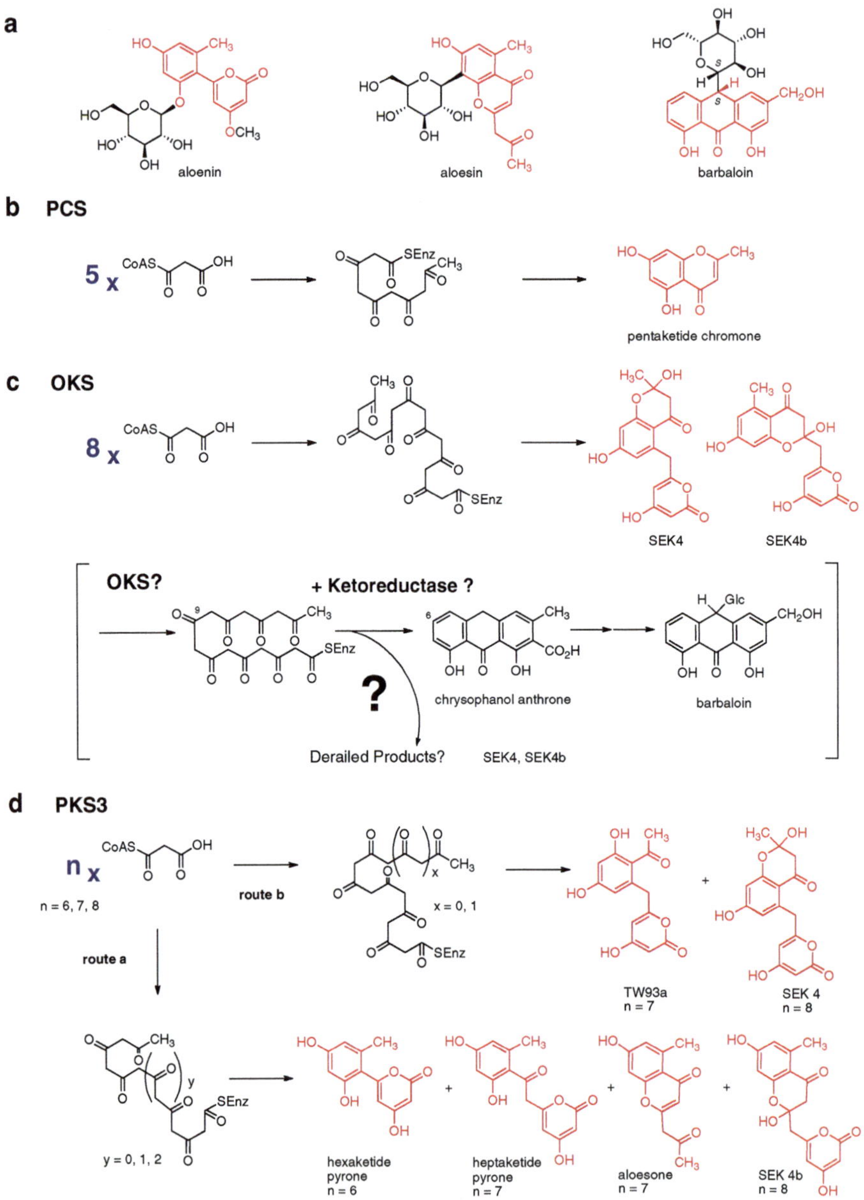

Fig. 4 (**a**) Structures of aromatic polyketides produced by *A. arborescens*. (**b**) Proposed enzyme reaction mechanism of PCS, (**c**) OKS, and (**d**) PKS3. A hypothetical scheme for the involvement of OKS and as yet unidentified ketoreductase in the biosynthesis of anthrones and anthraquinones is also included

active principles of the medicinal plant. Recently, another octaketides SEK4/SEK4b-producing type III PKS has been reported from *Hypericum perforatum* [37], whereas novel type III PKSs from *Drosophyllum lusitanicum* [38] and *Plumbago indica* [39] catalyzed sequential condensations of six molecules of malonyl-CoA to produce the hexaketide pyrone, the precursor of aloenin.

3 Sequence Analysis and Site-Directed Mutagenesis

Sequence analysis revealed that the catalytic triad (Cys164, His303, and Asn336) and most of the CHS active-site residues (Met137, Gly211, Gly216, Pro375, and the CHS "gatekeepers" Phe215 and Phe265) (numbering in *M. sativa* CHS) [4] are conserved in *A. arborescens* PCS and OKS (Fig. 3). One of the most characteristic features is that the conserved residues, Thr197, Gly256, and Ser338, are uniquely replaced with T197M/G256L/S338V in PCS and T197G/G256L/S338V in OKS, respectively [30, 31]. The three residues lining the active-site cavity are sterically altered in a number of functionally different type III PKSs; T197L/G256L/S338I in the triketide TAL-producing *Gerbera hybrida* 2PS [6] and T197A/G256L/S338T in the heptaketide-producing *R. palmatum* ALS [17]. On the basis of the crystal structures of *M. sativa* CHS and *G. hybrida* 2PS, Noel and coworkers have proposed that the three residues play a crucial role for controlling starter substrate selectivity and the numbers of malonyl-CoA condensations by steric modulation of the active-site cavity [6]. Thus, a T197L/G256L/S338I triple mutant of *M. sativa* CHS was shown to be functionally identical to 2PS, producing TAL from acetyl-CoA starter and two molecules of malonyl-CoA [6]. Another interesting feature is that both *A. arborescens* PCS and OKS retain a characteristic three-amino acid "PPE" insertion at 10 residues upstream from the catalytic His303 (Fig. 3). A similar six-amino acid "SIDSII" insertion at the same position has also been reported for coumaroyltriacetic acid synthase [40] (or stilebenecarboxylate synthase [41]) from *Hydrangea macrophylla*.

To investigate the structure-function relationship between the pentaketide-producing PCS and the octaketide-producing OKS, sharing the 91% amino acid sequence identity, we carried out site-directed mutagenesis on the active-site residue 207 (corresponding to Thr197 in *M. sativa* CHS) which is uniquely replaced with Met207 in PCS and with Gly207 in OKS, respectively (Fig. 3). We first constructed a PCS mutant in which Met207 was replaced with small Gly as in the case of OKS. To our surprise, there was a dramatic change in the enzyme activity; PCS M207G mutant no longer produced the pentaketide chromone from five molecules of malonyl-CoA, but instead efficiently catalyzed condensations of eight molecules of malonyl-CoA to produce a 1:4 mixture of SEK4/SEK4b, just as in the case of OKS (Fig. 5) [30]. The pentaketide-producing PCS was thus functionally converted to OKS by the single amino acid substitution. On the other hand, when Gly207 in OKS was substituted with bulky Met as in PCS, OKS G207M mutant produced a pentaketide 2,7-dihydroxy-5-methylchromone instead

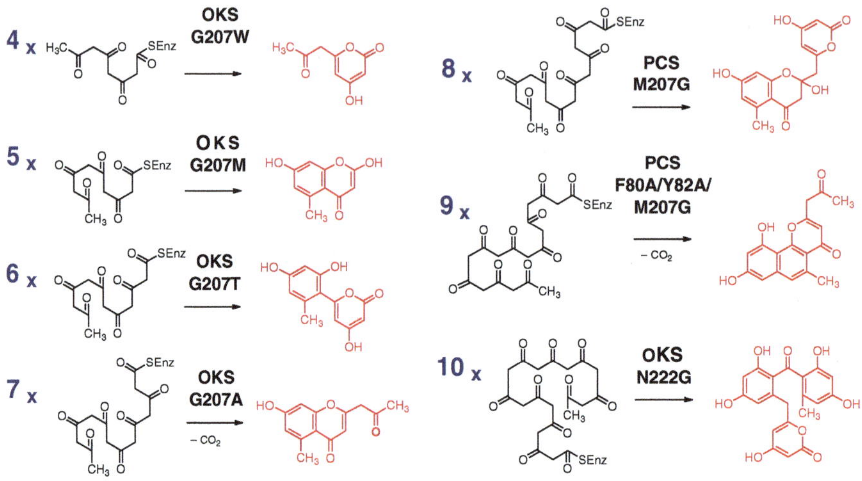

Fig. 5 Proposed mechanism of the formation of unnatural polyketides by site-directed mutants of *A. arborescens* PCS and OKS

of SEK4/SEK4b (Fig. 5) [31]. Thus, OKS was functionally converted to a PCS by the small-to-large amino acid substitution. Here it should be noted that the pentaketide product is a regioisomer of the PCS's pentaketide chromone. In the OKS mutant, a C-4/C-9 aldol-type cyclization yielded 2,7-dihydroxy-5-methylchromone, while in PCS a C-1/C-6 Claisen-type cyclization produces 5,7-dihydroxy-2-methylchromone. This suggests subtle structural differences of the active-site architecture between the two enzymes.

To test further a hypothesis that the single residue 207 lining the active-site cavity controls the polyketide chain length and the product specificity, we constructed a set of *A. arborescens* OKS mutants (G207A, G207T, G207L, G207F, and G207W) and investigated the mechanistic consequences of the point mutations (Fig. 5) [31]. As a result, OKS G207A mutant, in which Gly207 was replaced with Ala as in the case of the heptaketide-producing *A. arborescens* PKS3 [36] and *R. palmatum* ALS [16], indeed produced the heptaketide aloesone, the biosynthetic precursor of the anti-inflammatory aloesin, in addition to SEK4/SEK4b. On the other hand, when Gly207 was substituted with Thr as in CHS, the G207T mutation abolished the octaketide-forming activity and yielded the hexaketide pyrone, the precursor of the anti-histamic aloenin, from six molecules of malonyl-CoA. Furthermore, other bulky substitutions G207L and G207F yielded the pentaketide 2,7-dihydroxy-5-methylchromone, whereas the most bulky G207W only afforded a tetraketide and a triketide pyrone TAL, without formation of an aromatic ring system.

The small-to-large substitutions of the residue 207 (corresponding to Thr197 in *M. sativa* CHS) thus resulted in loss of the octaketide-producing activity and the concomitant formation of shorter chain length products, ranging from triketide to heptaketide, including the precursors of aloenin and aloesin. These results clearly established that the steric bulk of the chemically inert single residue lining the active-site cavity controls the numbers of malonyl-CoA condensations and the polyketide chain length [31]. Presumably, the single-amino acid replacement causes steric contraction of the active-site pocket that accommodates the growing polyketide intermediates, leading to the shortening of the product chain length. It is remarkable that such a simple steric modulation of the active-site cavity causes a dramatic change in the catalytic activity. Mechanistically, the enzymes are thought to catalyze the chain elongation and the initial aldol-type aromatic ring formation at the methyl end of the β-polyketo intermediates within the active-site cavity. After the C–S thioester bond cleavage from the catalytic Cys, the partially cyclized intermediates would then be released from the active-site and undergo spontaneous α-pyrone ring formations, thereby completing the construction of the fused-ring systems [31].

4 Crystallographic Analysis and Structure-Based Engineering

Crystallization and X-ray analysis of *A. arborescens* PCS, both the pentaketide-producing wild-type enzyme and the octaketide-producing M207G mutant, were carried out at 1.6 Å resolution (Fig. 6) [42, 43]. The overall folding structures of PCS are nearly identical to those of previously reported *M. sativa* CHS [4] and *G. hybrida* 2PS [6]. The Cys-His-Asn catalytic triad is buried deep inside, and sits at an intersection of the CoA binding tunnel and a large internal cavity, in a location and orientation very similar to those of the other plant type III PKSs. The CoA-binding tunnel connects to the protein surface, which enables the substrates to get into the catalytic center. Further, surprisingly, most of the active-site residues of *A. arborescens* PCS and *M. sativa* CHS are superimposable in nearly identical positions. The only exceptions are Cys143, Thr204, Met207, Leu266, and Val351 of *A. arborescens* PCS (corresponding to Ser133, Thr194, Thr197, Gly256, and Ser338 of *M. sativa* CHS, respectively), which causes steric contraction of the active-site cavity and loss of the coumaroyl-binding pocket. The three residues, Met207, Leu266, and Val351, sterically altered in a number of functionally divergent type III PKSs, are thought to be crucial in controlling the substrate and product specificities of the enzyme reactions (Fig. 6). Finally, a comparison of the structures with *M. sativa* CHS and *G. hybrida* 2PS revealed a significant backbone change in residues 296–309 of PCS (corresponding to residues 286–296 of *M. sativa* CHS and 291–301 of *G. hybrida* 2PS, respectively), which is due to the characteristic insertion of the "PPE" residues 304–306, and the addition of ten residues at the *N*-terminus [42].

On the other hand, the X-ray crystal structure of the PCS M207G mutant clearly demonstrated that location and orientation of the residues lining the active-site cavity in the mutant are almost perfectly conserved as in the wild-type enzyme. However, the cavity volume is dramatically increased because of newly formed additional novel pockets that extend into the floor of the active-site of the wild-type

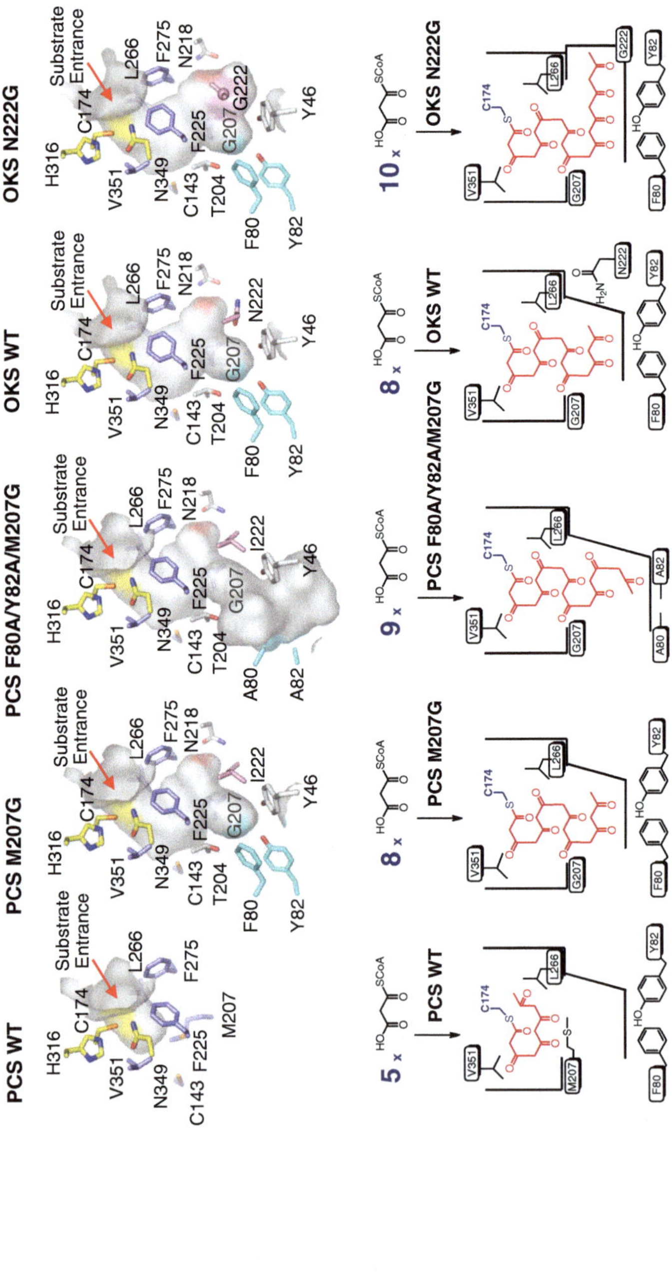

Fig. 6 Comparison and schematic representation of the active-site cavities of *A. arborescens* PCS and OKS, and their site-directed mutants

PCS (Fig. 6) [42]. As a result, the total cavity volume of the point mutant is 649 $Å^3$, which is 2.6-fold larger than that of the wild-type enzyme (247 $Å^3$), and almost as large as that of *M. sativa* CHS (754 $Å^3$). Thus, in the wild-type PCS, the bulky Met207 blocks the entrance of the pockets, whereas in the M207G mutant, the large-to-small amino acid substitution opens a gate to the buried pockets, thereby increasing the polyketide chain elongation up to eight condensations instead of five condensations of malonyl-CoA. The octaketide-producing activity of the M207G mutant is totally dependent upon the presence of the novel buried pockets and the growing linear polyketide chains are likely to extend into these pockets [42].

In addition to the "downward expanding" M207G substitution, the active-site architecture of PCS features a "horizontally restricting" Leu266 substitution (Fig. 6). The residue 266 lining the initiation and elongation cavity (corresponding to Gly256 in *M. sativa* CHS [4]) occupies a crucial position for the loading of the starter substrate, and is uniquely replaced with bulky Leu266 in both PCS and OKS, as in the case of *G. hybrida* 2PS [6]. As a result, both PCS and OKS no longer accept *p*-coumaroyl-CoA as a starter substrate to produce chalcone, but instead efficiently accept the smaller malonyl-CoA starter to initiate the iterative decarboxylative condensation reactions. In fact, OKS L266G mutant was shown to accept *p*-coumaroyl-CoA as a starter to produce a tetraketide pyrone, 4-coumaroyltriacetic acid lactone, after three condensations with malonyl-CoA [17]. It was thus confirmed that the 2PS-like G256L substitution indeed controls the starter substrate selectivity in *A. arborescens* PCS and OKS.

On the other hand, the active-site residue 351 (corresponding to Ser338 in *M. sativa* CHS), located in the neighborhood of the catalytic Cys at the "ceiling" of the active-site cavity (Fig. 6), also plays a crucial role in the polyketide chain elongation reactions [17, 44]. On the basis of recent structural analyses of *Pinus sylvestris* STS, Noel and Schröder proposed that a hydrogen bond network involving Ser338-H_2O-Thr132-Glu192 (numbering in *M. sativa* CHS) plays a critical role in balancing the competing cyclization specificities in CHS (Claisen-type) and STS (aldol-type) from a common tetraketide intermediate [7]. In *A. arborescens* PCS and OKS, however, the conserved Ser338 is uniquely substituted with hydrophobic Val351, and such hydrogen bond network and the nucleophilic water molecule are not observed in the crystal structures [42, 45]. In PCS and OKS, the hydrophobic Val351 is likely to guide the course of the chain extension so that the linear β-polyketo intermediate extends into the "downward expanding" pocket, thereby leading to the formation of the pentaketide and the octaketides, respectively.

Very interestingly, it has been recently demonstrated that a point mutant of *Scuterallia baicalensis* CHS in which Ser338 was replaced with Val as in the case of PCS and OKS produced a trace amount of the octaketides SEK4/SEK4b in addition to the triketide TAL [44]. Moreover, the octaketides-producing activity was dramatically increased in an OKS-like T197G/G256L/S338V triple mutant of *S. baicalensis* CHS [44]. Close examination of the crystal structure of *M. sativa* CHS [4] revealed that the presence of a similar buried pocket that extends into the "floor" of the conventional active-site cavity. It is likely that, just as in the case of PCS M207G mutant, the S338V mutation would have lead the polyketide intermediate chain to extend into this hidden pocket. Most of the chain elongations were however terminated at the triketide stage to yield TAL as a dominant product. It is remarkable that CHS, the ubiquitous plant type III PKS, can be easily engineered to produce the octaketides SEK4/SEK4b by such a simple steric modulation of the active-site cavity [44].

To manipulate further the pentaketide-producing PCS enzyme reaction, we constructed an F80A/Y82A/M207G triple mutant in which the active-site cavity of the octaketide-producing PCS M207G mutant was expanded by simultaneously substituting the two aromatic residues, Phe80 and Tyr82, forming the bottom face of the novel buried pocket, with small Ala (Fig. 6) [46]. A homology model predicted that the total cavity volume of the triple mutant (1030 $Å^3$) is increased four times larger than that of the wild-type enzyme (247 $Å^3$). The triple mutant was functionally expressed in *E. coli* at levels comparable with wild-type enzyme, and shown to produce an unnatural novel nonaketide naphthopyrone (0.2% yield, tenfold less efficient than that of the pentaketide formation by the wild-type enzyme) in addition to octaketides SEK4/SEK4b (Fig. 5) [46]. Interestingly, the structure of the nonaketide naphthopyrone showed close similarity to those of heptaketides aloesone and 6-hydroxymusizin from *R. palmatum* [47], and an octaketide eleutherinol isolated from another medicinal plant *Eleutherine bulbosa* [48], which suggested that these aromatic polyketides are also produced by closely related CHS-superfamily type III PKSs.

Here the mechanism of the formation of the nonaketide naphthopyrone involves iterative condensation of nine molecules of malonyl-CoA and a sequential C–C bond formation by intramolecular aldol-type condensations [46]. The enzyme is likely to catalyze the chain elongation and the first aromatic ring formation reaction from a linear nonaketide intermediate. After the thioester bond cleavage from the catalytic Cys, the partially cyclized aromatic intermediates would be released from the active-site and undergo subsequent cyclization, thereby producing the angular naphthopyrone skeleton with a fused tricyclic ring system. Most of the reactions were, however, terminated at the octaketide stage to afford SEK4/SEK4b as major products. Notably, Harris and coworkers have reported that β-polyketones having the two terminal carbonyl groups protected as ketals are efficiently and directly cyclized into aromatic polyketides including 6-hydroxymusizin and eleutherinol [49, 50]. On the other hand, despite the structural similarity to eleutherinol, formation of the "octaketide" naphthopyrone was not detected with either PCS M207G mutant or the triple mutant, suggesting that the naphthalene ring-forming activity was only attained by the F80A/Y82A/M207G triple mutation.

Crystallization and X-ray analysis of the octaketide-producing *A. arborescens* OKS has also been achieved [45]. The crystal structure at 2.6 Å resolution revealed that OKS shares almost identical active-site architecture with PCS M207G mutant catalyzing condensations of eight molecules of malonyl-CoA to produce the octaketides SEK4/SEK4b (Fig. 6). Thus, both OKS and PCS M207G mutant utilize the "downward expanding" pockets for the formation of SEK4/SEK4b. To manipulate the OKS enzyme reaction, a structure-based OKS N222G mutant, in which the active-site cavity was expanded by a large-to-small substitution of Asn222 located at the bottom of the "downward expanding" pocket (Fig. 6). As a result, OKS N222G mutant efficiently produced a C_{20} decaketide benzophenone SEK15 as a single product by sequential condensations of ten molecules of malonyl-CoA (Fig. 5) [51]. The mutant showed a K_M value of 55.0 μM and k_{cat} value of 2.7×10^{-3} min^{-1} for malonyl-CoA with a pH optimum at 7.0. The decaketide benzophenone has been previously reported as a product of genetically engineered type II PKSs [52], and is the longest polyketide produced by the structurally simple type III PKS. Very recently, the X-ray crystal structure of OKS N222G mutant was obtained, which clearly demonstrated that the active-site cavity was indeed expanded by the single amino acid substitution; the total cavity volume of the point mutant was increased to 693 $Å^3$ from 651 $Å^3$.

5 Substrate Specificity and Precursor Directed Biosynthesis

The CHS-superfamily type III PKSs exhibit unusually broad, promiscuous substrate specificities; the structurally simple homodimeric enzymes accept a variety of non-physiological substrates, including aromatic and aliphatic CoA thioesters, to produce an array of chemically and structurally divergent unnatural polyketides [53–62]. We have previously reported that *S. baicalensis* CHS, normally producing chalcone by sequential condensation of the C_6–C_3 unit of *p*-coumaroyl-CoA with three molecules of malonyl-CoA, shows remarkably broad substrate specificity (Fig. 7a) [53, 54]. For example, 4-fluorocinnamoyl-CoA was converted to a fluorochalcone along with a triketide and a tetraketide α-pyrone by-product, while other 4-substituted analogs (X=Cl, Br, and OCH_3) afforded only triketide and tetraketide pyrones. Further, it is remarkable that analogs in which the coumaroyl aromatic ring is replaced by furan or thiophene were efficiently converted into unnatural novel polyketides with the heteroaromatic ring system. The steric and/or electronic perturbations by the substituents alter the stability of the enzyme-bound intermediate or the optimally folded conformation in the cyclization pocket of the active-site cavity. In addition, CHS also accepts benzoyl-CoA and phenylacetyl-CoA to afford phlorobenzophenone and phlorobenzylketone, respectively [54]. Moreover, CHS accepts aliphatic CoA starters (isovaleryl, isobutyryl, *n*-hexanoyl, *n*-octanoyl, *n*-decanoyl, and *n*-dodecanoyl) to produce tetraketide phloroglucinols as well as triketide and tetraketide pyrones [58]. When incubated with methylmalonyl-CoA and *p*-coumaroyl-CoA, CHS afforded an unnatural C_6–C_5 aromatic polyketide, 1-(4-hydroxyphenyl)pent-1-en-3-one, formed by one-step decarboxylative condensation of the two substrates (Fig. 7b) [56]. On the other hand, CHS does not accept the shorter and the bulkier *N*-methylanthraniloyl-CoA as a starter. However, benzalacetone synthase (BAS) from *R. palmatum*, a type III PKS normally producing a diketide benzalacetone by one-step condensation with malonyl-CoA, efficiently catalyzes condensation of *N*-methylanthraniloyl-CoA (or anthraniloyl-CoA) and one molecule of malonyl-CoA (or methylmalonyl-CoA) to produce 4-hydroxy-2 (1*H*)-quinolones (Fig. 7c) [60]. The anthranilates are key intermediates in the

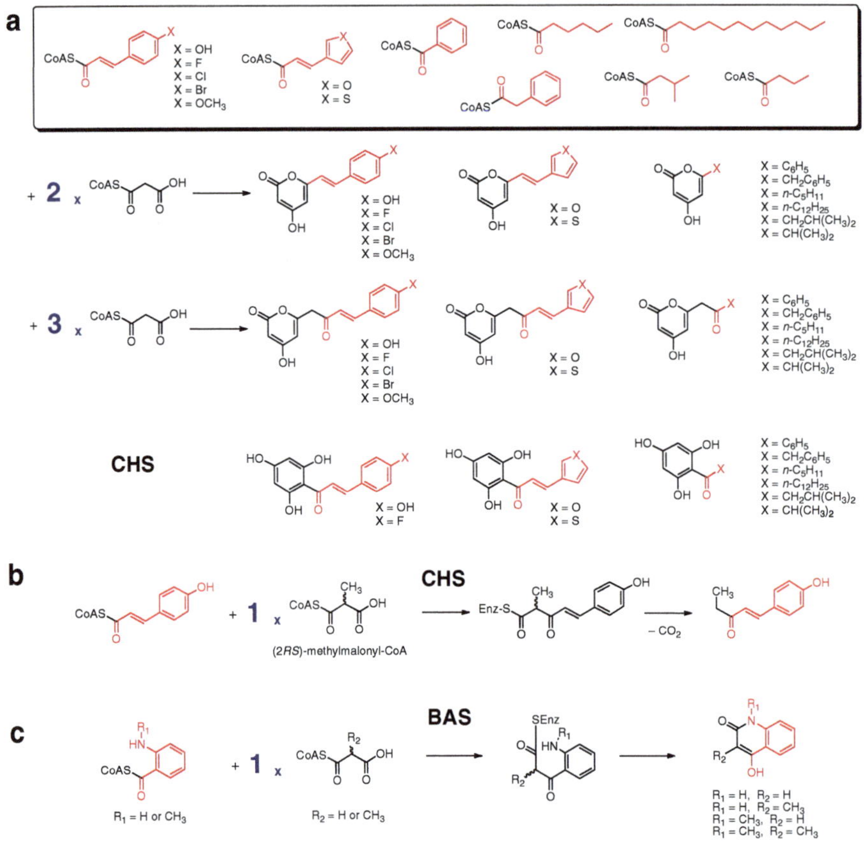

Fig. 7 Enzyme reaction products from (**a**) various starter substrate analogs, (**b**) extender substrate analog by CHS, and (**c**) enzymatic formation of quinolinones by BAS

biosynthesis of acridone alkaloids; acridone synthase is a plant-specific type III PKS that selects *N*-methylanthraniloyl-CoA as a starter to produce 1,3-dihydroxy-*N*-methylacridone after three condensations with malonyl-CoA (Fig. 1).

Like other type III PKSs, the octaketide-producing *A. arborescens* OKS exhibits significantly broad substrate specificities. The enzyme accepts disparate starter molecules, both aromatic and aliphatic CoA thioesters with different chain lengths, and efficiently performs sequential condensation and cyclization reactions to produce a series of unnatural polyketides [31]. For example, when OKS was incubated with *p*-coumaroyl-CoA and malonyl-CoA as substrates, most of the enzyme reactions were initiated by malonyl-CoA and the octaketides SEK4/SEK4b were obtained as major products. However, careful examination of the reaction products led to isolation of two less polar minor products; octaketide-producing OKS also catalyzed condensations of *p*-coumaroyl-CoA and five or six molecules of malonyl-CoA to produce an unnatural C_{21} heptaketide chalcone and a C_{19} hexaketide stilbene, novel polyketide scaffolds generated by the structurally simple type III PKS (Fig. 8a) [51]. Remarkably, the C_{21} chalcone-forming activity is dramatically increased in the structure-guided OKS N222G mutant that produces a C_{20} decaketide SEK15 from ten molecules of malonyl-CoA (Fig. 5) [51].

As described, in the CHS enzyme reaction, the C_6–C_3 coumaroyl starter is loaded onto the coumaroyl binding pocket; however, in OKS and PCS, the G256L and S338V substitutions causes the loss of the pocket from the active-site. As a result, OKS and PCS no longer accept *p*-coumaroyl-CoA as a starter substrate to produce the tetraketide chalcone, but instead, OKS utilize the novel buried pockets that extend into the traditional active-site cavity for the production of

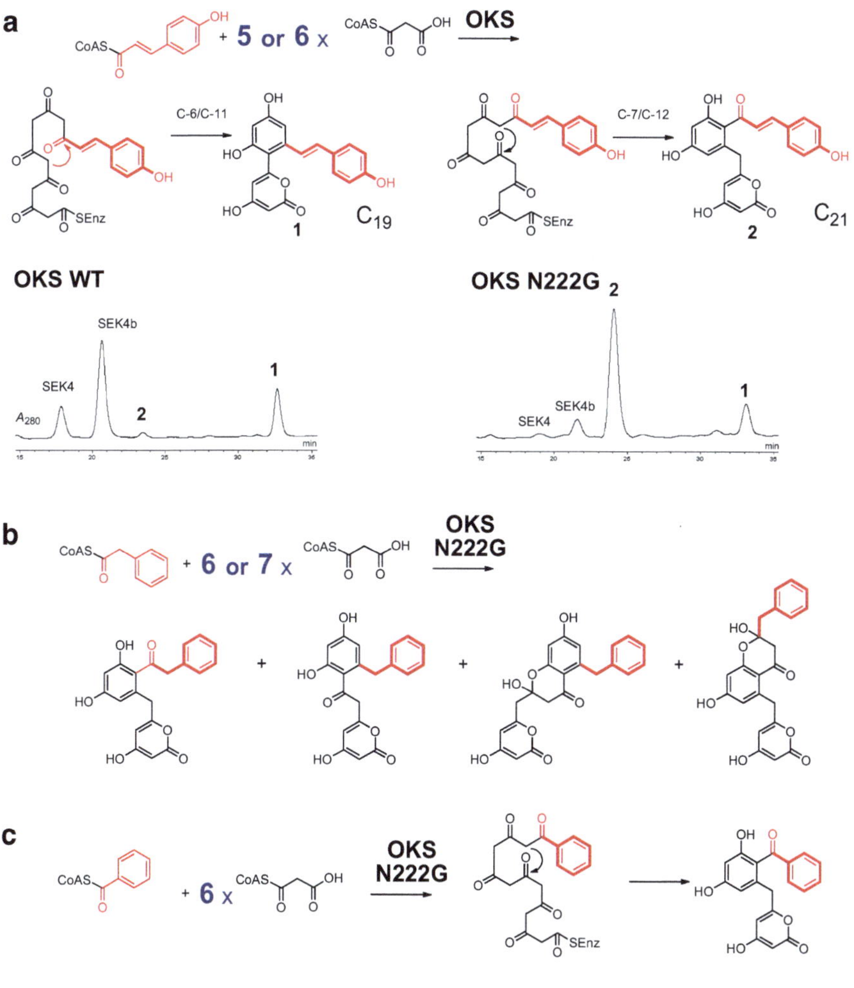

Fig. 8 Enzymatic formation of unnatural polyketides (and HPLC elution profiles) from (**a**) 4-coumaroyl-CoA, (**b**) phenylacetyl-CoA, and (**c**) benzoyl-CoA by *A. arborescens* wild-type OKS and N222G mutant

SEK4/SEK4b (Fig. 6) [42]. Therefore, the production of the coumaroyl-derived heptaketide/hexaketide is thought also to be dependent upon the presence of the novel buried pockets. Indeed, the structure-based OKS N222G mutant, in which the buried pockets is expanded by the large-to-small substitution of Asn222, efficiently produces the C_{21} heptaketide chalcone as a major product from the coumaroyl starter. Presumably, after the chain elongation, the enzyme catalyzes the first aromatic ring formation at the middle of the polyketide intermediate as in the case of the production of SEK4/SEK4b and SEK15. The C_{21} chalcone is thus generated by a C-7/C-12 aldol-type cyclization of a heptaketide intermediate, whereas a C-6/C-11 aldol-type cyclization of a hexaketide intermediate produces the C_{19} stilbene (Fig. 8a). The formation of the terminal α-pyrone ring could be an important process for the release of the polyketide products from the thioester-linked active-site Cys residue [51]. On the other hand, the structure-guided OKS N222G mutant produced novel polyketides including C_{22} aromatic octaketides from phenylacetyl-CoA (Fig. 8b), and a novel C_{19} heptaketide benzophenone from benzoyl-CoA (Fig. 8c) [63]. Notably, the C_6–C_2 phenylacetyl starter better fits into the active-site of the enzyme and more efficiently initiates the sequential condensation reactions than the C_6–C_1 benzoyl-CoA. The combination of the structure-based engineering and the precursor-directed biosynthesis can be a powerful approach for further production of chemically and structurally divergent unnatural novel polyketides.

6 Conclusion and Perspectives

The site-directed mutagenesis and X-ray crystallographic studies on the pentaketide-producing PCS and the octaketide-producing OKS clearly demonstrated that the chemically inert single residue lining the active-site cavity determines the polyketide chain length and the product specificity. It is remarkable that the functional diversity of the type III PKSs evolved from such a simple steric modulation of the active-site architecture. Further, the structure-based rationally designed mutant enzymes, PCS F80A/Y82A/M207G and OKS N222G mutant, dramatically extended the product chain length to produce an unnatural novel nonaketide and a decaketide, respectively, which are the longest polyketides generated by the structurally simple type III PKS so far. On the other hand, the CHS-superfamily type III PKSs exhibit unusually broad, promiscuous substrate specificities, accepting a variety of non-physiological substrates to produce an array of chemically and structurally divergent unnatural novel polyketides. These findings revolutionized our concept of the catalytic potential of the structurally simple type III PKSs and suggested possible involvement in the biosynthesis of structurally diverse plant polyketides such as anthrones and anthraquinones. Further analyses of the catalytic potential and plasticity of the functionally divergent type III PKSs promise to reveal intimate structural details of the enzyme-catalyzed processes.

Our results suggested that the engineered biosynthesis of plant polyketides would lead to further production of chemically and structurally divergent unnatural novel polyketides. Since we can now manage to control the starter substrate specificity and the number of the malonyl-CoA condensations, the next goal would be how to manipulate the mechanism of the cyclization reactions to generate polyketides with desired ring systems. On the other hand, the β-polyketo intermediates are highly reactive and readily react with amines to yield Schiff bases, which make it possible to introduce additional C–C or C–N bond forming chemistry such as the Mannich-type cyclization to generate more complex enzyme reaction products. Combination of the structure-based protein engineering and the precursor-directed biosynthesis with rationally designed nitrogen-containing synthetic analogs is expected to generate unnatural novel polyketide-alkaloid scaffolds with promising biological activities, which is now in progress in our laboratories.

Acknowledgments I would like to expresses sincere appreciation to an excellent group of coworkers whose contributions are cited in the text, in particular to Professor Hiroyuki Morita (The University of Tokyo), Dr. Toshiyuki Kohno (Mitsubishi Kagaku Institute of Life Sciences), and Professor Hiroshi Noguchi (University of Shizuoka). This work was supported by the PRESTO program from Japan Science and Technology Agency, Grant-in-Aid for Scientific Research from the Ministry of Education, Culture, Sports, Science and Technology, Japan, and Research Grants from the Uehara Memorial Foundation.

References

1. Schröder J (1999) The chalcone/stilbene-synthase family of condensing enzymes. In: Sankawa U (ed) Comprehensive natural products chemistry, vol 1. Pergamon, Oxford, pp 749–771
2. Austin MB, Noel JP (2003) The chalcone synthase superfamily of type III polyketide synthases. Nat Prod Rep 20:79–110
3. Abe I, Morita H (2010) Structure and function of the chalcone synthase superfamily of plant type III polyketide synthases. Nat Prod Rep 27: in press
4. Ferrer JL, Jez JM, Bowman ME, Dixon RA, Noel JP (1999) Structure of chalcone synthase and the molecular basis of plant polyketide biosynthesis. Nat Struct Biol 6:775–784
5. Eckermann S, Schröder G, Schmidt J, Strack D, Edrada RA, Helariutta Y, Elomaa P, Kotilainen M, Kilpeläinen I, Proksch P, Teeri TH, Schröder J (1998) New pathway to polyketides in plants. Nature 369:387–390
6. Jez JM, Austin MB, Ferrer J, Bowman ME, Schröder J, Noel JP (2000) Structural control of polyketide formation in plant-specific polyketide synthases. Chem Biol 7:919–930
7. Austin MB, Bowman ME, Ferrer J-L, Schröder J, Noel JP (2004) An aldol switch discovered in stilbene synthases mediates cyclization specificity of Type III polyketide synthases. Chem Biol 11:1179–1194
8. Shomura Y, Torayama I, Suh DY, Xiang T, Kita A, Sankawa U, Miki K (2005) Crystal structure of stilbene synthase from *Arachis hypogaea*. Proteins 60:803–806
9. Lukacin R, Springob K, Urbanke C, Ernwein C, Schröder G, Schröder J, Matern U (1999) Native acridone synthases I and II from *Rura graveolens* L. form homodimers. FEBS Lett 448:135–140

10. Wanibuchi K, Zhang P, Abe T, Morita H, Kohno T, Chen G, Noguchi H (2007) An acridone-producing novel multifunctional type III polyketide synthase from *Huperzia serrata*. FEBS J 274:1073–1082
11. Morita H, Kondo S, Kato R, Wanibuchi K, Noguchi H, Sugio S, Abe I, Kohno T (2007) Crystallization and preliminary crystallographic analysis of an acridone-producing novel multifunctional type III polyketide synthase from *Huperzia serrata*. Acta Crystallograph Sect F Struct Biol Cryst Commun 63:576–587
12. Abe I, Takahashi Y, Morita H, Noguchi H (2001) Benzalacetone synthase: a novel polyketide synthase that plays a crucial role in the biosynthesis of phenylbutanones in *Rheum palmatum*. Eur J Biochem 268:3354–3359
13. Abe I, Sano Y, Takahashi Y, Noguchi H (2003) Site-directed mutagenesis of benzalacetone synthase: the role of Phe215 in plant type III polyketide synthases. J Biol Chem 278:25218–25226
14. Abe T, Morita H, Noma H, Kohno T, Noguchi H, Abe I (2007) Structure function analysis of benzalacetone synthase from *Rheum palmatum*. Bioorg Med Chem Lett 17:3161–3166
15. Morita H, Tanio M, Kondo S, Kato R, Wanibuchi K, Noguchi H, Sugio S, Abe I, Kohno T (2008) Crystallization and preliminary crystallographic analysis of a plant type III polyketide synthase that produces benzalacetone. Acta Crystallograph Sect F Struct Biol Cryst Commun 64:304–306
16. Abe I, Utsumi Y, Oguro S, Noguchi H (2004) The first plant type III polyketide synthase catalyzing formation of aromatic heptaketide. FEBS Lett 562:171–176
17. Abe I, Watanabe T, Lou W, Noguchi H (2006) Active site residues governing substrate selectivity and polyketide chain length in aloesone synthase. FEBS J 273:208–218
18. Katsuyama Y, Matsuzawa M, Funa N, Horinouchi S (2007) In vitro synthesis of curcuminoids by type III polyketide synthase from *Oryza sativa*. J Biol Chem 282:37702–37709
19. Katsuyama Y, Kita T, Funa N, Horinouchi S (2009) Curcuminoid biosynthesis by two type III polyketide synthases in the herb *Curcuma longa*. J Biol Chem 284:11160–11170
20. Katsuyama Y, Kita T, Horinouchi S (2009) Identification and characterization of multiple curcumin synthases from the herb *Curcuma longa*. FEBS Lett 583:2799–2803
21. Funa N, Ohnishi Y, Fujii I, Shibuya M, Ebizuka Y, Horinouchi S (1999) A new pathway for polyketide synthesis in microorganisms. Nature 400:897–899
22. Austin MB, Izumikawa M, Bowman ME, Udwary DW, Ferrer JL, Moore BS, Noel JP (2004) Crystal structure of a bacterial type III polyketide synthase and enzymatic control of reactive polyketide intermediates. J Biol Chem 279:45162–45174
23. Jez JM, Ferrer JL, Bowman ME, Dixon RA, Noel JP (2000) Dissection of malonyl-coenzyme A decarboxylation from polyketide formation in the reaction mechanism of a plant polyketide synthase. Biochemistry 39:890–902
24. Lanz T, Tropf S, Marner FJ, Schröder J, Schröder G (1991) The role of cysteines in polyketide synthases. Site-directed mutagenesis of resveratrol and chalcone synthases, two key enzymes in different plant-specific pathways. J Biol Chem 266:9971–9976
25. Jez JM, Noel JP (2000) Mechanism of chalcone synthase. pKa of the catalytic cysteine and the role of the conserved histidine in a plant polyketide synthase. J Biol Chem 275:39640–39646
26. Jez JM, Bowman ME, Noel JP (2001) Structure-guided programming of polyketide chain-length determination in chalcone synthase. Biochemistry 40:14829–14838
27. Tropf S, Kärcher B, Schröder G, Schröder J (1995) Reaction mechanisms of homodimeric plant polyketide syntahses (stilbene synthase and chalcone synthase). J Biol Chem 270: 7922–7928
28. Suh DY, Fukuma K, Kagami J, Yamazaki Y, Shibuya M, Ebizuka Y, Sankawa U (2000) Identification of amino acid residues important in the cyclization reactions of chalcone and stilbene synthases. Biochem J 350:229–235
29. Suh DY, Kagami J, Fukuma K, Sankawa U (2000) Evidence for catalytic cysteine-histidine dyad in chalcone synthase. Biochem Biophys Res Commun 275:725–730

30. Abe I, Utsumi Y, Oguro S, Morita H, Sano Y, Noguchi H (2005) A plant Type III polyketide synthase that produces pentaketide chromone. J Am Chem Soc 127:1362–1363
31. Abe I, Oguro S, Utsumi Y, Sano Y, Noguchi H (2005) Engineered biosynthesis of plant polyketides: chain length control in an octaketide-producing plant Type III polyketide synthase. J Am Chem Soc 127:12709–12716
32. Abe I (2007) Engineered biosynthesis of plant polyketides. ACS Symp Ser 955:105–127
33. Abe I (2008) Engineering of plant polyketide biosynthesis. Chem Pharm Bull 56:1505–1514
34. Fu H, Ebert-Khosla S, Hopwood DA, Khosla C (1994) Engineered biosynthesis of novel polyketides: dissection of the catalytic specificity of the *act* ketoreductase. J Am Chem Soc 116:4166–4170
35. Fu H, Hopwood DA, Khosla C (1994) Engineered biosynthesis of novel polyketides: evidence for temporal, but not regiospecific, control of cyclization of an aromatic polyketide precursor. Chem Biol 1:205–210
36. Mizuuchi Y, Shi S-P, Wanibuchi K, Kojima A, Morita H, Noguchi H, Abe I (2009) Novel type III polyketide synthases from *Aloe arborescens*. FEBS J 276:2391–2401
37. Karppinen K, Hokkanen J, Mattila S, Neubauer P, Hohtola A (2008) Octaketide-producing type III polyketide synthase from *Hypericum perforatum* is expressed in dark glands accumulating hypericins. FEBS J 275:4329–4342
38. Jindaprasert A, Springob K, Schmidt J, De-Eknamkul W, Kutchan TM (2008) Pyrone polyketides synthesized by a type III polyketide synthase from *Drosophyllum lusitanicum*. Phytochemistry 69:3043–3053
39. Springob K, Samappito S, Jindaprasert A, Schmidt J, Page JE, De-Eknamkul W, Kutchan TM (2007) A polyketide synthase of *Plumbago indica* that catalyzes the formation of hexaketide pyrones. FEBS J 274:406–417
40. Akiyama T, Shibuya M, Liu HM, Ebizuka Y (1999) *p*-Coumaroyltriacetic acid synthase, a new homologue of chalcone synthase, from *Hydrangea macrophylla* var. *thunbergii*. Eur J Biochem 263:834–839
41. Eckermanna C, Schröder G, Eckermann S, Strack D, Schmidt J, Schneider B, Schröder J (2003) Stilbenecarboxylate biosynthesis: a new function in the family of chalcone synthase-related proteins. Phytochemistry 62:271–286
42. Morita H, Kondo S, Oguro S, Noguchi H, Sugio S, Abe I, Kohno T (2007) Structural insight into chain-length control and product specificity of pentaketide chromone synthase from *Aloe arborescens*. Chem Biol 14:359–369
43. Morita H, Kondo S, Abe T, Noguchi H, Sugio S, Abe I, Kohno T (2006) Crystallization and preliminary crystallographic analysis of a novel plant type III polyketide synthase that produces pentaketide chromone. Acta Crystallograph Sect F Struct Biol Cryst Commun 62:899–901
44. Abe I, Watanabe T, Morita H, Kohno T, Noguchi H (2006) Engineered biosynthesis of plant polyketides: manipulation of chalcone synthase. Org Lett 8:499–502
45. Morita H, Kondo S, Kato R, Wanibuchi K, Noguchi H, Sugio S, Abe I, Kohno T (2007) Crystallization and preliminary crystallographic analysis of an octaketide-producing plant type III polyketide synthase. Acta Crystallograph Sect F Struct Biol Cryst Commun 63:947–949
46. Abe I, Morita H, Oguro S, Noma H, Wanibuchi K, Kawahara N, Goda Y, Noguchi H, Kohno T (2007) Structure-based engineering of a plant type III polyketide synthase: formation of an unnatural nonaketide naphthopyron. J Am Chem Soc 129:5976–5980
47. Tsuboi M, Minami M, Nonaka G, Nishioka I (1977) 1977) Studies on rhubarb (*Rhei rhizoma)*. IV. Chem Pharm Bull 25:2708–2712
48. Birch AJ, Donovan FW (1953) Studies in relation to biosynthesis. III. Structure of eleutherinol. Aust J Chem 6:373–378
49. Harris TM, Wittek PJ (1975) Biogenetic-type syntheses of polycyclic polyketide metabolites using partially protected β-hexa- and β-heptaketones. 6-Hydroxymusizin, barakol, emodin, and eleutherinol. J Am Chem Soc 97:3270–3271

50. Harris TM, Harris CM (1986) Biomimetic syntheses of aromatic polyketide metabolites. Pure Appl Chem 58:283–294
51. Shi S-P, Wanibuchi K, Morita H, Endo K, Noguchi H, Abe I (2009) Enzymatic formation of unnatural novel chalcone, stilbene, and benzophenone scaffolds by plant type III polyketide synthase. Org Lett 11:551–554
52. Yu T-W, Shen Y, McDaniel R, Floss HG, Khosla C, Hopwood DA, Moore BS (1998) Engineered biosynthesis of novel polyketides from *Streptomyces* spore pigment polyketide synthases. J Am Chem Soc 120:7749–7759
53. Abe I, Morita H, Nomura A, Noguchi H (2000) Substrate specificity of chalcone synthase: enzymatic formation of unnatural polyketides from synthetic cinnamoyl-CoA analogues. J Am Chem Soc 122:11242–11243
54. Morita H, Takahashi Y, Noguchi H, Abe I (2000) Enzymatic formation of unnatural aromatic polyketides by chalcone synthase. Biochem Biophys Res Commun 279:190–195
55. Morita H, Noguchi H, Schröder J, Abe I (2001) Novel polyketides synthesized with a higher plant stilbene synthase. Eur J Biochem 268:3759–3766
56. Abe I, Takahashi Y, Noguchi H (2002) Enzymatic formation of an unnatural C_6–C_5 aromatic polyketide by plant type III polyketide synthases. Org Lett 4:3623–3626
57. Abe I, Takahashi Y, Lou W, Noguchi H (2003) Enzymatic formation of unnatural novel polyketides from alternate starter and non-physiological extension substrate by chalcone synthase. Org Lett 5:1277–1280
58. Abe I, Watanabe T, Noguchi H (2004) Enzymatic formation of long-chain polyketide pyrones by plant type III polyketide synthases. Phytochemistry 65:2447–2453
59. Oguro S, Akashi T, Ayabe S, Noguchi H, Abe I (2004) Probing biosynthesis of plant polyketides with synthetic *N*-acetylcysteamine thioesters. Biochem Biophys Res Commun 325:561–567
60. Abe I, Abe T, Wanibuchi K, Noguchi H (2006) Enzymatic formation of quinolone alkaloids by a plant type III polyketide synthase. Org Lett 8:6063–6065
61. Abe T, Noma H, Noguchi H, Abe I (2006) Enzymatic formation of an unnatural methylated triketide by plant type III polyketide synthases. Tetrahedron Lett 47:8727–8730
62. Jez JM, Bowman ME, Noel JP (2002) Expanding the biosynthetic repertoire of plant type III polyketide synthases by altering starter molecule specificity. Proc Natl Acad Sci USA 99:5319–5324
63. Shi S-P, Wanibuchi K, Morita H, Noguchi H, Abe I (2009) Enzymatic formation of unnatural novel polyketide scaffolds by plant-specific type III polyketide synthase. Tetrahedron Lett 50:2150–2153

Top Curr Chem (2010) 297: 67–103
DOI: 10.1007/128_2010_80

Published online: 13 July 2010

Enzymatic and Chemo-Enzymatic Approaches Towards Natural and Non-Natural Alkaloids: Indoles, Isoquinolines, and Others

Joachim Stöckigt, Zhong Chen, and Martin Ruppert

Abstract The multi-step enzyme catalysed biosyntheses of monoterpenoid indole and isoquinoline alkaloids are described. Special emphasis is placed on those pathways leading to alkaloids of pharmacological and medicinal significance which have been fully elucidated at the enzyme level. The successful identification and cloning of cDNAs of single enzymes and their application provides great opportunities to develop novel strategies for both *in vitro* and *in vivo* alkaloid production in whole plants or tissue cultures, as well as in microbial systems such as *Escherichia coli* and yeast.

Enzyme crystallisation, 3D analyses and site-directed mutation allowed rational engineering of enzyme substrate acceptance, which in turn can be used for re-programming *in vivo* alkaloid biosynthesis and for the design of biomimetic alkaloid syntheses. These strategies broaden structural diversity and allow the creation of large libraries of unnatural alkaloid with expected optimised or novel biological activities. The chemo-enzymatic syntheses of the above-mentioned alkaloid groups and their precursors (in addition to selected examples of other alkaloid families) provides an overview of how enzyme reactions are integrated into the development of total chemical syntheses.

Keywords Alkaloids · Chemo-enzymatic synthesis · Enzymatic formation · Generation of diversity · Genetic approaches

---

J. Stöckigt (✉) and Z. Chen
College of Pharmaceutical Sciences, Zhejiang University, Zijingang Campus, 388 Yu Hang Tang Road, Hangzhou 310058, P. R. China
e-mail: joesto2000@yahoo.com

M. Ruppert
Institute of Pharmacy, Johannes Gutenberg-University, Staudinger Weg 5, 55128, Mainz, Germany

Contents

1 Introduction

> Once organic chemistry has succeeded in a more complicated research field, the proteins, then the chemical riddle of life can be solved.
>
> (E. Fischer, in his speech after he received his Nobel Price Award)

The use of enzymes in organic chemistry dates back over 100 years, when the hydrolysis of racemic mandelic acid by liver esterase was investigated [1, 2]. The application of enzymes to the synthesis of natural products reached breakthrough for some groups of secondary metabolites, such as alkaloids, in the last two decades. With a few exceptions, most of the literature relating to enzymatic synthesis or alkaloid transformation has been published in the last decade. An increase in the number of papers in very recent years indicates a strong growing interest in the field of chemo-enzymatic alkaloid syntheses. However, it is important to distinguish between the two topics (1) biosynthesis of alkaloids at the enzyme level, which allows enzyme-catalysed synthesis of alkaloids and their derivatives, and (2) chemo-enzymatic approaches using a combination of enzymatic and chemical reactions for alkaloid synthesis.

This review does not focus on the former topic, though detection of the individual enzymes involved in each pathway through biosynthesis related research provides the prerequisites for understanding the basic biochemistry of biosynthetic routes, ultimately delivering the tools for genetic and synthetic application of these

---

Dedicated to Professor Handong Sun on the occasion of his 70th birthday.

catalysts. Rather, major examples are then selected to illustrate the latter topic and are described in order to provide an overview of how to generate various alkaloid structures by combination of classical and modern enzymatic reactions.

Within the natural products field, the investigation of complete biosynthetic pathways at the enzyme level has been especially successful for plant alkaloids of the monoterpenoid indole alkaloid family generated from the monoterpene glucoside secologanin and decarboxylation product of tryptophan, tryptamine [3–5]. The most comprehensive enzymatic data are now available for the alkaloids ajmalicine (raubasine) from *Catharanthus roseus,* and for ajmaline from Indian *Rauvolfia serpentina* [6]; the latter alkaloid with a six-membered ring system bearing nine chiral carbon atoms. Entymatic data exsist also for vindoline, the vincaleucoblastin (VLB) precursor for which some studies on enzymatic coupling of vindoline and catharanthine exist in order to get the so-called dimeric *Catharanthus* indole-alkaloids VLB or vincristine [7–9] with pronounced anti-cancer activity [10, 11].

For most of this work, optimised *in vitro* plant cell suspension cultures were used, especially from *R. serpentina* and *C. roseus*, and in the latter case, differentiated tissue (seedling) was also successfully used for investigation of various aspects of vindoline biosynthesis [12, 13].

Biosynthetic research relating to the isoquinoline family was extremely successful, with such important members as morphine [3, 14], codeine [3, 15] or berberine [3, 14, 16–18]. Extensive efforts have provided details pertaining to multiple sets of enzymes participating in the biosynthesis of the alkaloids above, in many cases with the help of plant cell suspension culture techniques. Since 1988, when the breakthrough in cloning of cDNA from alkaloid biosynthesis occurred [19, 20], a significant number of enzymes known from the indoles' and isoquinolines' biosynthesis have been isolated, their biochemical properties described and the majority of their corresponding cDNAs cloned and functionally over-expressed in non-plant hosts such as *Escherichia coli*, yeast or insect cells.

Moreover, significant knowledge has accumulated relating to the pathway leading to the diterpenoid alkaloid taxol, particularly in the isolation and characterisation of a whole set of technically challenging cytochrome P450 enzymes in addition to the enzymatic cyclisation of the diterpene unit, several acyltransferases and for cDNA cloning [21, 22]. Significant progress has also been made in unravelling the key reactions in the biosynthesis of tropine and pseudotropine alkaloids [23, 24]. The chapter will not place particular focus on these alkaloid groups.

2 Enzymatic Formation of Monoterpenoid Indole Alkaloids

2.1 *Heteroyohimbine/Sarpagine/Ajmaline Type Alkaloids*

Ajmalicine (=raubasine) is a heteroyohimbine alkaloid with positive effects on postischemic hypoxia and cerebral protection [25–27] and has anti-hypertensive, sedative and oxygen supplying activities in brain tissue. It can be enzymatically obtained along with its isomers (structures not shown) 19-*epi*-ajmalicine and tetrahydroalstonine. Plant cell suspension cultures producing a broad range of different classes of monoterpene indole alkaloids, but optimised for the production of ajmalicine, were the key source for the individual enzymes required for heteroyohimbine biosynthesis [28, 29]. This was one of the earliest alkaloid pathways to be characterised at the enzymatic level (summarised in Fig. 1) and also allowed the first enzymatic synthesis of the heteroyohimbines, including most of the intermediates [30, 31]. The first reaction is catalysed by strictosidine synthase (STR), the first known enzyme to catalyse a Pictet–Spengler reaction [32, 33], the central reaction required for the formation of the entire alkaloid family, which consists of about 2,000 naturally occurring members. The second enzyme in the pathway is strictosidine glucosidase, which delivers the highly

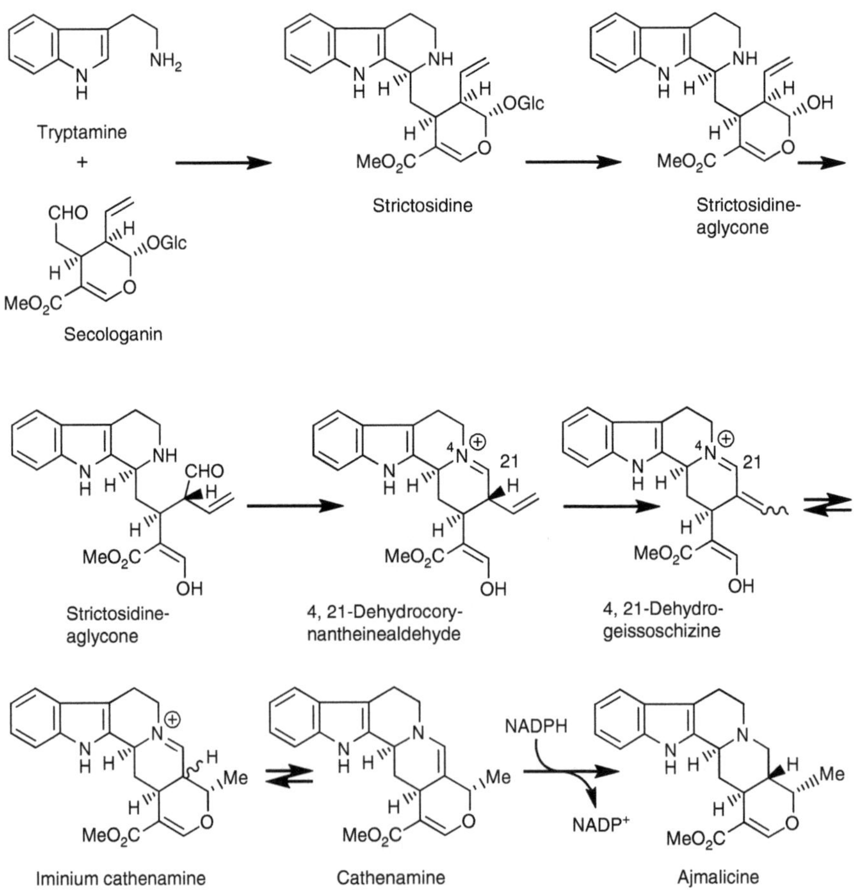

Fig. 1 First example of a complete biosynthetic pathway to an alkaloid (ajmalicine = raubasine) based on enzymatic reactions elucidated by isolation of the single enzymes from *Catharanthus roseus* cell suspension cultures

reactive aglycone that, through intermediate dehydrocorynantheine aldehyde, forms the tetracyclic dehydrogeissoschizine [31]. Enzyme assisted and pH-dependent spontaneous ring closure by intra-molecular Michael addition then leads to the five-ring heteroyohimbine system (cathenamine), which is finally reduced enzymatically in the presence of NADPH as a cofactor, a reaction step which finalises the enzymatic formation of, e.g. ajmalicine [31, 34]. It is this pathway which offers the enzymes and some of the appropriate key genes (cDNAs) required to develop advanced methods for future biological alkaloid synthesis (see Genetic Approaches 2.3 and 3.2).

Synthesis of the anti-arrhythmic drug ajmaline from *Rauvolfia* plants, an efficient source of several alkaloid types used in therapy, represents an even more advanced study of the total enzymatic synthesis of terpenoid indole alkaloids. Ajmaline is a class I anti-arrhythmic alkaloid because of its activity as heart muscle sodium channels antagonist [35].

Ajmaline and some of its derivatives are used for pharmaceutical intervention in Europe [36]. Its enzymatic synthesis is depicted in Fig. 2 and starts, as described for the *Catharanthus* alkaloid ajmalicine, with the enzymatic Pictet–Spengler reaction performed by STR1 [19, 20, 47], followed by deglucosylation of the glucoalkaloid strictosidine by strictosidine glucosidase (SG) [48–50]. Both enzymes are not identical to the *Catharanthus* enzymes. Although they catalyse the same reactions, they differ in their amino acid sequences, resulting in slight changes in their biochemical properties.

Through the dehydrogeissoschizine stage, known from the enzymatic formation of ajmalicine (see Fig. 1), the sarpagan bridge enzyme (SBE) yields the first sarpagan-type alkaloid, polyneuridine aldehyde, formed on the way to the ajmalan-type [51]. A highly substrate specific esterase, polyneuridine aldehyde esterase (PNAE) [52, 53], leads, by ester hydrolysis and decarboxylation, to 16-*epi*-vellosimine which is transformed in turn by vinorine synthase (VS) in the presence of acetyl-CoA as a cofactor to the six-ring system, the alkaloid vinorine [54, 55]. On the way to ajmaline, all the reactions which follow occur at the periphery of the alkaloid structure by cytochrome P450 hydroxylation (vomilenine formation) [56], followed by four enzyme-catalysed reactions involving two NADPH-dependent reductases, an acetylesterase and an *N*-methyltransferase [57–60]. In summary, tryptamine and secologanin provide the skeletons for a pathway involving ten enzymes belonging to the synthase, glucosidase, esterase, hydroxylase, reductase and methyltransferase enzyme families, ultimately giving rise to the complex structure of ajmaline. All these enzymes have been characterised in detail, and half of them can now be produced heterologously and very efficiently in *E. coli*.

The multiple side-reactions of this pathway result in a complex metabolic network of *Rauvolfia* alkaloids [5]. In Fig. 2 only two of the enzymatic side steps are mentioned: the deglucosylation of the gluco-alkaloid raucaffricine and the NADPH-dependent reduction of perakine into raucaffrinoline. In recent years (2005–2009), six of the enzymes mentioned have been crystallised following synthesis in *E. coli* (summarised in [61, 62]), opening novel and significant

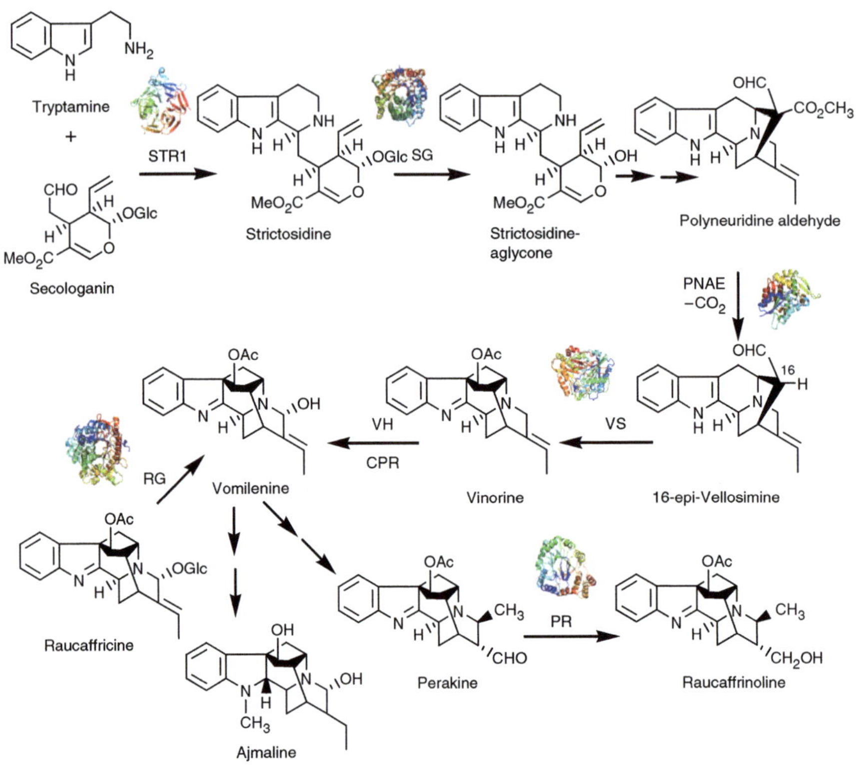

Fig. 2 Part of the biosynthetic network of indole alkaloids in the Indian medical plant *Rauvolfia serpentina*; elucidated by isolation and characterisation of individual enzymes, cloning of corresponding cDNAs by reverse genetics and crystallisation and 3D-X-ray analysis of the major catalysts. (*STR1* strictosidine synthase; *SG* strictosidine glucosidase; *SBE* sarpagan bridge enzyme; *PNAE* polyneuridine acetyl esterase; *RG* Raucaffricine glucosidase; *PR* perakine reductase).STR1 is the enzyme from *R. serpentina*, STR is the *C. roseus* enzyme. 3D structures were obtained during recent years for: STR1 [37–39]; SG [40, 41]; PNAE [42]; VS [43, 44]; RG [45]; PR [46]

opportunities for exploiting these enzymes in the development of new strategies to generate new alkaloids by rational enzyme engineering (see Sect. 2.4).

2.2 *Indole Alkaloid Precursors: Tryptophan and Strictosidine*

A significant number of tryptophan and tryptamine chemical syntheses are known: this section describes their biological synthesis *in vivo* (whole cells), the *in vitro* enzymatically catalysed synthesis and synthesis through chemo-enzymatic approaches.

Selected tryptophan "syntheses" will first be described because tryptophan is easily converted into its decarboxylation product tryptamine. Major emphasis will be placed on tryptophan analogues with a variety of substitution patterns at the aromatic ring system. These derivatives could be of significance for producing new tryptophan derived alkaloids.

Two recently published procedures should be highlighted. Development of simple and scalable one-step biotransformation of 4.5.6.7-halogenated (F, Cl, Br) and methylated indoles into the appropriate tryptophan analogues (Fig. 3) is based on 3-days incubation of a bacterial lysate together with L-serine at 37°C [63].

The lysate was obtained from *E. coli* cells which were pre-transformed with a plasmid (pSTB7) expressing tryptophan synthase from *Salmonella enterica*. This *E. coli* strain is commercially available (ATCC 37845). Reported yields on tryptophans ranged from 3% to 63%, but, in several cases, could be significantly enhanced up to 100% using the *E. coli* lysate, separated from the reactant by a dialysis tube.

The enzyme-mediated tryptophan synthase dependent approach is in theory well established but has the disadvantage of requiring the prior purification of tryptophan synthase [64, 65] and was therefore not widely employed until recent development of the ATCC 37845 *E. coli* line.

Since tryptophan synthase does not (or poorly) tolerate the substitution of indoles by large groups (iodo- and nitro-groups), such cases have lead to the development of a facial chemical synthesis of racemic *N*-α-acetyltryptophan derivatives from 5- and 6-monosubstituted indoles and L-serine, followed by an enzymatic resolution step using acylase "Amano" resulting in substituted L-tryptophans in high yields and enantiomeric purity of 91–100% ee [66] (Fig. 4).

Fig. 3 Enzymatic synthesis of L-tryptophans from indoles, L-serine and lysates of recombinant *E. coli* overexpressing tryptophan synthase (X = F, Cl, Br at carbon 4,5,6,7)

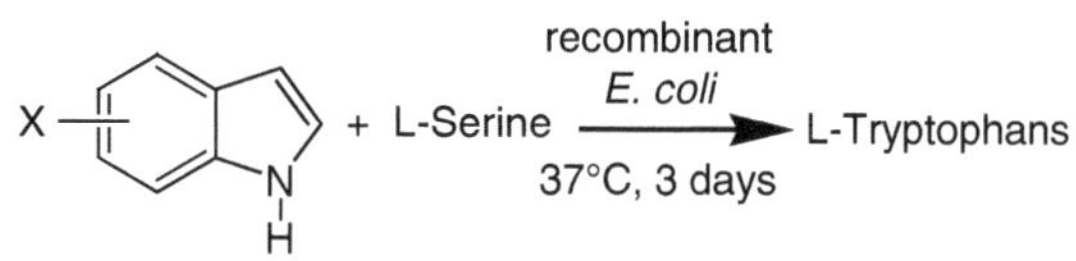

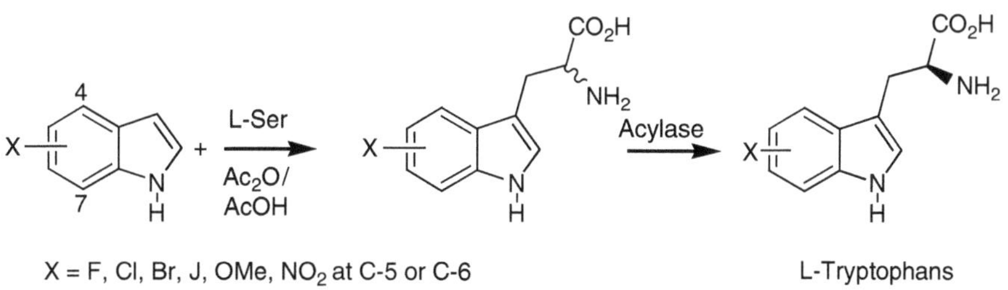

Fig. 4 Chemical synthesis of substituted tryptophanes followed by enzymatic resolution of the racemates

With reference to this, a recent summary describing biological halogenation, in particular that catalysed by $FADH_2$-dependent halogenases, should be mentioned [67]. These enzymes exhibit the conserved "flavin binding site" (GlyxGlyxxGly) and represent a family possessing 4, 5, 6 and 7 tryptophan-halogenase activity. These enzymes provide the possibility to generate tryptophan-derivatives starting with indole, L-serine and tryptophan synthase followed by halogenation with different tryptophan halogenases as the last reaction step, resulting in a selection of halogenated tryptophans which can be employed for further synthetic use.

Recently, an entire series of novel dimethylallyltryptophan synthases was detected, although the first described enzyme of this enzyme type dates back to 1971, when dimethylallyltryptophan synthase (DMATS; E.C.2.5.1.34) was detected [68]. DMATS prenylates L-tryptophan at position 4. Analysis of genome sequences from species of the *Claviceps, Aspergillus and Neosartorya* genera lead to the identification of a significant number of putative indole prenyltransferase genes. From a single organism alone, *C. purpurea,* up to ten prenyltransferase genes were detected [69, 70]. The gene products exhibit 24–62% amino acid sequence similarity to DMATS isolated from different fungi.

In total, ten prenyltransferase genes have been cloned and overexpressed in *E. coli* and yeast and the corresponding prenyltransfer reactions onto various substrates (tryptophan derivatives, indoles and tryptophan-containing cyclic dipeptides) investigated [71]. Depending on the presence of DMAPP as a prenyl donor, these prenyltransferases are soluble, metal ion-independent enzymes. Having been overproduced as His_6-tagged enzymes, and purified to near homogeneity, these can now be easily used as tools for the production of prenylated indoles. The major advantage of these enzymes is that they possess a relatively broad substrate acceptance, and can thus be generally applied to the chemo-enzymatic synthesis of a whole range of prenylindole derivatives [72, 73] leading to remarkably high yields, especially when "long-term" (overnight) incubations are carried out.

However, there are also examples of prenyltransferase in which substrates are only poorly accepted with low rate constants (K_{cat}/K_m), especially when their structures differ significantly from the natural substrates of the enzymes. There are two ways to avoid such drawbacks in future: (1) enzyme optimisation by site-directed mutation and (2) significantly increased catalyst overexpression, so that the amount of enzyme is not the limiting factor.

In conclusion, current achievements in the field of chemo-enzymatic application of these fungal prenyltransferases is remarkable, and future work must show whether preparative application for higher gram scales will be possible. Future identification and characterisation of novel prenyltransferases will allow expansion of their application with regards to substrate spectrum and position of prenylation. This is chiefly due to the fact that there are in excess of 100 fungal genome sequences now available (www.ncbi.nlm.nih.gov) consisting of approximately 100 homologues of dimethylallyltryptophan synthase. For most of these genes the functions are still waiting to be discovered [71]. Figure 5 represents selected

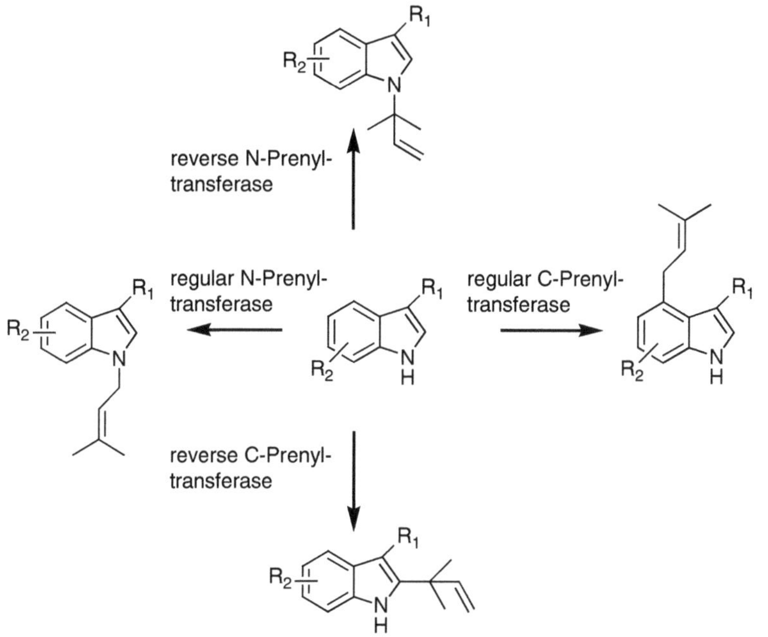

Fig. 5 Regio-selective prenylation of substituted indoles by novel, heterologously expressed His_6-tagged prenyltransferases [71]

examples for prenylation and overviews, present regio-selective enzymatic prenylations by various DMAPP-transferring proteins.

The successful application of enzymes to the preparation of structurally more advanced intermediates of alkaloids belonging to the monoterpenoid indole alkaloid family was convincingly demonstrated by the pioneering work of Zenk's research group at Munich University, approximately 25 years ago [74, 75]. They used pre-purified strictosidine synthase (EC 4.3.3.2) isolated from cell suspension cultures of the plant *C. roseus*. The enzyme was coupled to CNBr-activated Sepharose. At the immobilised stage the thermo-stability of the synthase was significantly enhanced (about 300 times compared to the enzyme in solution). Under these conditions, the biocatalyst was used for the gram-wise production of stereochemically pure 3(α)*S*-strictosidine, simply by passing the substrate tryptamine and secologanin (ratio 1:1) through a column containing the immobilised enzyme followed by freeze drying of the effluent. The half-life of the enzyme at 37°C was also enhanced up to 68 days compared with the soluble enzyme of $T_{1/2} \sim 5$ h, so that the strictosidine preparation column could be used for more than half a year without serious loss of production capacity. This enzymatic synthesis will be of future significance if strictosidine analogues are needed for the generation of novel alkaloid libraries (see Sect. 2.4).

2.3 *Genetic Approaches Towards Monoterpenoid Indoles*

2.3.1 Engineered *C. roseus* "Hairy Roots"

The biochemical characterisation of strictosidine synthase has been made possible by cDNA isolation and heterologous expression [19, 20, 47] followed by elucidation of its three-dimensional structure and its binding pocket [37–39], paving the way for the creation of rationally designed STR mutants [76, 77]. These mutants (e.g. Val214Met) show a broader substrate acceptance compared to the wild-type enzyme. Genetically engineered plants expressing this mutant (which preferentially accepts 5-substituted tryptamines and forms new, unnatural strictosidines (e.g. 5-halo-strictosidines)) were achieved through genomic transformation of *C. roseus* "hairy roots" by the pCMABIA plasmid [78]. Feeding of the transgenic root lines with 5-subsituted tryptamines (Cl, Me, Br) for 1 week, followed by phytochemical analysis of the root material, allowed the detection of a range of novel, unnatural alkaloids such as 5-substituted heteroyohimbine alkaloids (ajmalicine, serpentine), aspidosperma-type (tabersonine) and iboga-type alkaloids such as catharanthine. This strategy of re-programming alkaloid biosynthesis *in vivo* may generate whole sets of novel alkaloid analogues as formation of about 15 intermediates can be expected from strictosidine to tabersonine or catharanthine. The fact that the technique of cultivating plant organs such as the "hairy roots" described here has been established and optimised for some time and been developed for large scale production suggests that there is potential for further biosynthetic engineering in related systems such as cell suspension cultures, a promising tool for the future alkaloid diversification (Fig. 6).

2.3.2 Engineered *E. coli* and Yeast Cells

A dual vector containing the cDNA of STR1 and cDNA of strictosidine glucosidase (SG, EC 3.2.1.105) from the Indian medicinal plant *R. serpentina* was designed and, following transformation of an *E. coli* M15 line, functionally expressed [79]. Enzyme extracts of the bacteria could be used to convert tryptamine and secologanin through strictosidine into cathenamine. Chemical reduction by $NaBH_4$ then resulted in the synthesis of the naturally occurring tetrahydroalstonine and finished this short chemo-enzymatic route (Fig. 7). This new vector should allow insertion of cDNA libraries designed and produced from mutated STR1 and SG for generation of both diverse unnatural tetrahydroalstonines, as well as insertion of multiple additional available cDNAs, so that the vector may in future become an important tool for multi-step *in vivo* or *in vitro* enzymatic or chemo-enzymatic production strategies for structurally diverse and pharmacologically promising monoterpenoid indole alkaloids [79].

There is significant debate as to whether *E. coli* or yeast constitutes the better host. The significantly faster replication rate of the former is clearly superior to the

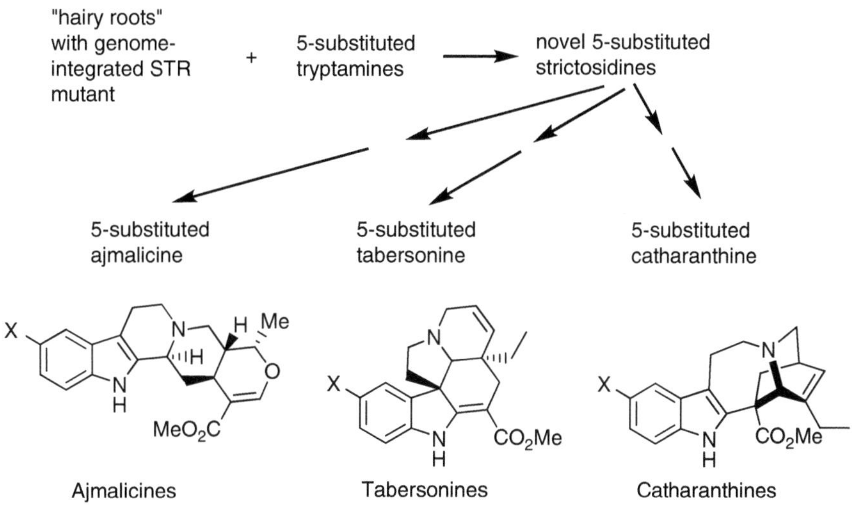

Fig. 6 In vivo reprogramming of alkaloid biosynthesis in "hairy roots" of *C. roseus* by introduction of a mutant cDNA of the key enzyme strictosidine synthase (STR) with broader, unnatural substrate specificity leading to diversification of alkaloid content in roots following long-term feeding with 5-substituted tryptamines (X = Cl, Br, Me) [78]

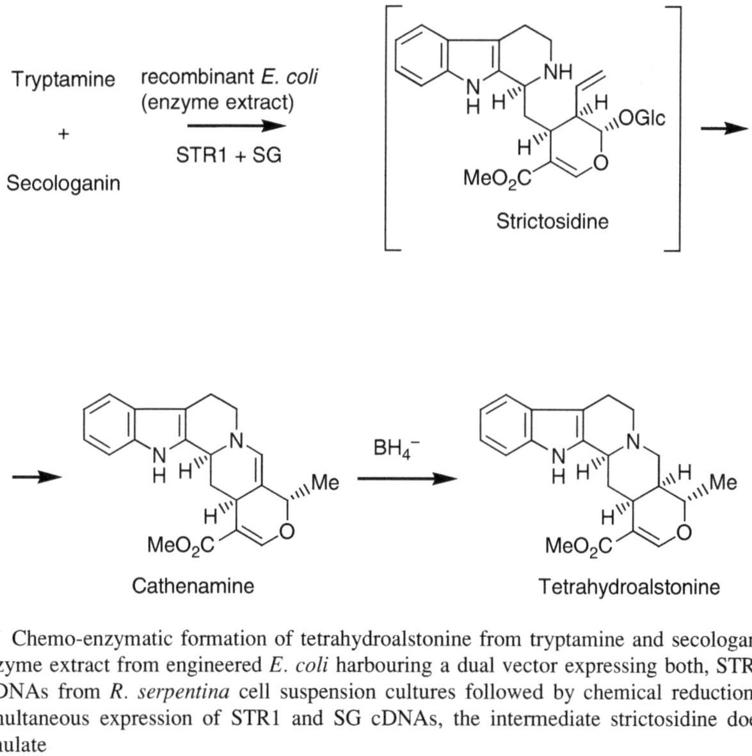

Fig. 7 Chemo-enzymatic formation of tetrahydroalstonine from tryptamine and secologanin by an enzyme extract from engineered *E. coli* harbouring a dual vector expressing both, STR1 and SG cDNAs from *R. serpentina* cell suspension cultures followed by chemical reduction. Due to simultaneous expression of STR1 and SG cDNAs, the intermediate strictosidine does not accumulate

second though posttranslational enzyme modifications can only be achieved in yeast. The choice as to which host is more useful for the genetic approaches described here must therefore be decided on a "case by case" basis.

In a similar approach to the one just described, two plasmids, each harbouring the cDNA of STR and SG from *C. roseus*, were used for transformation of *Saccharomyes cerevisiae* strain YPH500. After 2 days of growth of the recombinant yeast strain in the presence of 2 mM tryptamine and secologanin, strictosidine accumulated in the surrounding medium in yields as high as 2 g/L. The deglucosylation product cathenamine was, however, detected inside the yeast cells and, after permeabilisation of the cells, major extracellular strictosidine was hydrolysed to cathenamine by this short two-step method [80]; this methodology should allow the production of preparative amounts of both alkaloid intermediates.

2.4 Non-Natural Alkaloid Libraries by Re-Designed Enzymes

The power of engineered enzymes in the synthesis of novel alkaloids, to generate structural diversity and establish new alkaloid libraries, is best represented by the enzyme strictosidine synthase (STR1).

The crystal structure of STR1 together with its complexes of substrates (tryptamine and secologanin), enzyme product (strictosidine) and an inhibitor, not only provided a first deep insight into the enzyme mechanism [37–39, 81] of this central catalyst involved in the biosynthesis of ~2,000 alkaloids, but also provided a view into the binding pocket. Three-dimensional analysis also provides significant opportunities for the rational re-design and expansion of STR's substrate specificity. The aromatic moiety of the substrate tryptamine is very versatile, in particular when targeting development of combinatorial biomimetic approaches.

Observations made in the past demonstrated that in presence of an excess of nucleophiles (e.g. primary amines), the "normal" enzymatic biosynthesis of the heteroyohimbine system is interrupted by formation of E-ring analogues, resulting in a range of novel *N*-heteroyohimbines after chemical reduction. With the re-designed STRs [76, 77] this biomimetic approach might now be extended by use of rationally designed STR mutants such as Val2008Ala and Val214Met, which can now be used to generate A- and E-ring substituted heteroyohimbine libraries by combination with the above "nucleophiles" approach (Figs. 8 and 9).

2.5 Chemo-Enzymatic Synthesis of Monoterpenoid Indole Alkaloids

Chemo-enzymatic approaches to monoterpenoid indole alkaloids have attracted much interest since the beginning of the 1990s. At that time, synthetically "early"

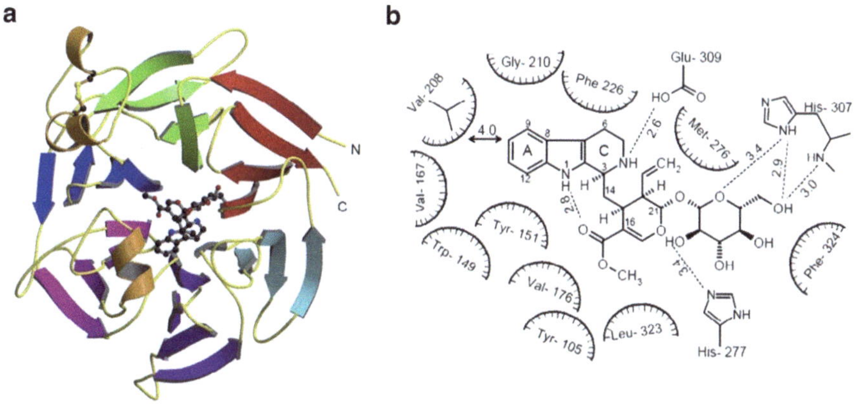

Fig. 8 (**a**) Complex of STR1 with the product strictosidine (*ribbon diagram*, strictosidine in *ball and stick presentation*). (**b**) 2D-representation of the complex, pointing to Val208 residue and hydrogen bond network (<4.1 Å); figures were taken from [76]

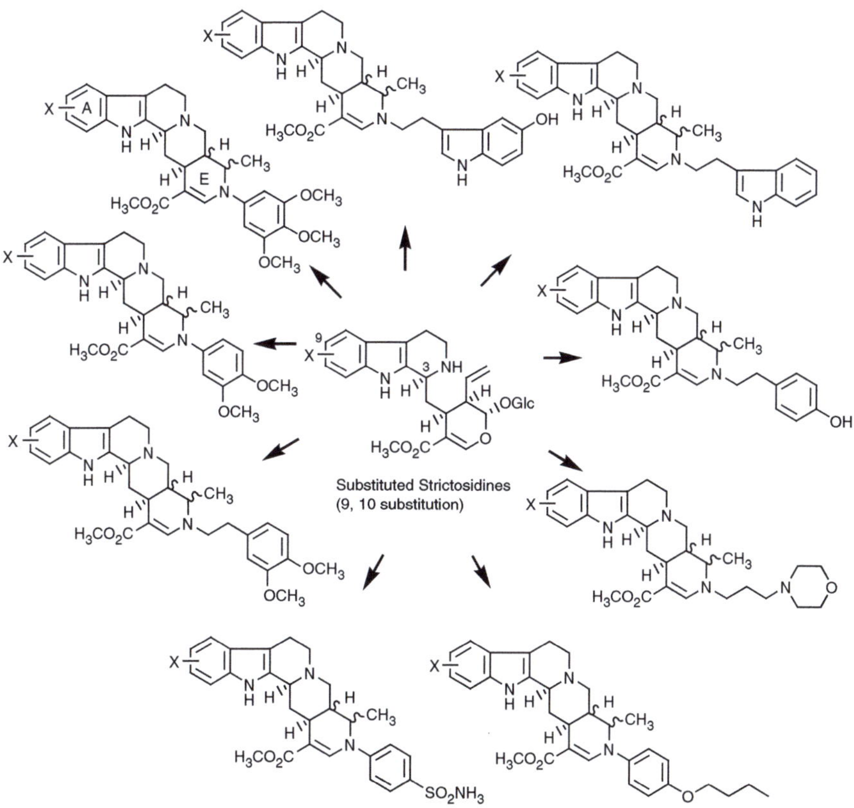

Fig. 9 Heteroyohimbine analogues that can be obtained by a combinatorial approach. Strictosidine is the precursor of the compounds displayed (X = H). Based on engineered mutants of STR1, the approach can now be extended to various substituted strictosidines

and simple building blocks such as pure enantiomers were used in order to develop strategies to generate, e.g. two tetracyclic indole alkaloids (−)-antirhine [82], (−)-akagerine [83], the pentacyclic (−)-allo-yohimbane [84] and (+)-meroquinene [85]. The latter is a degradation product of Cinchona alkaloids such as cinchonine. Key intermediates of the first three alkaloids are the enzymatically prepared (1*S*,2*R*)-cyclohex-4-ene-dimethanol monoacetate.

It became apparent early on that compounds of the 1,2-disubstituted cyclohexenes allow exploitation of a wide range of synthetic opportunities, leading to their frequent use in natural product synthesis; the application of chemo-synthetic routes for selected monoterpenoid indole alkaloid alone are summarised here [86].

Chiral *cis*-1,2-disubstituted cyclohex-4-ene precursors could in principle mimic the carbon framework of the monoterpenoid secologanin, leading to powerful chiral synthons, from readily available *meso*-forms. Moderate results were obtained with lipases such as porcine liver esterase (PLE), delivering the (1*S*,2*R*)-mono-acetate at a yield of 78% with 96% ee (Fig. 10), though preparation of the enzyme seemed to be crucial for the appropriate result. In contrast, pig pancreatic lipase (PPL) was significantly more efficient in forming the (−)-(1*R*,2*S*) enantiomer at a yield of 96% and 98% ee [86, 87].

An optimised approach to (1*S*,2*R*)-monoacetate, consisting of the addition of PPL to a solution of the diol in dry EtOAc, resulted in excellent >99% ee acetylation, but with only 88% conversion [83].

A retrosynthetic synthesis of (−)antirhine (Fig. 11) was performed through the monoacetate (1) and lactone (2) which was condensed with tryptamine. Product 3 was then finally converted, through a number of intermediates, to the tetracyclic (−)-antirhine, resulting in a remarkable overall yield of 13% (starting from the monoacetate 1), together with the 3-epimer (+)-antirhine at a yield of 6%. This strategy allowed the chemo-enzymatic synthesis of both (−) – and (+)-antirhines [82].

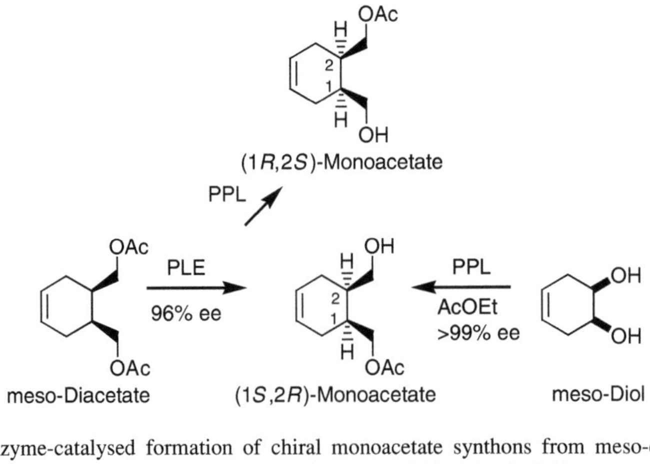

Fig. 10 Enzyme-catalysed formation of chiral monoacetate synthons from meso-diacetate or meso-diol for chemo-enzymatic synthesis of monoterpenoid indole alkaloids

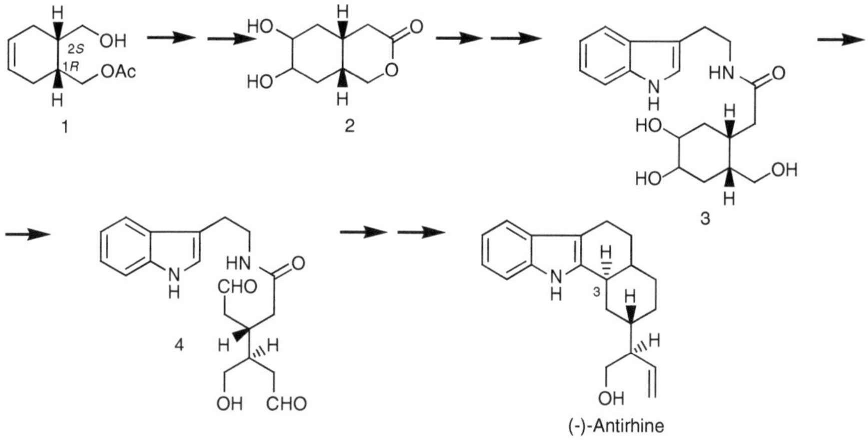

Fig. 11 Complete synthesis of the major alkaloid (−)-antirhine known from the plant *Antirhea putaminosa* (except for formation of 1, all other reaction steps were not enzyme-catalysed)

In a previously described approach, a largely similar strategy was employed in order to generate enantio-selectively (−)-allo-yohimbane [84]. Yohimbane alkaloids such as reserpine and yohimbine were of long-standing interest from pharmacological and synthetic points of view, and the first enantio-selective (−)-alloyohimban synthesis was reported some time ago by Isobe et al. [88].

As summarised here for the chemo-enzymatic approach of (−)-allo-yohimbane, synthon was once again the key lactone 1 (derived enzymatically from the above-mentioned meso-diacetate). Through condensation with tryptamine, a lactam was obtained which was then sequentially converted into (−)-allo-yohimban, indicating the high potential of this chemo-enzymatic approach for production of optically pure alkaloids of the terpenoid indoles.

Preparation of (−)-akagerine should be mentioned as the third example using this strategy of enantio-selective chemo-enzymatic indole alkaloid synthesis [83]. The alkaloid is a typical member of tetracyclic Strychnos alkaloids and exhibits convulsive activity [89]. Moreover, akagerine displays an unusual tetracyclic structure: a perhydroazepine ring is coupled to the tetrahydro-β-carboline moiety. Its first enantio-selective total synthesis summarised here was based on the lactone 3, generated from the 1*S*,2*R*-monoacetate 2 (Fig. 12). In a 17-step synthesis with an overall yield of 19.3%, pure (−)-akagerine was obtained, showing once again the remarkable power of this approach [83].

Extension of the approach described resulted in a short synthesis of (+)-meroquinene (Fig. 13). Formation of this alkaloid also synthetically connects the indole with the quinoline alkaloids, a link which has been suggested from the biosynthetic point of view for multiple decades: quinine alkaloids are expected to be enzymatically formed from the indole framework. Meroquinene is a simple piperidine derivative which can be obtained by degradation of cinchonine and other Cinchona alkaloids under acidic conditions. It also plays a key role in the chemical synthesis

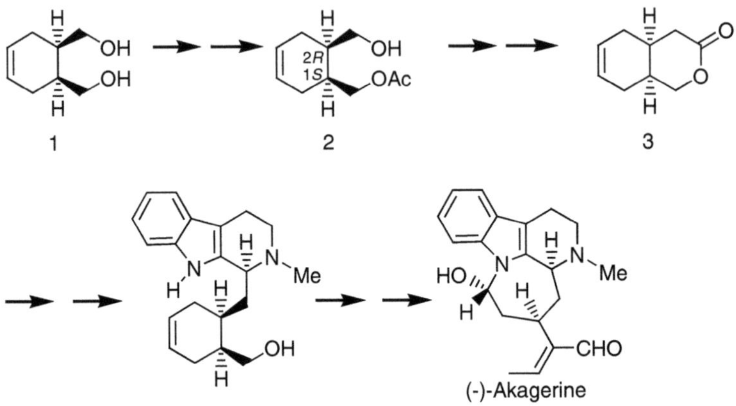

Fig. 12 First enantio-selective, chemo-enzymatic synthesis of the monoterpenoid indole alkaloid (−)-akagerine, member of the family of *Strychnos* alkaloids

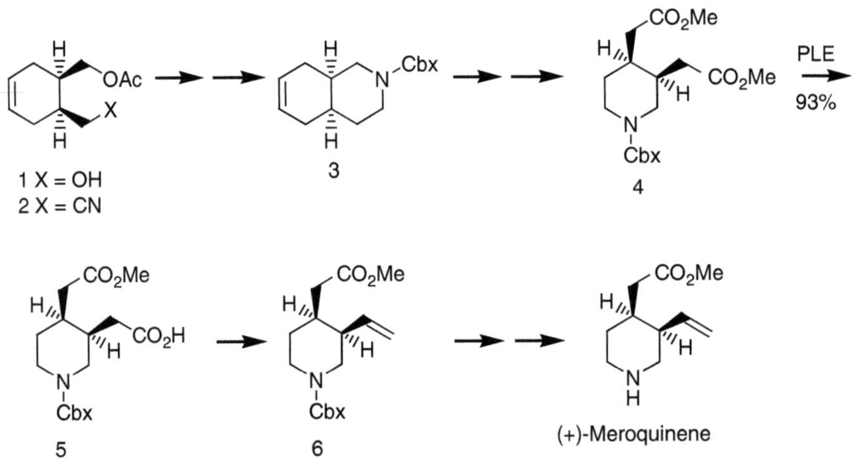

Fig. 13 Chemo-enzymatic synthesis of (+)-meroquinene

of several Chinchona alkaloids [90]. The monoacetate (1) was converted to its nitrile (2), its *cis*-octahydroquinoline (3) and followed by several chemical steps, followed by a highly efficient enzymatic reaction catalysed by PLE, which delivered by regio-selective hydrolysis intermediate 5 in surprisingly high yield of 93% and resulted finally in formation of (+)-meroquinene [85].

As a final example in this series of alkaloids, the enantio-selective synthesis of (+)-tacamonine is illustrated, which is also based on enzymatic generation of a chiral lactone related to the approaches described above.

The key lactone became available first by diol (1) efficiently converted to 2*R*-2-(3-butenyl)-1.3-propanediol acetate (2) by Amano PS Lipase as catalyst with 98%

yield (>99% ee) and second by enzymatic transesterification of diol 1 by PPL, leading to 2(*R*)-diol monoacetate 2, which was transformed to the lactone by ozonolysis at a yield of 97%. The lactone served as the key precursor for the (+)-tacamonine synthesis. The chemo-enzymatic strategy is outlined in Fig. 14.

The enzymatic conversion of a 9-methoxycamptothecin derivative to 9-methoxy-20-(*S*)-mappicine should be mentioned as it is a short, simple and unusual process which uses microwave irradiation and baker's yeast as catalyst [91].

The Indian tree *Nothapodytes foetida* (syn. *Mappia foetida*) is an important source of camptothecin, the DNA topoisomerase I inhibitor possessing anti-cancer and anti-HIV activity of significant therapeutic importance. One of the minor components of the plant has been identified as 9-methoxy-20-(*S*)-mappicine, a tetracyclic derivative of the pentacyclic camptothecin. When the latter compound was treated for a short-time (7 min) with microwave irradiation, it was converted into 9-methoxymappicine ketone with an exceptionally high yield of 95%. This ketone was then easily transformed by baker's yeast into the target compound 9-methoxy-20-(*S*)-mappicine as illustrated in Fig. 15 [91].

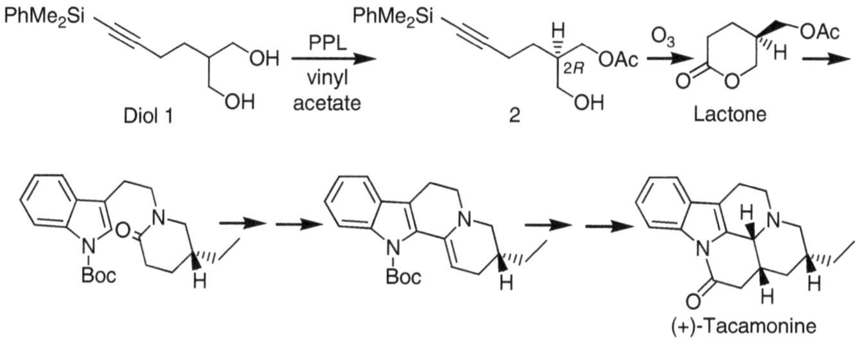

Fig. 14 Chemo-enzymatic formation of (+)-tacamonine [86]

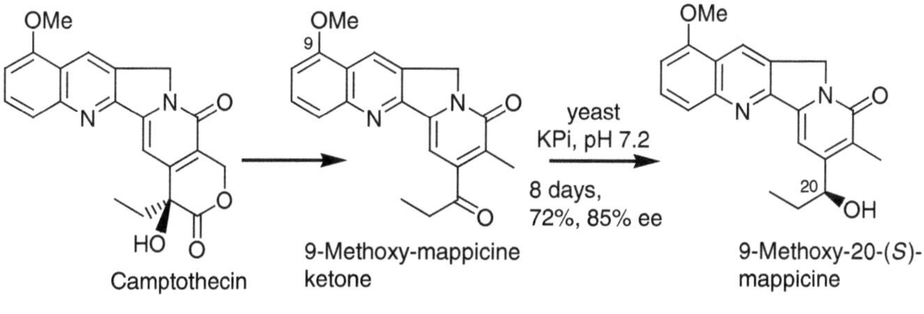

Fig. 15 The two-step formation of 9-methoxy-20-(*S*)-mappicine from 9-methoxycamptothecin by microwave irradiation and *Saccharomyces cerevisiae*

3 Enzymatic Formation of Isoquinoline Alkaloids

3.1 Enzymes of Morphine and Berberine Synthesis

Morphine is probably the most exciting alkaloid for a number of reasons, including its very early detection as a plant natural product, its excellent therapeutic value as an extremely strong pain killer, its complex chemical structure (and resulting complexity of synthesis), its occurrence not only in plants but also in human beings but primarily due to the challenge of understanding the complexity of its natural enzymatic synthesis. Following approximately two decades of work, Zenk's laboratories in Munich and Halle provided the single enzymes from poppy plants, catalysing the multi-step synthesis with 16 reaction steps starting from two precursors, dopamine and *p*-hydroxyphenylacetaldehyde (Fig. 16) [3].

The demethylation reactions which take place at the end of the pathway are the only remaining steps for which the corresponding enzymes have not yet been described [92]. The remaining enzymes are well known and many have been purified to homogeneity meaning that the prerequisites for complete synthesis of morphine at the enzyme level is imminent. Moreover, more than a half of these enzymes can now be generated by cloning of their cDNAs followed by their heterologous expression in *E. coli*, yeast or insect cells ([92] and cited literature).

Morphine as a typical plant natural product has also been identified many times in animals and humans, initially leading to the belief that its occurrence is of a dietary origin or of de novo synthesis [93]. However, surprisingly, feeding experiments with labelled precursors to neuroblastome cells have recently provided clear evidence of the presence of endogenous morphine formation in human cells [94]. Moreover, the finding that some of the morphine biosynthesis-specific enzymes are present in some of these cells provides the first, surprising, clear-cut evidence for its enzyme catalysed formation in non-plant cells [95, 96].

The alkaloidal intermediates (*S*)- and (*R*)-reticuline occupy a crucial position in the enzyme-catalysed formation of various isoquinoline alkaloids. On one hand, as shown in Fig. 16, they are precursors of codeine and morphine; on the other hand, 12-step biosynthesis from (*S*)-reticuline leads to the benzophenanthridines sanguinarine, chelirubine and macarpine, enzymatic pathways which have been extensively investigated [97] but that are not illustrated here. In addition, as depicted in Fig. 17, the (*S*)-precursor also leads to a short and comprehensively elucidated, vesicle-harboured [98] pathway to berberine, through intermediates (*S*)-scoulerine, (*S*)-tetrahydrocolumbanine and columbamine. Although there is extensive knowledge relating to a multitude of novel enzymes catalysing all of the single steps for several isoquinoline alkaloids of pharmacological importance, there has been no direct synthetic application or use of these proteins in synthetic chemistry so far. This situation is, however, in sharp contrast to the biological use of the corresponding cDNAs (genes) which, in general, led to the development of the research field "synthetic biology", also applicable to alkaloids [99]. Such cDNAs allow, e.g. the production of isoquinoline alkaloids in microbial systems.

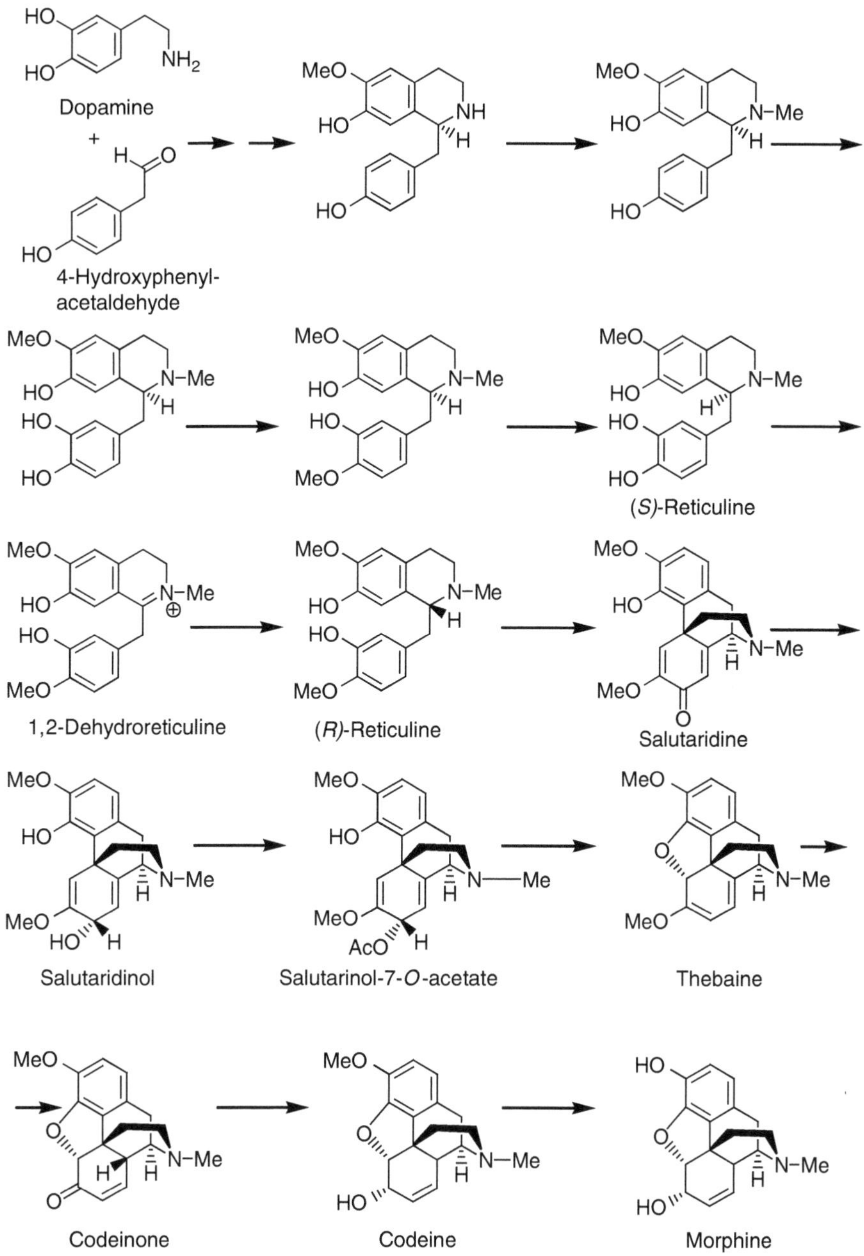

Fig. 16 The multi-step enzymatic formation of morphine from dopamine and 4-hydroxyphenylacetaldehyde

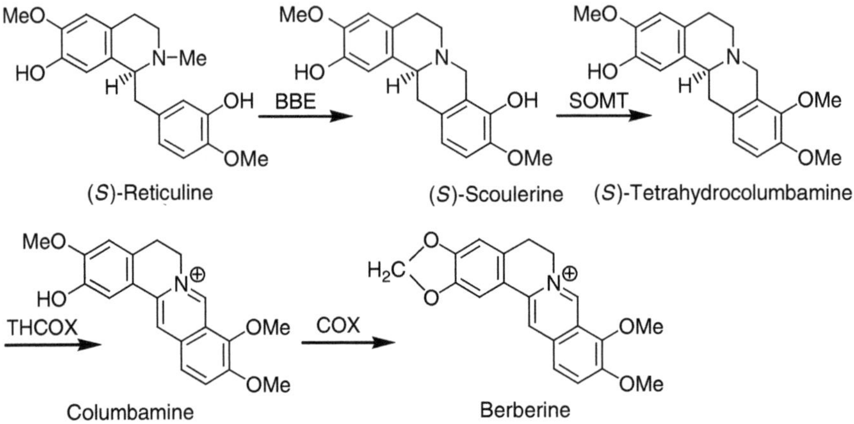

Fig. 17 Sequential enzymatic formation of berberine from (*S*)-reticuline (*BBE* berberine bridge enzyme; *SOMT* scoulerine *O*-methyltransferase; *THCOX* tetrahydrocolumbamine oxidase; *COX* columbamine oxidase)

Last but not least, the wide range of pharmacological effects and activities coupled to systematic research on plant cell suspension cultures of the genus *Berberis* and *Coptis* boosted successful investigation of the enzymes involved in the biosynthesis of the isoquinoline alkaloid berberine. The alkaloid displays anti-inflammatory activity, which might be based in part on its inhibition of DNA-synthesis as shown in activated lymphocytes [100, 101]. The alkaloid also has immunotherapeutic activity because of its induction of interleukin-12 production in mouse macrophages [102]. In combination with *Panax ginseng* saponins, berberine has an effect on hemodynamics, on the concentration of plasma brain natriuretic peptide and on calcium-ion concentration in experimental rats with congestive heart failure [103]. Unsurprisingly, the alkaloid also shows anti-cancer activity [104], it acts as a potent telomerase inhibitor and is known to have anti-oxidant activity [105].

Elucidation of the enzymatic synthesis of berberine has thus been a long-standing goal (Fig. 17) [3, 106].

The first step in the berberine pathway is the conversion of the branch-point intermediate (*S*)-reticuline, which also leads as indicated above to sanguinarine/macarpine and codeine/morphine biosynthesis. The appropriate enzyme for berberine formation is called berberine bridge enzyme (BBE), and it catalyses a unique reaction [107]. The N-methyl group of (*S*)-reticuline is oxidised by BBE and by oxygen under H_2O_2 formation followed by ring closure delivering C-8 of (*S*)-scoulerine. The BBE product is then methylated by a specific SAM-dependent methyltransferase, leading to (*S*)-tetrahydrocolumbamine, which in turn is oxidised by (*S*)-stylopine oxidase. The columbamine formed is transformed by a methylendioxy-bridge forming enzyme into the target alkaloid berberine.

As observed in a different cell system (*Coptis japonica* cell suspension cultures), an alternative step in berberine synthesis is the formation of the methylendioxy group at the stage of tetrahydrocolumbamine [108]. The cytochrome P450 enzyme

CYP719A1 (canadine synthase) [109] is responsible for (*S*)-tetrahydroberberine production, which is then enzymatically oxidised to berberine.

3.2 Genetic Approaches Towards Isoquinoline

3.2.1 Plant Metabolic Engineering

The isolation of the cDNAs encoding the enzymes involved in diverse isoquinoline alkaloid formation in plants and microorganisms allowed the first metabolic engineering routes to be developed and paved the way for new ways of future production of isoquinoline alkaloids. For instance, transgenic opium poppy plants were created in which codeinone reductase was suppressed by RNAi, resulting in the substitution of morphine synthesis with the non-narcotic precursor reticuline [110]. In a similar approach, RNAi suppression or overexpression of salutaridinol 7-*O*-acetyltransferase in opium poppy led to accumulation of salutaridine or increase of morphine, codeine and thebaine content [111], suppression of the BBE led to accumulation of berberine in California poppy cells [112].

In contrast, when an "early gene" (CYP80B) of morphine biosynthesis or a late gene (the cDNA coding for codeinone reductase) were overexpressed in poppy plants, the yield of morphine significantly increased (4.5-fold and 20% increase, respectively) [113, 114]. These results are testament to the power of metabolic engineering of medicinal plants in the field of isoquinoline alkaloids, through characterisation of genes (cDNAs) and their protein product function. However, the extremely slow growth and regeneration time of the plant system compared to microorganisms remains a major drawback. It appears therefore that for future alkaloid production, more significant emphasis is placed on the appropriate engineering of bacteria and yeast systems.

However, one should keep in mind that by classical mutagenesis of seeds, poppy plants were obtained which produced 3% thebaine instead of morphine and codeine, providing a simple supply of thebaine [115].

3.2.2 Microbial Formation of Plant Isoquinolines

Microbial formation of plant isoquinolines became an attractive issue over recent years [116]. Magnoflorine was of particular interest thanks to its pharmacological activities. On one hand, it was found that this aporphine alkaloid protects human high density lipoprotein (HDL) against lipid peroxidation but it also exhibits anti-HIV activity [117–119]. New production strategies for this alkaloid will be vital in the future.

In a remarkable approach involving both microbial and plant genes, reticuline was produced by transgenic *E. coli* (Fig. 18) [120]. This resulting strain, expressing the cDNAs of the five enzymes monoamine oxidase (MAO) from *Micrococcus*

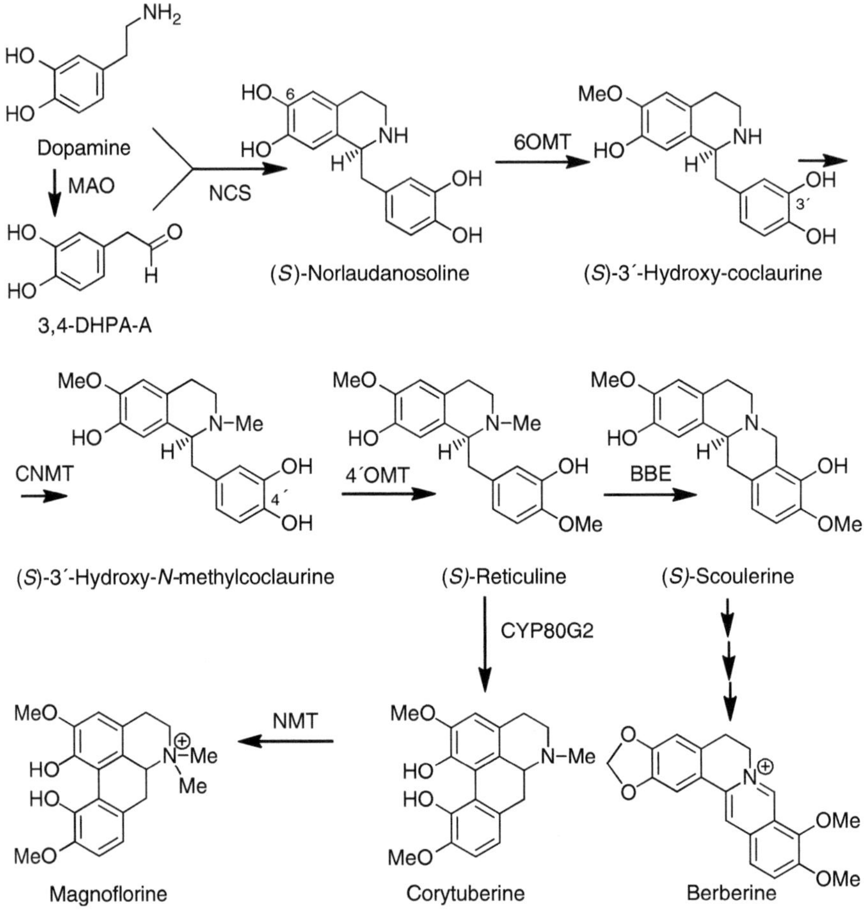

Fig. 18 Reconstructed isoquinoline alkaloid biosynthetic pathway in microbes (*E. coli* and *S. cerevisiae*) [120]. Accession numbers of the genes can be taken from Gene Bank. (*3,4-DHPA-A* 3,4-dihydroxyphenylacetaldehyde; *NCS* norcoclaurine synthase; *6OMT* 6-*O*-methyltransferase; *CNMT* coclaurine-*N*-methyltransferase; *4′OMT* 4′-*O*-methyltransferase; *BBE* berberine bridge enzyme; *NMT* *N*-methyltransferase)

luteus, and all others from the plant cell of *Coptis japonica*, such as norcoclaurine synthase (NCS), norcoclaurine 6-*O*-methyltransferase, coclaurine-*N*-methyltransferase and 3′-hydroxy-*N*-methyl-coclaurine-4′-*O*-methyltransferase, produced up to 11 mg of racemic reticuline in 1 L culture (yield 2.9%). When crude enzymes of the engineered *E. coli* cells were used for reticuline production, 22 mg/L (14.4% yield) (*S*)-reticuline was obtained, which increased up to 55 mg/L in the presence of optimal levels of dopamine [120].

Using a combination of transgenic *E. coli* and *S. cerevisiae* engineered with the appropriate expression vectors, the benzylisoquinoline alkaloid reticuline, the protoberberine type alkaloid scoulerine, and the aporphine alkaloid magnoflorine were synthesised *in vivo* (Fig. 18) [120].

In a different but closely related process, recombinant yeast strains were used for the production of a range of known benzylisoquinolines. Yeast represents an ideal organism for the study of metabolic engineering. A remarkable example of this approach was published in the steroid field a few years ago, by way of a yeast strain harbouring nine foreign cDNAs and deletion of four host genes. The resulting strain produced hydrocortisone through generation of its own substrate, a process that could be the future basis for technical production of hydrocortisone by a one-step fermentation [121, 122].

Yeast-based production of benzylisoquinolines represents a consistent and challenging continuation in the development of transgenic *Saccharomyces* for generation of complex plant-derived alkaloids [123]. In an ambitious approach, several yeast strains have been successfully designed containing for instance:

1. The three cDNAs of norcoclaurine-6-*O*-methyltransferase, coclaurine-*N*-methyltransferase and 3′-hydroxy-*N*-methylcoclaurine 4′-*O*-methyltransferase converting (*R*, *S*)-norlaudanosoline into (*R*, *S*)-reticuline with a 5–10% yield
2. When strain 1 was transformed with cDNA of BBE, 80% of produced (*S*)-reticuline was converted to (*S*)-scoulerine
3. About 20 mg salutaridine was formed from 4 mM norlaudanosoline in strain 1 which additionally expressed a human cytochrome P450 reductase

Moreover, transgenic yeast cells formed (*S*)-tetrahydroberberine (~30 mg/L) amounting to a ~1–2% total conversion of norlaudanosoline in the seven-step pathway, and demonstrating the efficiency of the system, which most probably can be further optimised in future [123].

3.3 Chemo-Enzymatic Approach to Isoquinoline Alkaloids

3.3.1 Simple Isoquinolines

The 1.2.3.4-tetrahydroisoquinoline skeleton represents the framework found in many isoquinoline alkaloid derivatives and not only from plants. Some derivatives attracted much interest because of their anti-cancer activity [124], which has prompted many groups to invest in their chemical synthesis. The Pictet–Spengler reaction has become an important method in the preparation of this alkaloid type, and has often been described with phenylalanine derivatives and pyruvates as starting materials. Synthesis of appropriate tetrahydroisoquinoline-3 and the corresponding tetrahydroisoquinoline-1-carboxylic acid has been the key target [125].

Efficient and rapid synthesis of chiral (*S*)-6-hydroxy-tetrahydroisoquinoline-3-carboxylic acid with a yield of 30% has been reported [126]. The route starts with separation of (*S*)-*m*-tyrosine from the (*R*,*S*)-racemate by resolution with immobilised D-amino acid oxidase from *Trisonopsis variabilis* (EC 1.4.3.3). Catalase from bovine liver is added for destruction of the H_2O_2 formed during transformation of (*R*)-*m*-tyrosine to *m*-hydroxyphenylpyruvate. The mixture is incubated at 30°C (180 rpm) for 60 h. The formed crude *m*-hydroxyphenylpyruvate is easily separated (86% yield) from the (*S*)-*m*-tyrosine formed (88% yield) and the latter is transformed into the target carboxylic acid at a notable yield of 30% (see Fig. 19).

3.3.2 Morphinan Skeleton

A total of more than 20 total syntheses have been described which aim to generate the most important members of the morphinan-type alkaloids, morphine and codeine. These long-standing efforts in alkaloid synthesis have been primarily due to the exceptional pharmacological importance of both compounds, as the most efficacious therapies for pain and cough, respectively.

A comprehensive and elegant strategy for application of chemo-enzymatic methods to the synthesis of such alkaloids has been elucidated by the group of Hudlicky for many years, with the aim of providing efficient and practical ways of syntheses by including biocatalysis into traditional synthetic protocols. In this respect, the earlier investigations of Gibson [127] on the metabolism of arenes as a carbon source for soil bacteria constituted ground-breaking explorations because they opened the way to whole cell and enzymatic transformation of simple aromatics for synthetically valuable intermediates as the examples summarised here show; for a review see [128].

In particular, dihydroarene diols were extremely useful for the design of syntheses of natural products, such as the examples of isoquinolines summarised here. The homochiral diol in (Fig. 20) was available from β-bromo-ethylbenzene by enzymatic oxidation [129]. The whole-cell enzymatic oxidation of β-bromoethylbenzene with *E. coli* strain JM109 (carried a plasmid expressing a dioxygenase)

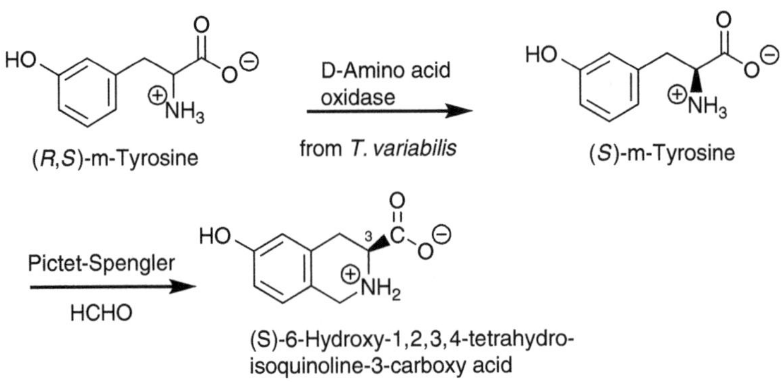

Fig. 19 Chemo-enzymatic formation of chiral 6-hydroxy-1,2,3,4-tetrahydroisoquinoline-3-carboxylic acid [126]

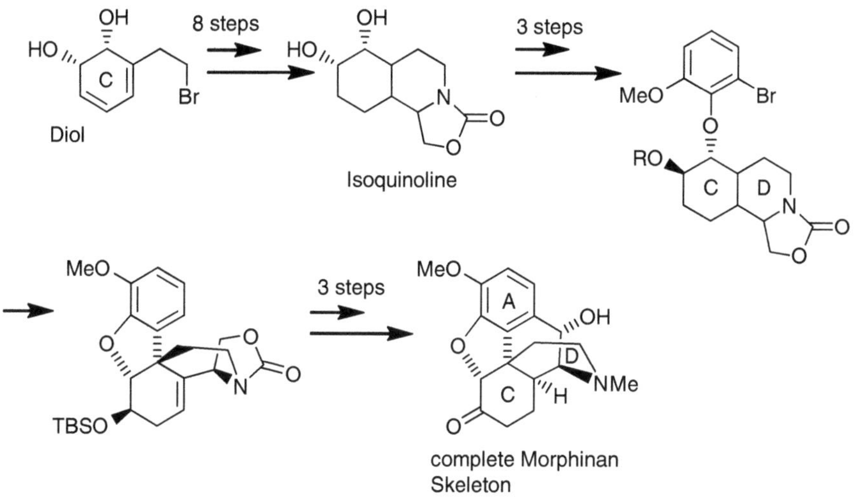

Fig. 20 Fourteen-step synthesis of the complete morphinan skeleton of morphine and codeine, starting with the enzymatically formed diol providing ring C of the carbon frame work of these alkaloids [129, 130]

resulted in excellent yields of 10–15 g/L/h production of the homochiral diol and was applied in an only 13-step synthesis of (+)-codeine by sequential Heck cyclisations and 14-step synthesis of the complete morphinan skeleton [129, 130].

4 Chemo-Enzymatic Synthesis of Other Alkaloids: Selected Examples

4.1 Enzyme Assisted Synthesis of (+)-Pilocarpine

(+)-Pilocarpine can be regarded as an "old" alkaloid but is becoming increasingly interesting because of its pharmacological importance not only in the well-known treatment of glaucoma, but also in the therapy of xerostomia and Sjoegren's syndrome. Many syntheses have been previously developed for that alkaloid but have usually yielded the racemic pilocarpine mixture and have necessitated long multi-step routes. In the 1930s, the first syntheses of (+)-pilocarpine were published and many others followed where the resolution of appropriate racemates at the stage of (±)-pilopic acid or (±)-iso-pilopic acid were ultimately frequently necessary. Chiral approaches gave (+)-pilocarpine and (+)-isopilocarpine mixtures and even the patented industrial synthesis [131, 132] delivered racemic pilocarpine in addition to two formal chemical total syntheses [133].

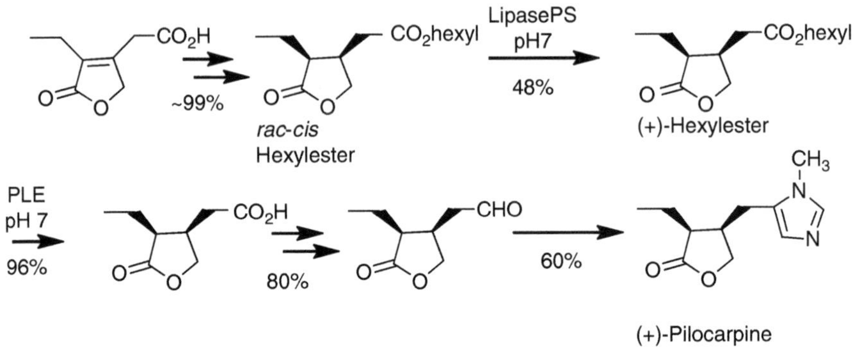

Fig. 21 Short chemo-enzymatic approach to enantiomerically pure (+)-pilocarpine from an easy available precursor using two lipase-catalysed reactions (PS (from *AmanoEnzyme Inc.*); *PLE* porcine liver esterase)

A short chemo-enzymatic approach to the enantiomerically pure alkaloid, published in 2008 is notable, because it avoids long sequences with low overall yields, mixtures of isomers and at the end racemic alkaloid which is outlined in Fig. 21 ([134], also includes citations for earlier syntheses).

4.2 *Methanomorphanthridine Alkaloids by Chemo-Enzymatic Processes*

4.2.1 Synthesis of Natural and Unnatural Montanine Alkaloids

Montanine alkaloids are members of the alkaloids of the plant family of *Amaryllidaceae*. They represent a relatively small class of compounds with the typical methanomorphanthridine skeleton exhibiting some anxiolytic, anti-depressive and anti-convulsive activities [135, 136]. The total syntheses of the natural (−)-montanine enantiomers are only mentioned, not described, here (for these (−)-enantiomers see, e.g. literature cited in [137]).

The total chemo-enzymatic synthesis of three of the (+)-montanine alkaloids, namely (+)-brunsvigine [138], (+)-nangustine [139] and (+)-montabuphine [140], has been reported in the last 3 years. These non-natural enantiomers have been prepared from monochiral 3-halo-*cis*-1,2-dihydrocatechols precursors. A total of over 17–20 reaction steps are needed for their syntheses (Fig. 22).

Important to note is that the precursors can be generated very efficiently on the multi-gram scale by whole-cell biotransformation of the corresponding halobenzene with genetically engineered microorganism over-expressing toluene dioxygenase. For a recent review on generating *cis*-1,2-dihydrocatechol and their extensive synthetic application as synthons (see [128]).

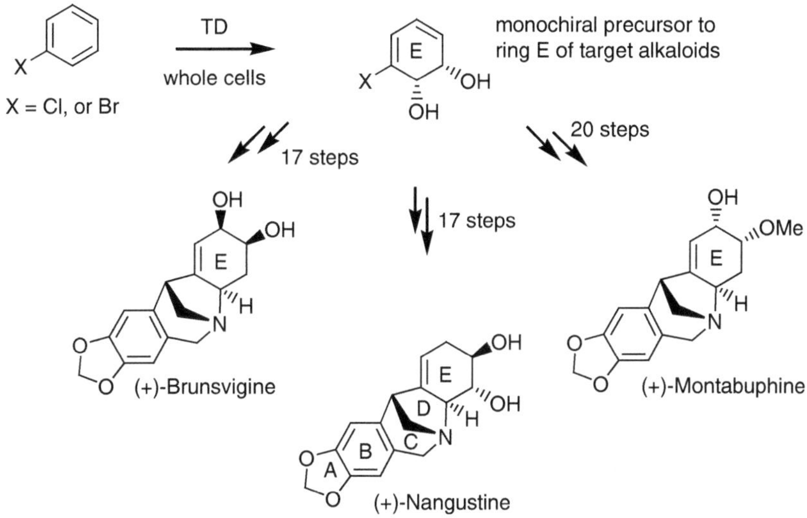

Fig. 22 Chemo-enzymatic, enantio-selective total synthesis of three non-natural montanine type alkaloids, (+)-brunsvigine, (+)-nangustine and (+)-montabuphine (*TD* toluene dioxygenase)

The synthetic route to the three alkaloids is summarised in Fig. 22, in which generation of the catechol is the crucial step based on enzyme-mediated transformation.

4.3 Chemo-Enzymatic Formation of Lycoricidines

A different structural type in the *Amaryllidaceae* alkaloid family belongs to the lycorine-skeleton. There are three alkaloid examples (+)-lycoricidine, (+)-narciclasine and (+)-pancratistatin, which have been known for many years following their isolation from *Amaryllidaceae* plants such as daffodils and narcissi. Potent pharmacological effects such as anti-viral and anti-cancer activity of these alkaloids [141] has resulted in recent developments in practical synthetic approaches. For instance, (+)-pancratistatin exhibits promising properties against various cancer cell lines. Here, the synthesis of the non-natural enantiomer (−)-*ent*-lycoricidine, of the natural (+)-lycoricidine is summarised together with the 3-*epi*-*ent*-lycoricidine and its 4-deoxy derivative [142]. These routes once again take advantage of the well known readily available chiral synthon, *cis*-1.2-dihydrocatechol, which has been so successfully introduced to chemo-enzymatic alkaloid syntheses by Hudlicky's group nearly a decade ago [143, 144]. These authors have also made available the natural (+)-enantiomers mentioned above in elegant synthetic routes.

As Fig. 23 illustrates, the enzymatically produced brominated key compound was the starting compound for the chemo-enzymatic formation of the three (+)-enantiomers [142].

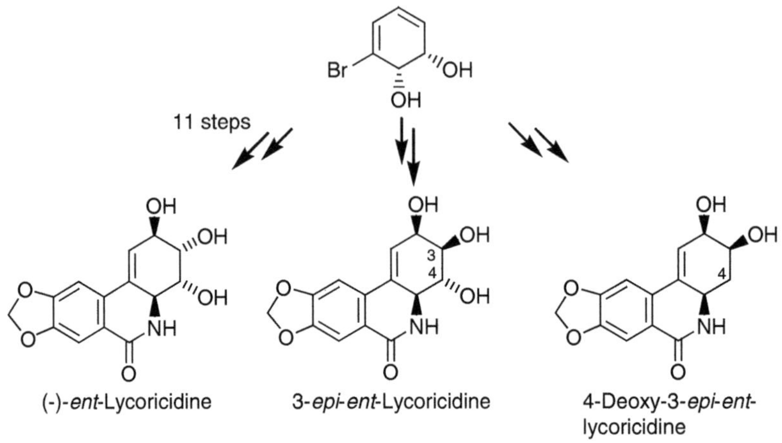

Fig. 23 Enzyme-assisted synthesis of lycoricidines

4.4 *Pyrrolizidines and Indolizidine Examples*

4.4.1 Pyrrolizidines

Alkaloids of the pyrrolizidine type possessing the common bicyclo skeleton, 1-azabicyclo-[3.3.0]octane, are well known from, e.g. plants of the genera *Castanospermum* and *Alexa*. Despite their extensive and diverse hydroxylation pattern, they are divided into three groups, dependent on a further hydroxymethyl group at C-1 (necines), at C-3 (alexines) or at C-3 and C-5 (hyacinthacines). The inhibitory effects of the latter group of alkaloids towards enzymes of the glycosidase- and glycosyltransferase-type have long been known, in addition to their anti-cancer and anti-viral activities [145–148].

By an enzyme-assisted short synthesis, several tetrahydroxylated pyrrolizidines (3-*epi*-australine, australine, 7-*epi*-alexine) have been directly obtained by applying an enzyme-catalysed aldo-reaction, bypassing the necessity for tedious efforts in the protecting OH-groups as demonstrated by Romero and Wong [149] (Fig. 24).

A few years ago, hyacinthacines A_1 and A_2, which are also hydroxylated pyrrolizidine alkaloids, were isolated from bulbs of *Muscaria armeniacum* (*Hyacinthaceae*) [146]. They were found to be good inhibitors of rat intestinal lactase, α-L-fucosidase and amyloglucosidase. With the intention of optimising their biological activities and their selectivity, the generation of configurational diversity at stereogenic centres soon became an important synthetic aim.

More effort was therefore invested in the application of synthetic methodologies for these alkaloids and some straightforward chemo-enzymatic approaches were recently developed [150]. An enzyme-catalysed aldol reaction was again a crucial step in that synthetic route and is strongly reminiscent of Wong's research strategy relating to the chemo-enzymatic synthesis of pyrrolizidines mentioned earlier.

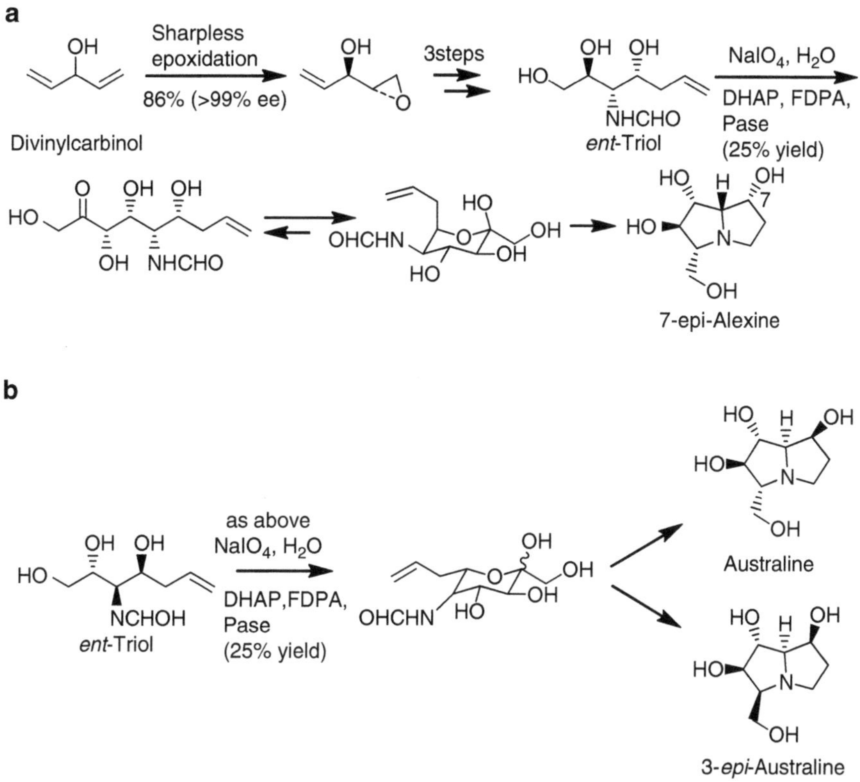

Fig. 24 (**a, b**) Chemo-enzymatic process for synthesis of tetrahydroxylated pyrrolizidines 7-*epi*-alexine, australine and 3-*epi*-australine utilising dihydroxyacetone phosphate (DHAP), stereospecific aldol reaction catalysed by fructose-1.6-diphosphate aldolase (FDPA) and acid phosphatase (Pase) [149]

The unnatural enantiomers (−)-hyacinthacine A_2 (natural enantiomer is (+)-hyacinthacine A_2), 7-deoxy-2-*epi*-alexine (enantiomer of 3-*epi*-hyacinthacine A2), *ent*-7-deoxyalexine (enantiomer of 7-deoxy-alexine) and 2-*epi*-hyacinthacine A_2 were synthesised [150]. Interestingly, (−)-hyacinthacine A_2 exhibited an α-D-glucosidase inhibitory effect which was not observed for the naturally occurring (+) enantiomer. This fact clearly indicates the urgent need for additional syntheses of non-natural enantiomers in the natural products field.

For the described approach, it is important to note that aldolases of different origin were tested and that, in contrast to L-rhamnulose-1-phosphate aldolase (RhuA), the D-fructose-1,6-biphosphate aldolase from rabbit muscle and L-fuculose-1-phosphate aldolase from *E. coli* were not active for DHAP/(*R*)-*N*- and (*S*)-*N*-Cbz-prolinal condensation. Since RhuA accepts both, (*S*)-N- and (*R*)-N-Cbz prolinals, the chemo-enzymatic synthesis of both, hyacinthacines A_1 and A_2 isomers could be achieved. In conclusion, the origin and the particular enzyme itself are of high importance for consideration of the best enzyme for the particular reaction step. Moreover, it is also necessary to keep in mind that the described aldol-condensation reactions were more efficient at 4°C (very unusual for enzyme reactions) instead of 25°C, although reaction times at low temperature are significantly longer. For synthetic details we refer to Calveras et al. [150].

4.4.2 Indolizidines

A recent example in the alkaloid group of indolizidines is the chemo-enzymatic enantio-selective route which delivers the alkaloid (−)-gephyrotoxin-223 [151]. This toxin is a 3,5-disubstituted indolizidine which was detected in amphibian skin, and is suggested to originate from ants, dietaries of the amphibians. (−)-Gephyrotoxin-223 was first found in the dendrobatid frog *Dendrobates histrionicus*. About 15 alkaloids from amphibian skin belong to these 3.5-disubstituted indolizidines and, immediately after detection, their synthesis and their structural determination was a great challenge for synthetic chemists. The chemo-enzymatic synthesis of gephyrotoxin approaches are based on two enantiomeric monoacetates with (+)-(2*R*,6*S*)- and (−)-(2*S*,6*R*)- configuration. These chiral intermediates occupied a crucial role in the synthesis of several enantiomeric pairs of indolizidiines, of which the gephyrotoxin synthesis was performed as shown in Fig. 25.

The meso-diol was enzymatically acetylated, which afforded the mono-acetate with 2*S*,6*R*- configuration. Hydrolysis by lipase reaction (PLE) gave an intermediate through which four reaction steps resulted in the N-protected (Cbz-group) ethyl ester, which could be finally transformed to the (−)-enantiomer of the target alkaloid (−)-gephyrotoxin.

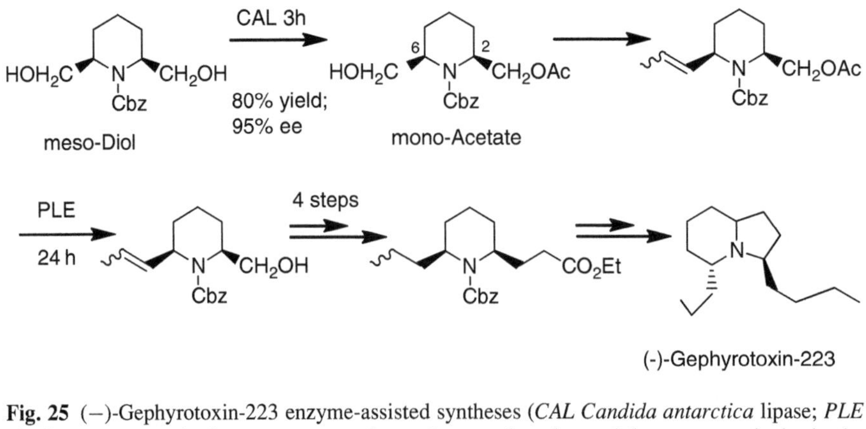

Fig. 25 (−)-Gephyrotoxin-223 enzyme-assisted syntheses (*CAL Candida antarctica* lipase; *PLE* pig liver esterase; vinyl acetate was used as solvent and as the acyl-donor, respectively, in the enzymatic acetylation of the meso-diol)

5 Complex Alkaloids Formed by Spontaneous Assembling?

Are structurally complex natural products also assembled spontaneously without the help of enzymes?

It has been generally accepted that natural products are nearly exclusively formed by enzyme catalysed reactions because of usually complete stereoselectivity and complete or high stereo- and regio-specificity of the enzymes. They can be regarded as "high-speed" transformations, compared to "low-speed" biomimetic reactions. There are, however, sporadic discussions in the literature as to whether this accounts for all natural products. In a recent microreview by Gravel and Poupon [152] a selection of examples is described including various classes and structural types of secondary metabolites which are assumed to be generated spontaneously without enzyme assistance. Mechanisms of intra-molecular cyclisations of reactive linear molecules, of photo-induced reactions, spontaneous dimerisations and so-called multi-component reactions are summarised and discussed as being formed through non-enzymatic means [152]. The following two remarkable examples from the alkaloid field are illustrated by a light-induced conversion (Fig. 26) and by a spontaneous dimerisation (Fig. 27) leading to indole alkaloids.

Brevianamide A, an indole alkaloid from *Penicillium*, was isolated first and later its derivatives C and D were detected. In a simple experiment it could be demonstrated that C and D were formed when the fungus was grown in the light but were not found under light-exclusion conditions. This observation clearly indicates a

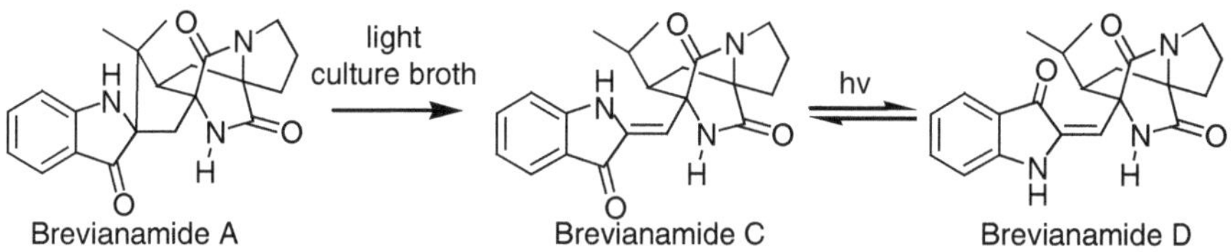

Fig. 26 Brevianamide A conversion into its isomers brevianamide C and D by exposure to light

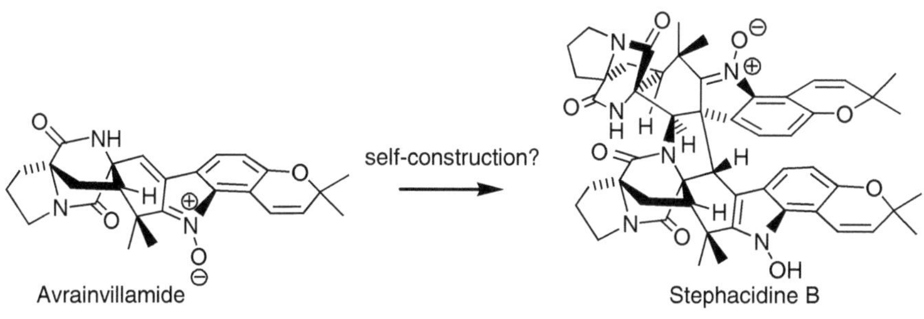

Fig. 27 Suggested stephacidine B formation by artificial not enzyme-dependent dimerisation of avrainvillamide

photo-induced rearrangement of compound A into C. In this context one should remember that one of the basic biochemical reactions in plants, the recognition of red and blue light, is based on light-dependent isomerisation of phytochromes 660 and 730, which suggests that such a process may occur more frequently than might be expected.

More complicated alkaloid structures like stephacidine B exhibiting the impressive number of 15 rings and 9 stereogenic centres may have originated by spontaneous, not enzyme-catalysed dimerisation of avrainvillamide. The efficient bio-mimetic conversion of the precursor molecule avrainvillamide into its dimer stephacidine B under diverse experimental conditions by two Michael additions and additionally the retro-dimerisation of the dimer would support a self-assembly mechanism for formation of the dimer B [153–155].

6 Conclusions and Perspectives

During the last two decades, the use of enzyme-catalysed reactions in the total chemical synthesis and in the biological formation of alkaloids has been rapidly developed and a large number of enzymes and genes (cDNAs) involved in the biosynthesis of alkaloids have become available.

Based on these achievements, various modern concepts of alkaloid "synthesis" have been developed. The chemo-enzymatic approaches to alkaloids applied by synthetic chemists are in general all very similar, as shown by the examples discussed; organic chemists create chiral synthons by applying enzymes *in vitro* and *in vivo*. These synthons are used as starting materials for divergent synthetic routes. They must be available in high gram-scales (10–50 g), otherwise such an approach is unlikely to be of interest to synthetic chemists. The use of whole cells (recombinant microorganisms) is becoming more important and there is likely to be an increasingly significant demand for using such commercially available systems in organic chemistry. In contrast, enzymes from biosynthetic pathways have been largely neglected by alkaloid chemistry so far, and is unlikely that there will be significant progress as long as there are no efficient and simple techniques available for use. Why should a chemist try to apply, e.g. a special reductase he needs in a synthetic sequence if it cannot be selected from a catalogue or its cDNA cannot be overexpressed over the weekend with a commercially available kit and the reaction be performed on Monday?

Such developments are still likely to require at least a decade. Production of natural products by enzymatic means alone by biological chemists will presently meet similar problems and might become successful only in special cases. The most promising strategy in future alkaloid production might be the so-called "synthetic biology". An increasing number of genome analyses for pharmaceutically interesting microorganisms and useful medicinal plants is expected to be performed in the near future and will deliver an excess of novel genes (cDNAs) involved in natural product biosynthesis.

The achievements in using recombinant *E. coli* and yeast strains for production of natural metabolites have been impressive in the last decade. The biological hydrocortisone synthesis by multiple-engineered yeast [121, 122] exemplifies a recently established new platform for the development of alkaloid production strategies for future demands and represents an attractive case of "synthetic biology". There is the need however for these platforms to be optimised and rendered more efficient. As an example, from the point of view of productivity, berberine type alkaloids so far generated at milligram-level/litre of recombinant *E. coli* or yeast, must in future compete with about 400 times higher production observed in plant cell cultures (3.5 g/L) more than two decades ago [156]. From a commercial point of view these amounts were still not high enough. Where generation of alkaloid diversity is concerned, the combination of synthetic biology with the search and application of novel genes and rationally designed and engineered enzymes, should provide a bright future not only in the field of alkaloids but also in natural product research in general. The importance of modern search and creation of enzymes applicable in organic chemistry has recently been outlined in detail [157].

Acknowledgements Support by Fonds der Chemischen Industrie (Frankfurt/Main, Germany), K.P. Chao's High-Tech Foundation at Zhejiang University (Hangzhou, China) and National Basic Research of China, 973 Program (2009CB521906) is appreciated.

References

1. Dakin HD (1903) Proc Chem Soc 19:161
2. Dakin HD (1905) J Physiol 32:199
3. Zenk MH, Juenger M (2007) Phytochemistry 68:2757
4. O'Connor SE, Maresh JJ (2006) Nat Prod Rep 23:532
5. Ruppert M, Ma X, Stöckigt J (2005) Curr Org Chem 9:1431
6. Barleben L, Stöckigt J (2008) Sci Cult 74:71
7. Sottomayor M, Lopes Cardoso I, Pereira LG, Ros Barcelo A (2004) Phytochem Rev 3:159
8. Kumar S, Dutta A, Sinha AK, Sen J (2007) FEBS J 274:1290
9. Costa MM, Hilliou F, Duarte P, Pereira LG, Almeida I, Leech M, Memelink J, Barceló AR, Sottomayor M (2008) Plant Physiol 146:403
10. Cordell GA (1998) The alkaloids: chemistry and biology, vol 50. Academic Press, San Diego, p 260
11. Islam MN, Iskander MN (2004) Minirev Med Chem 4:1077
12. DeLuca V (2003) Recent Adv Phytochem 37:181
13. Murata J, Roepke J, Gordon H, DeLuca V (2008) Plant Cell 20:524
14. Kutchan TM (1998) In: Cordell GA (ed) The alkaloids: chemistry and biology, vol 50. Academic Press, San Diego, p 257
15. Weid M, Ziegler J, Kutchan TM (2004) Proc Natl Acad Sci USA 101:13957
16. Galneder E, Rüffer M, Wanner G, Tabata M, Zenk MH (1988) Plant Cell Rep 7:1
17. Sato F, Takeshita N, Fitchen JH, Fujiwara H, Yamada Y (1993) Phytochemistry 32:659
18. Sato F, Tsujita T, Katagiri Y, Yoshida S, Yamada Y (1994) Eur J Biochem 225:125
19. Kutchan TM, Hampp N, Lottspeich F, Beyreuther K, Zenk MH (1988) FEBS Lett 237:40
20. Kutchan TM (1989) FEBS Lett 257:127

21. Hezari M, Croteau R (1997) Planta Med 63:291
22. Jennewein S, Wildung MR, Chau MD, Walker K, Croteau R (2004) Proc Natl Acad Sci USA 101:9149
23. Dräger B (2004) Nat Prod Rep 21:211
24. Biastoff S, Dräger B (2007) In: Cordell GA (ed) The alkaloids, chemistry, biology, vol 64. Academic Press, New York, p 49
25. Chan P (1995) Eur Neurol 35(Suppl 1):23
26. Li S, Long J, Ma Z, Xu Z, Li J, Zhang Z (2004) Curr Med Res Opin 20:409
27. Li S (1998) Eur Neurol 93:26
28. Stöckigt J, Soll HJ (1980) Planta Med 40:22
29. Zenk MH, El-Shagi H, Arens H, Stöckigt J, Weiler EW, Deus B (1977) In: Barz W, Reinhard E, Zenk MH (eds) Plant tissue culture and its biotechnological application. Springer, Berlin, p 27
30. Zenk MH (1980) J Nat Prod 43:438
31. Stöckigt J, Hemscheidt T, Höfle G, Heinstein P, Formacek V (1983) Biochemistry 22:3448
32. Stöckigt J, Zenk MH (1977) J Chem Soc Chem Commun 646
33. Stöckigt J, Ruppert M (1999) In: Barton DHR, Nakanishi K (eds) Comprehensive natural products chemistry, vol 4. Elsevier, Oxford, p 109
34. Stöckigt J (1980) In: Philipson JD, Zenk MH (eds) Indole and biogenetically related alkaloids. Academic Press, London, pp 113–141
35. Rolf S, Bruns H-J, Wichter T, Kirchhof P, Ribbing M, Wasmer K, Paul M, Breithardt G, Haverkamp W, Eckhardt L (2003) Eur Heart J 24:1104
36. Aniszewski T (2007) Alkaloids – secrets of life alkaloid chemistry. Biological significance: applications and ecological role. Elsevier, Amsterdam, p 181
37. Ma X, Koepke J, Fritzsch G, Diem R, Kutchan TM, Michel H, Stöckigt J (2004) Biochim Biophys Acta 1702:121
38. Ma X, Panjikar S, Koepke J, Loris E, Stöckigt J (2006) Plant Cell 18:907
39. Stöckigt J, Barleben L, Panjikar S, Loris E (2008) Plant Physiol Biochem 46:340
40. Barleben L, Ma X, Koepke J, Peng G, Michel H, Stöckigt J (2005) Biochim Biophys Acta 1747:89
41. Barleben L, Panjikar S, Ruppert M, Koepke J, Stöckigt J (2007) Plant Cell 19:2886
42. Yang L, Hill M, Wang M, Panjikar S, Stöckigt J (2009) Angew Chem Int Ed 48:5211
43. Ma X, Koepke J, Bayer A, Linhard V, Fritzsch G, Zhang B, Michel H, Stöckigt J (2004) Biochim Biophys Acta 1701:129
44. Ma X, Koepke J, Panjikar S, Fritzsch G, Stöckigt J (2005) J Biol Chem 280:13576
45. Ruppert M, Panjikar S, Barleben L, Stöckigt J (2006) Acta Crystallogr F 62:257
46. Rosenthal C, Mueller U, Panjikar S, Sun L, Ruppert M, Zhao Y, Stöckigt J (2006) Acta Crystallogr F 62:1286
47. Kutchan TM (1993) Phytochemistry 32:493
48. Hemscheidt T, Zenk MH (1980) FEBS Lett 110:187
49. Geerlings A, Ibanez MM, Memelink J, van der Heiden, Verpoorte R (2000) J Biol Chem 275:3051
50. Gerasimenko I, Sheludko Y, Ma X, Stöckigt J (2002) Eur J Biochem 269:2204
51. Schmidt D, Stöckigt J (1995) Planta Med 61:254
52. Dogru E, Warzecha H, Seibel F, Haebel S, Lottspeich F, Stöckigt J (2000) Eur J Biochem 267:1397
53. Mattern-Dogru E, Ma X, Hartmann J, Decker H, Stöckigt J (2002) Eur J Biochem 269:2889
54. Pfitzner A, Polz L, Stöckigt J (1986) Z Naturforsch 41c:103
55. Bayer A, Ma X, Stöckigt J (2004) Bioorg Med Chem 12:2787
56. Falkenhagen H, Stöckigt J (1995) Z Naturforsch 50c:45
57. Gao S, von Schumann G, Stöckigt J (2002) Planta Med 68:906
58. von Schumann G, Gao S, Stöckigt J (2002) Bioorg Med Chem 10:1913
59. Ruppert M, Woll J, Giritch A, Genady E, Ma X, Stöckigt J (2005) Planta 222:888

60. Stöckigt J, Pfitzner A, Keller PI (1983) THL 24:2485
61. Stöckigt J, Panjikar S, Ruppert M, Barleben L, Ma X, Loris E, Hill M (2007) Phytochem Rev 6:15
62. Stöckigt J, Panjikar S (2007) Nat Prod Rep 24:1382
63. Goss RJM, Newill PLA (2006) Chem Commun 4924
64. Lee M, Philips RS (1992) Bioorg Med Chem Lett 2:1563
65. Philips RS (2004) Tetrahedron Asymmetry 15:2787
66. Blaser G, Sanderson JM, Batsanov AS, Howard JAK (2008) Tetrahedron Lett 49:2795
67. Ngnigha ARK, Muffler K, Ernyei A, van Pee K-H, Ulber R (2008) Chemie Ingenieur Technik 80:783
68. Heinstein PF, Lee SI, Floss HG (1971) Biochem Biophys Res Commun 44:1244
69. Li S-M Phytochemistry 70:1746
70. Steffan N, Grundmann A, Yin W-B, Kremer A, Li S-M (2009) Curr Med Chem 16:218
71. Li S-M (2009) Appl Microbiol Biotechnol 84:631
72. Kremer A, Li S-M (2008) Appl Microbiol Biotechnol 79:951
73. Ruan H-L, Yin W-B, Wu Y-Z, Li S-M (2008) Chembiochem 9:1044
74. Pfitzner U, Zenk MH (1982) Planta Med 46:10
75. Pfitzner U, Zenk MH (1987) Methods Enzymol 136:342
76. Loris EA, Panjikar S, Ruppert M, Barleben L, Unger M, Schübel H, Stöckigt J (2007) Chem Biol 14:979
77. Bernhardt P, McCoy E, O'Connor SE (2007) Chem Biol 14:888
78. Runguphan W, O'Connor SE (2009) Nat Chem Biol 5:151
79. Stöckigt J, Hammes B, Ruppert M (2010) Nat Prod Res 24:759
80. Geerlings A, Redondo FJ, Contin A, Memelink J, van Der Heijden R, Verpoorte R (2001) Appl Microbiol Biotechnol 56:420
81. Maresh IJ, Giddings L-A, Friedrich A, Loris EA, Panjikar S, Trout BL, Stöckigt J, Peters B, O'Connor SE (2008) J Am Chem Soc 130:710
82. Danieli B, Lesma G, Mauro M, Palmisano G, Passarella D (1994) Tetrahedron 50:8837
83. Danieli B, Lesma G, Mauro M, Palmisano G, Passarella D (1995) J Org Chem 60:2506
84. Riva R, Banfi L, Danieli B, Guanti G, Lesma G, Palmisano G (1987) J Chem Soc Chem Commun 299
85. Danieli B, Lesma G, Mauro M, Palmisano G, Passarella D (1990) Tetrahedron Asymmetry 1:793
86. Danieli B, Lesma G, Passarella D, Silvani A (2000) Curr Org Chem 4:231
87. Laumen K, Schneider M (1985) Tetrahedron Lett 26:2073
88. Isobe M, Fukami N, Goto T (1985) Chem Lett 14:71
89. Schuelzer GH, Gurib-Fa-kim (2008) Plant resources of tropical Africa 11(1). Medicinal plants. PROTA Foundation/Backhuys Publishers/CTA/Wageningen 568
90. Dalton DR (1979) The alkaloids. Marcel Dekker, New York, p 509
91. Das B, Madhusudhan P (1999) Nat Prod Res 14:135
92. Ziegler J, Voigtlaender S, Schmidt J, Kramel R, Miersch O, Ammer C, Gesell A, Kutchan TM (2006) Plant J 48:177
93. Weitz CJ, Faull KF, Goldstein A (1987) Nature 330:674
94. Poeaknapo C, Schmidt J, Brandsch M, Draeger B, Zenk MH (2005) Proc Natl Acad Sci USA 101:14091
95. Boettcher C, Fellermeier M, Boettcher C, Draeger B, Zenk MH (2005) Proc Natl Acad Sci USA 102:8495
96. Grobe N, Zhang B, Fisinger U, Kutchan TM, Zenk MH, Guengerich FP (2009) J Biol Chem 284:24425
97. Zenk MH (1994) Pure Appl Chem 66:2023
98. Bock A, Wanner G, Zenk MH (2002) Planta 216:57
99. Keasling J (2005) Bridge Natl Acad Eng 35:18
100. Ckless K, Schlottfeld JL, Pasqual M, Moyna P, Henriques JH, Wajner M (1995) J Pharm Pharmacol 47:1029

101. Kuo CL, Chi CW, Liu TY (2004) Cancer Lett 203:127
102. Kang BY, Chung SW, Cho D, Kim TS (2002) Biochem Pharmacol 63:1901
103. Li Y, Chen X, Liu H, Luo F, Li G (2009) Zhongguo Zhong Yao Za Zhi 34:324
104. Mantena SK, Sharma SD, Katiyar SK (2006) Carcinogenesis 27:2018
105. Maiti M, Kumar GS (2006) Med Res Rev 27:p683
106. Ziegler J, Facchini PJ (2008) Annu Rev Plant Biol 59:735
107. Winkler A, Kutchan TM, Macheroux P (2007) J Biol Chem 282:24437
108. Kato N, Dubouzet E, Kokabu Y, Yoshida S, Taniguchi Y, Dubouzet JG, Yazaki K, Sato F (2007) Plant Cell Physiol 48:8
109. Ikezawa N, Tanaka M, Nagayoshi M, Shinkyo R, Sakaki T, Inouye K, Sato F (2003) J Biol Chem 278:38557
110. Allen RS, Millgate AG, Chitty JA, Thisleton J, Miller JA, Fist AJ, Gerlach WL, Larkin PJ (2004) Nat Biotechnol 22:1559
111. Allen RS, Miller JA, Chitty JA, Fist AJ, Gerlach WL, Larkin PJ (2008) Plant Biotechnol J 6:22
112. Fuijii N, Inui T, Iwasa K, Morishige T, Sato F (2007) Transgenic Res 16:363
113. Frick S, Kramell R, Kutchan TM (2007) Metabol Eng 9:169
114. Larkin PJ, Miller JA, Allen RS, Chitty JA, Gerlach WL, Frick S, Kutchan TM, Fist AJ (2007) Plant Biotechnol J 5:26
115. Millgate AG, Pogson BJ, Wilson IW, Kutchan TM, Zenk MH, Gerlach WL, Fist AJ, Larkin PJ (2004) Nature 431:413
116. Sato F, Inui T, Takemura T (2007) Curr Pharm Biotechnol 8:211
117. Hung TM, Li JP, Min BS, Choi JS, Na M, Zhang X, Ngoc TM, Lee I, Bae K (2007) Biol Pharm Bull 30:1157
118. Hung TM, Na M, Min BS, Zhang X, Lee I, Ngoc TM, Thuong PT, Sok DE, Bae K (2007) Planta Med 73:1281
119. Rashid MA, Gustafson KR, Kashman Y, Cardellina JH II, McMahon JB, Boyd MR (1995) Nat Prod Res 6:153
120. Minami H, Kim J-S, Ikezawa N, Takemura T, Katayama T, Kumagai H, Sato F (2008) Proc Natl Acad Sci USA 105:7393
121. Szczebara FM, Chandelier C, Villeret C, Masurel A, Bourot S, Duport C et al (2003) Nat Biotechnol 21:143
122. Dumas B, Brocard-Masson C, Assemat-Lebrun K, Achstetter T (2006) Biotechnol J 1:299
123. Hawkins KM, Smolke CD (2008) Nat Chem Biol 4:564
124. Scott JD, Williams RM (2002) Chem Rev 102:1669
125. Kathawala FG, Coppola GM, Schuster HF (eds) (1990) The chemistry of heterocyclic compounds, vol 28. Wiley, New York
126. Benz P, Wohlgemuth R (2007) J Chem Technol Biotechnol 82:1082
127. Gibson DT, Koch JR, Schuld CI, Kallio RE (1968) Biochemistry 7:3795
128. Johnson RA (2004) Org React 63:117
129. Omori AT, Finn KJ, Leisch H, Caroll RJ, Hudlicky T (2007) Synlett 18:2859
130. Zezula J, Rice KC, Hudlicky T (2007) Synlett 18:2863
131. Reimann E (1995) European Patent Application EP 647640
132. Reimann E (1995) Chem Abstr 123:33492
133. Lei A, He M, Zhang X (2002) J Am Chem Soc 124:8198
134. Csuk R, Woeste B (2008) Tetrahedron 64:9384
135. Viladomat R, Bastida J, Codina C, Campbell WE, Mathee S (1995) Phytochemistry 40:307
136. da Silva AF, de Andrade JP, Bevilaqua LR, de Souza MM, Izquierdo I, Henriques AT, Zuanazzi JA (2006) Pharmacol Biochem Behav 85:148
137. Banwell MG, Coster MJ, Harvey MJ, Moraes J (2003) J Org Chem 68:613
138. Banwell MG, Kokas OJ, Willis AC (2007) Org Lett 9:3503
139. Kokas OJ, Banwell MG, Willis AC (2008) Tetrahedron 64:6444
140. Matveenko M, Banwell MG, Willis AC (2008) Org Lett 10:4693

141. Pettit GR, Eastham SA, Melody N, Orr B, Herald DL, McGregor J, Knight JC, Doubek DL, Pettit GR III, Garner LC, Bell JA (2006) J Nat Prod 69:7
142. Matveenko M, Kokas OJ, Banwell MG, Willis AC (2007) Org Lett 9:3683
143. Hudlicky T (2000) J Heterocycl Chem 37:535
144. Rinner M, Hudlicky T (2005) Syn Lett 365
145. Kato A, Adachi I, Miyauchi M, Ikeda K, Komae T, Kizu H, Kameda Y, Watson AA, Nash RJ, Wormald MR, Fleet GWJ, Asano N (1999) Carbohydr Res 316:95
146. Asano N, Kuroi H, Ikeda K, Kizu Y, Kameda Y, Kato A, Adachi I, Watson AA, Nash RJ, Fleet GWJ (2000) Tetrahedron Asymmetry 11:1
147. Oikonomakos NG, Tiraidis C, Leonidas DD, Zographos SE, Kristiansen M, Jessen CU, Norskov-Lauritsen L, Agius L (2006) J Med Chem 49:5687
148. Watson AA, Fleet GWJ, Asano N, Molyneux RJ, Nash RJ (2001) Phytochemistry 56:265
149. Romero A, Wong CH (2000) J Org Chem 65:8264
150. Calveras J, Casas J, Parella T, Joglar J, Clapes P (2007) Adv Synth Catal 349:1661
151. Ghodsi MZ, Chenevert R, Badiei AR (2006) Iran J Chem Eng 25:31
152. Gravel E, Poupon E (2008) Eur J Org Chem 2008:27
153. Herzon SB, Myers AG (2005) J Am Chem Soc 127:5342
154. Baran PS, Hafensteiner BD, Ambhaikar NB, Guerrero CA, Gallaghaer JD (2006) J Am Chem Soc 128:8678
155. Artmann GD III, Grubbs AW, Williams RM (2007) J Am Chem Soc 129:6336
156. Matsubara J, Kitani S, Yoshioka T, Morimoto T, Fujita Y, Yamada Y (1989) Chem Tech Biotechnol 46:61
157. Reetz MT (2009) J Org Chem 74:5767

Top Curr Chem (2010) 297: 105–148
DOI: 10.1007/128_2010_78

Published online: 13 July 2010

Chemoenzymatic and Bioenzymatic Synthesis of Carbohydrate Containing Natural Products

Bohdan Ostash, Xiaohui Yan, Victor Fedorenko, and Andreas Bechthold

Abstract The domain of bioactive natural products contains many oligosaccharides and aglycones decorated with various sugars. Glycan moieties influence essential aspects of biology of small molecules, such as mode of action, target recognition, pharmacokinetics, stability, and others. Methods of generation of novel glycosylated natural products are therefore of great value, as they, for example, may help fight human diseases more efficiently or provide healthier diet. This review covers the existing literature published mainly over the last decade that deals with biology-based approaches to novel glycoforms. Both genetic manipulations of biosynthesis of glycoconjugates and chemoenzymatic synthesis of novel "sweet" molecules are reviewed here. Wherever available, relationships between carbohydrate portions of the natural products and their biological activities are highlighted.

Keywords Natural products · Carbohydrate · Glycosyltransferase · Chemoenzymatic synthesis · Biosynthesis

Contents

B. Ostash and V. Fedorenko
Department of Genetics and Biotechnology, Ivan Franko National University of L'viv, L'viv 79005, Ukraine

X. Yan and A. Bechthold (✉)
Albert-Ludwigs-Universität Freiburg, Institut für Pharmazeutische Wissenschaften, Pharmazeutische Biologie und Biotechnologie, Freiburg 79104, Germany
e-mail: andreas.bechthold@pharmazie.uni-freiburg.de

Abbreviations

GT(s)	Glycosyltransferases
NP(s)	Natural product(s)
Ntfs	Nucleotidylyltransferases
Susy	Sucrose synthase

1 Introduction

The essentiality of carbohydrates for life is indisputable. These macromolecules are found across a wide range of pivotal biochemical processes, like storage of energy/information and cell signaling. Unlike nucleic acids and proteins, sugars can form branching structures. This, in combination with various sugar-tailoring reactions (e.g., deoxygenation, methylation, amination, epimerization, etc.), accounts for the unmatched structural diversity of carbohydrates and their rich biological capacities.

Carbohydrates are often found in natural products (NPs) – small molecules (usually less than 2 kDa) synthesized mainly by bacteria, fungi, and lower and flowering plants. NPs are thought to be "nonessential," or "secondary" metabolites. Consider, for example, gentamicin A_2 **1**, an aminoglycoside antibiotic produced by several actinomycetes belonging to genus *Micromonospora* [1], and Lipid II **2**, a ubiquitous primary metabolite involved in bacterial cell wall biosynthesis. While aminoglycoside-nonproducing mutants can be readily isolated, mutations in genes for Lipid II biosynthesis are lethal [2]. The reasons why genes for production of NPs persist through evolution remain debatable [3–5]. Nevertheless, it is likely that, under certain (natural) conditions, at least some NPs provide(d) selective advantages to the producer. NPs exert valuable biological activities (e.g., anticancer, antibacterial, antifungal, cholesterol-lowering, insecticidal, immunosuppressive, etc.), making them a focus of several industries, most notably those involved with pharmaceuticals.

Structural diversity of glycan moieties of carbohydrate-containing NPs is astonishing and greatly surpasses that found in primary metabolism, reflecting myriads of bioactivities/targets of glyco-NPs. In some cases sugars are directly responsible for interaction of NP with a target. For instance, hydroxyl and amino groups of deoxysugars of gentamicin **1** contact 16S rRNA and thus inhibit ribosome, triggering aminoglycoside-associated bacterial cell death [6]. Taking vancomycin **3** and aranciamycin **4** as an example, glycosylation influences activity and solubility [7]. Judicious alteration of glycan moieties of NPs may improve their activity or other pharmacological properties. Therefore, development of methods of manipulation of carbohydrate structures is of keen interest. Such methods have for a long time been the domain of organic chemists. Recently, progress in understanding the biosynthetic pathways of glyco-NPs has opened the door for biology-driven approaches towards manipulation of NPs and glycan moieties in particular. These approaches hinge on identification and exploitation of carbohydrate biosynthetic genes for enzymes showing substrate ambiguity (performing the same chemical transformation on different substrates) and/or catalytic promiscuity (catalyze different transformations on similar substrates). Broad or "relaxed" enzymatic properties were recently shown to be a rather common trait of the biology of prokaryotes that creates the basis for metabolic innovation [8]. In the case of secondary metabolism, isolation of such genes is facilitated by the fact that genetic determinants of biosynthesis of glyco-NPs are usually clustered. Two convergent but methodologically distinct ways can be followed to generate novel glycoforms. In one approach, genetic manipulations are used to create novel biosynthetic pathways for *in vivo* production of novel molecules. This approach we refer to as "bioenzymatic," underscoring the use of the cells as biological factories to make an NP. In the second approach, known as "chemoenzymatic," purified enzymes are used *in vitro* to produce a desired compound in the presence of appropriate substrates and cofactors. The inherent advantages of bioenzymatic approach are: (1) use of cellular machinery to produce complex aglycon and glycan portions of NP; (2) good potential to increase the production of novel compound of interest on industrial scale; (3) simplicity (in most cases) of genetic manipulations of NP-producing bacteria as compared to *in vitro* reconstitution of enzymatic reaction. At the same time, the bioenzymatic approach is, essentially, a metabolic "black box," that usually gives a complex mixture of compounds requiring tedious purification and thorough chemical analysis. *In vitro* systems are more flexible than *in vivo* ones in terms of supply of substrates and control of reaction environment. For example, it is possible to carry out *in vitro* reaction in nonaqueous media [9], which is impossible *in vivo*. In real life, both approaches evolve and intertwine, blurring the boundary between them. They are believed to offer ultimately economically sound, faster, and greener solutions for generation of novel glycoconjugates.

To date, great numbers of gene clusters for glyco-NPs have been identified and characterized. In this review we summarize major achievements in the field of diversification of carbohydrate moieties of NPs over the last decade. We start with a

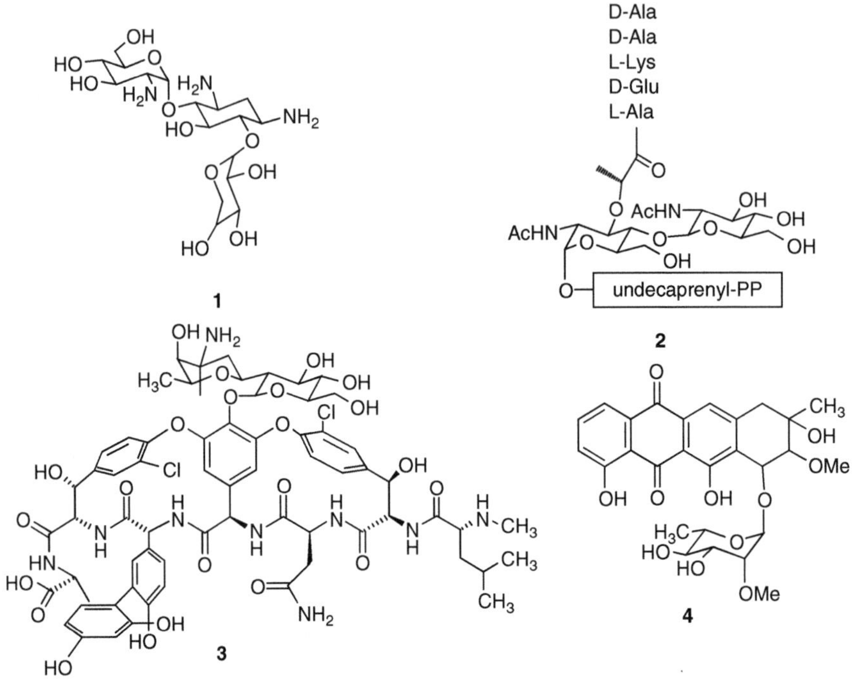

Fig. 1 Structures of gentamicin, lipid II, vancomycin, and aranciamycin

description of *in vivo* and *in vitro* strategies towards NDP-hexoses, an activated form of sugars which are necessary cosubstrates during biosynthesis of glyco-NPs. Then we describe different approaches for generation of novel glycoforms within structurally and biogenetically different classes of NPs. Whenever such information exists, we shall also mention how changes in the carbohydrate portion of given classes of NPs influence their biological activities. Finally, a general outline of main issues in the field of glycodiversification as well as future research directions will conclude the review (Fig. 1).

2 NDP-Sugars as Donor Substrates in Glycosylation Reactions: Strategies Toward Their Generation

Glycosylation requires nucleotide diphosphate activated sugars (NDP-sugars). The phosphonucleotidyl moiety serves as both a recognition element for enzymes involved in the biosynthetic pathway and a leaving group for glycosyltransfer reaction [10]. Several GTs have been shown to be very promiscuous with respect to their NDP-sugar donors [11]. Thus, a limiting factor in employing this promiscuity to produce new NPs is the lack of the NDP-sugars [12]. Although synthetic routes to these complex NDP-sugars are feasible [13–15], they are often plagued by laborious processes.

In contrast, enzymatic routes to NDP-sugars could be conducted in a relatively short time by stereo- and regioselective reactions. For this reason, currently NDP-sugars are mainly produced *in vitro* by chemoenzymatic strategies or *in vivo* by bioenzymatic strategies [16]. Chemoenzymatic strategies generally utilize two purified enzymes, an anomeric sugar kinase, and a nucleotidylyltransferase (Ntf) to generate NDP-sugars; bioenzymatic strategies require expression of sugar biosynthetic genes in a suitable host. Once these NDP-sugars are formed, combinations of different reactions, such as deoxygenation, epimerization, oxidation/reduction, transamination, alkylation, and decarboxylation, diversify them into a large array of usual and unusual NDP-deoxysugars [17].

A remarkable advantage in chemoenzymatic methods is that they allow for the activation of sugars with "reactive handles" (e.g., azide, thiols, ketones, aminooxy substituents) which can be further modified to enhance the diversity of the final glycol-NPs [18]. However, the general application of chemoenzymatic approaches is limited by expensive reagents in the reactions and by the difficulty in establishing conditions for multienzyme reactions [19]. Bioenzymatic routes are advantageous in that they are scalable and can use cheap substrates [20]. But their application is hampered by the background activities from the host organism and by the difficulty in controlling the reaction conditions (Fig. 2).

2.1 Generation of NDP-Sugars by In Vitro *Glycorandomization*

In vitro glycorandomization (IVG) is a chemoenzymatic method that uses substrate-flexible anomeric sugar kinases and Ntfs to convert diverse sugars into their NDP-activated forms [21]. The ability to enhance the promiscuity of these two

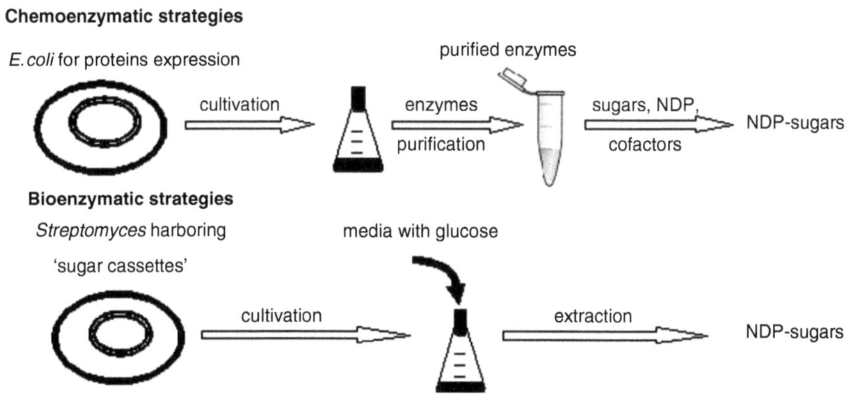

Fig. 2 Chemo- and bioenzymatic strategies toward NDP-sugars

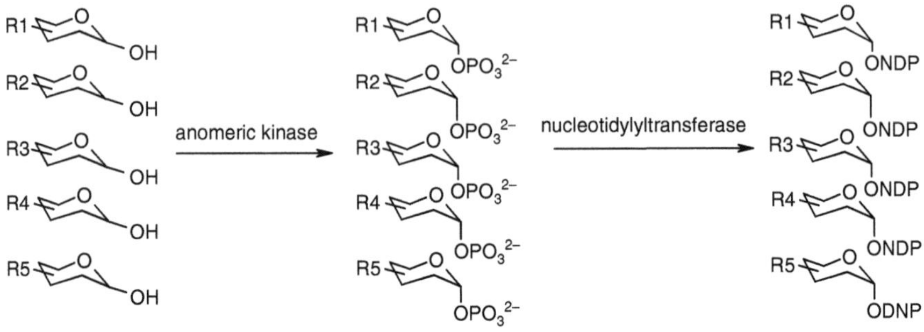

Fig. 3 Schematic for NDP-sugars generation by *in vitro* glycorandomization

enzyme families is critical for this approach. Although application of evolutionary methods together with a screen for enzymes with desired properties is feasible [22, 23], the efficiency of such an approach is still low. Thorson and coworkers applied the structure-based engineering method to enhance the promiscuity of the anomeric sugar kinase GalK and nucleotidylyltransferase E_p, and generated numerous NDP-sugars with the engineered enzymes (Fig. 3).

2.1.1 Production of Sugar-1-Phosphates by Engineered Anomeric Sugar Kinase GalK

Sugar-1-phosphates are the starting materials in the sugar activation processes. Thus, the ability to form sugar-1-phosphate directly influences the synthesis of NDP-sugar. Many synthetic routes to sugar phosphates already exist; these methods are often tedious due to multistep synthetic reactions [12, 16, 24]. Phosphorylation of sugars by kinase would be a better choice, as it is a one-step process. Formation of α-D-galactose-1-phosphate (Gal-1-P) **5**, for example, could be catalyzed by the galactokinase (GalK) from galactose **6** and ATP. However, all the naturally occurring kinases were shown to have limited substrate scopes [25–28]. Therefore, in order to apply these kinases to generate sugar-1-phosphates, their substrate flexibilities toward monosaccharides must first be enhanced.

Utilizing the well characterized *E. coli* GalK [29] as a model system, Thorson and coworkers combined a directed evolution approach with a high-throughput multisugar colorimetric assay [28] to enhance the catalytic capabilities of this enzyme [30]. A mutant (Y371H) exhibited kinase activity toward D-galacturonic acid **7**, D-talose **8**, L-altrose **9**, L-glucose **10**, and 6-amino-D-galactose **11**, all of which cannot be recognized by the wild-type GalK [30]. Afterward, based on a homology model with the crystal structure of *L. lactis* GalK [31], Thorson and coworkers developed a structure–activity model [32] and generated another *E. coli* GalK mutant (M173L) with enhanced promiscuity [33]. The double mutant enzyme

(M173L-Y371H) displayed kinase activity toward at least 22 sugars, which cannot be utilized by the wild-type GalK [32].

2.1.2 Production of NDP-Sugars by Engineered Nucleotidylyltransferases E_p

Ntfs catalyze the condensation of sugar-1-phosphates with nucleotide triphosphates (NTPs), yielding NDP-sugars and pyrophosphate. However, Ntfs involved in secondary metabolism are often substrate stringent [34, 35]. Out of the numerous Ntfs, Thorson and coworkers started their structures-based engineering work with the uniquely promiscuous *rmlA*-encoded α-D-glucopyranosyl phosphate thymidylyltransferase (E_p) from *Salmonella enterica* LT2 [36]. E_p catalyzes the conversion of α-D-glucopyranosyl phosphate (Glc-1-P) and dTTP to dTDP-α-D-glucose and pyrophosphate [37]. Unique among Ntfs is that this enzyme displays unusual flexibility toward both sugar phosphates and NTPs [38]. Specifically, E_p converts a broad spectrum of α-D-hexopyranosyl- and α-D-pentopyranosyl phosphates to their dTDP- and UDP-sugars [39]. The structure of E_p [40], as well as its homologs in *Pseudomonas aeruginosa* (RmlA) [37] and *E. coli* (G1p-TT) [41], revealed the molecular details of substrate recognition and substrate specificity of E_p.

Using a structure-based engineering approach, Nikolov and coworkers created several E_p variants capable of using sugar phosphates that were not accepted by the wild-type enzyme [42]. The W224H mutant accommodated bulkier substitutions at C-6 [40], and the T201A mutation allowed larger functional groups at C-2 and C-3. Notably, the L89T mutant displayed enhanced activity toward sugar phosphate substrates with unique C(2), C(3), and C(4) substitutions. Application of such E_p variants has led to formation of about 40 NDP-sugars [39].

2.1.3 Chemoenzymatic Synthesis of NDP-Sugars by Sucrose Synthase

Sucrose synthase (UDPG: D-fructose-2-glucosyltransferase, Susy) is a unique plant GT catalyzing the reversible cleavage of sucrose with NDP to generate activated glucose and D-fructose. It was reported that Susy from rice grains [43] and recombinant Susy1 from potato [44–46] were able to synthesize at least five different NDP-glucoses (UDP-, dUDP-, dTDP-, ADP-, and CDP) [46]. Different from the IVG method, Elling and coworkers followed an alternative route using Susy to synthesize NDP-sugars from sucrose and NDPs. Production of dTDP-glucose from sucrose and dTDP was first conducted in a continuous mode in an enzyme membrane reactor (EMR) [47]. In this EMR, 90% conversion after 2 h with a space–time-yield of 98 g L^{-1} d^{-1} was achieved. Furthermore, they set up a Susy module in which recombinant NMP kinase [48] from yeast, pyruvate kinase and recombinant Susy1 from potato were combined. This module enables production of NDP-glucose from the economic substrates sucrose and NMP. As the presence of 300–500 mM sucrose prolongs the stability of these enzymes, this module can be used in a repetitive batch mode [49].

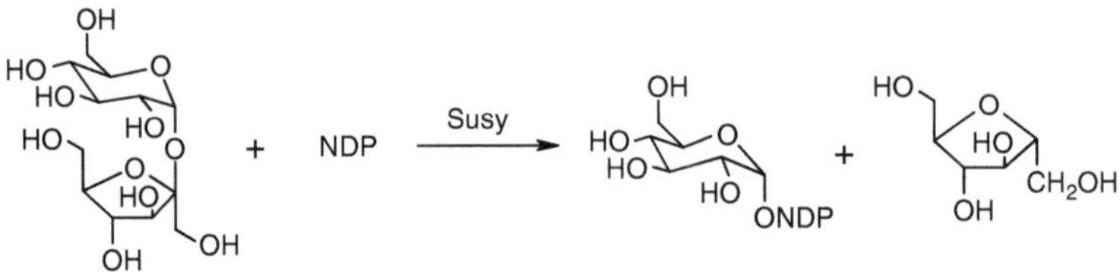

Fig. 4 Synthesis of nucleotide sugars from sucrose and nucleoside diphosphates (NDPs) by a sucrose synthase

Addition of dTDP-Glc-4,6-dehydratase (RmlB) to this module opens the door toward the diverse dTDP-deoxysugars. This enzyme catalyzes the formation of NDP-4-keto-6-deoxysugars, the common intermediate in almost all deoxyhexose biosynthetic pathways [50–52] (Fig. 4).

2.1.4 Generation of NDP-Sugars by Archaeal Nucleotidylyltransferases

Archaea contain many extremophiles and therefore can be a source of heat-stable enzymes for industrial applications [53]. A few archaeal Ntfs have recently been shown to be highly flexible toward their substrates. For instance, ST0452 of the *Sulfolobus tokodaii* accepts dTTP, dATP, dCTP, dGTP, and UDP in the presence of glucose-1-phosphate (Glc-1-P) **12**. It also accepts *N*-acetylglucosamine-1-phosphate (GlcNAc1P) **13** in the presence of UTP and dTTP [54]. A uridyltransferase (*Tca* U_p) from the *Thermus thermophilus* activates GlcNAc1P, Gal1P, Man1P **14**, and Glc1P in the presence of UTP, and utilizes ATP, GTP, CTP, and TTP in the presence of Glc1P [55].

Pohl and coworkers identified a bifunctional mannose-6-phosphate isomerase/GDP-mannose pyrophosphorylase (ManC) from the thermophile *Pyrococcus furiosus* DSM 3638 [56]. This enzyme could catalyze the formation of GDP-mannose from mannose-1-phosphate and GTP. GDP-mannose is an essential metabolic intermediate for the biosynthesis of other GDP-sugars, such as GDP-fucose **15** [57], GDP-colitose **16** [58], GDP-talose **17** [59], GDP-perosamine **18** [60], and GDP-D-rhamnose **19** [61]. Purified ManC showed surprising thermo- and pH stability. Substrate promiscuity tests showed that ManC accepted all five major NTPs (GTP, ATP, CTP, UTP, and dTTP) in the presence of mannose-1-phosphate. It also showed a relatively high degree of acceptance to various sugar-1-phosphates, such as, Man1P, Glc1P, GlcN1P, and GlcNAc1P. Overall, this enzyme was used for the synthesis of 17 different nucleotide sugars from the commercially available sugar-1-phosphates and NTPs [58]. Pohl and coworkers also studied the substrate tolerance of the UDP-α-D-glucose pyrophosphorylase (UDPG-PPase) from *P. furiosus*. This enzyme showed maximum activity at 99°C, and had little loss of activity at 110°C in phosphate buffer with glycerol. Unlike the ManC, this enzyme accepted Glc1P, Man1P, Gal1P, Fuc1P, GlcN1P, GalN1P, and GlcNAc1P with UTP and dTTP [62]. A conversion rate of 92% for Man1P with UTP was found when the reaction was incubated for 60 min with only 0.01 units of this enzyme (Fig. 5).

2.2 In Vivo *Production of NDP-Deoxysugars by "Sugar Cassettes"*

Other than the chemoenzymatic strategies, NDP-activated deoxysugars can be produced *in vivo* by expressing the plasmids containing sugar biosynthetic genes in suitable hosts [63, 64]. Based on the knowledge of sugar biosynthesis genes, Salas and coworkers designed a plasmid-based sugar-synthesizing system ("cassette plasmids" or "sugar cassettes") that was used to synthesize various NDP-deoxysugars in different streptomycetes. These plasmids replicate in both *E. coli* (origin of replication of pUC19) and *Streptomyces* (ori pIJ101), and contain one or two copies of the strong constitutive promoter *ermE* to control the expression of deoxysugar biosynthetic genes. These genes were amplified by PCR and flanked by unique restriction sites. In this way, each "cassette gene" can be easily moved among the plasmids, thereby creating pathways for biosynthesis of different NDP-deoxysugars [17, 65, 66].

The first of these "sugar cassette" plasmids, pLN2 [67], contains seven genes (*oleVWUYLSE*) from the oleandomycin pathway of *S. antibioticus* involved in the biosynthesis of NDP-L-oleandrose **20**. In order to facilitate the expression and exchange of genes on the "cassette" plasmid, another plasmid, pFL942, which

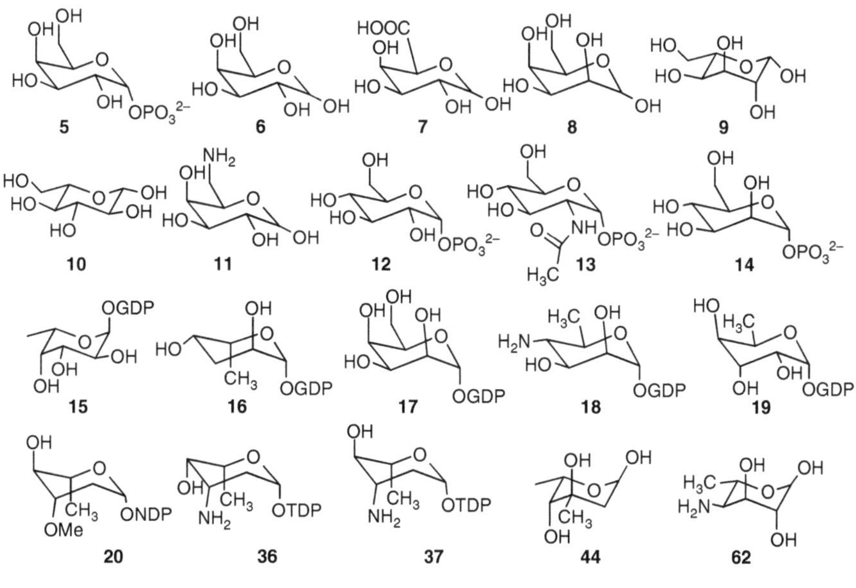

Fig. 5 Structures of selected sugars

contains two divergent *ermEp* promoters to control the genes for NDP-L-mycarose **26** biosynthesis, was also constructed. Using pLN2 and pFL942 as starting plasmids, a series of "cassette plasmids" to direct the biosynthesis of activated dideoxysugars were constructed [68, 69] (Table 1).

Table 1 The sugar cassette plasmids derived from pLN2 and pFL942

Plasmid	Gene cassettes	Deoxysugar	Structure
pLN2	*oleVWUYoleLSE*	NDP-L-olivose **21**	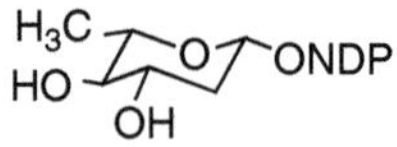
pLN2b	*oleUYoleLSE*	NDP-L-rhamnose **22**	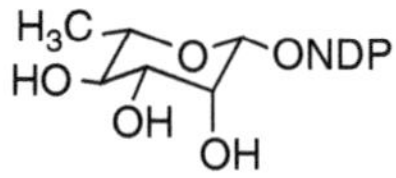
pLNR	*oleVWurdRoleYoleLSE*	NDP-D-olivose **23**	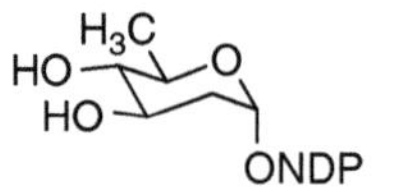
pLNBIV	*oleVWeryBIVoleYloeLSE*	NDP-L-digitoxose **24**	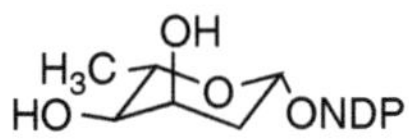
pLNRHO	*oleVWurdZ3oleYoleLSEurdQ*	NDP-L-rhodinose **25**	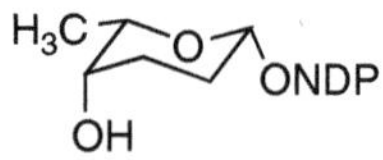
pFL942	*mtmDEoleVeryBIIeryBIVBIIIBVII*	NDP-L-mycarose **26**	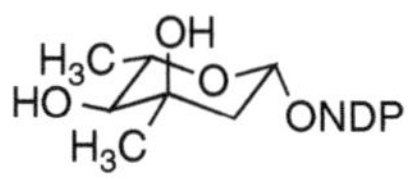
pFL844	*oleVWeryBIVoleYoleLSEurdQ*	NDP-L-amicetose **27**	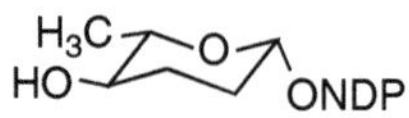
pFL845	*oleVWurdRoleYoleLSEurdQ*	NDP-D-amicetose **28**	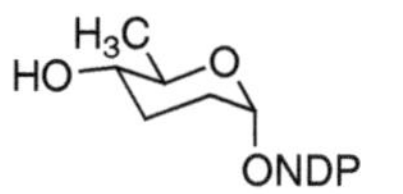
pFL947	*mtmDEoleVWeryBIVmtmCeryBVII*	NDP-L-chromose B **29**	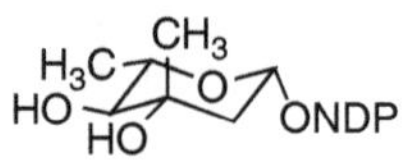
pMP1*UII	*mtmDEoleVWcmmUIIoleY*	NDP-D-oliose **30**	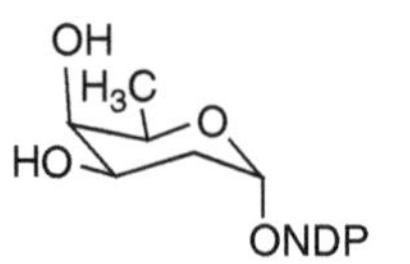
pMP3*BII	*mtmDEoleVeryBIIurdRoleY*	NDP-D-digitoxose **31**	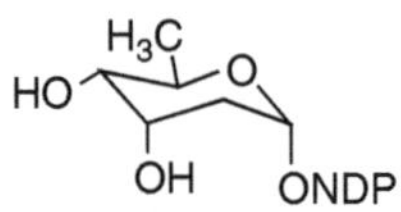
pMP1*BII	*mtmDEoleVeryBIIoleUoleY*	NDP-D-boivinose **32**	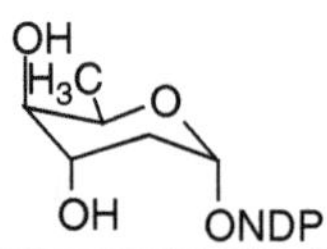

3 Altering Carbohydrate Moieties of NPs Via Bioenzymatic Approaches

Historically, genetic engineering of NP metabolism in bacteria was the first approach towards novel glycosylated compounds. Two basic techniques are available to change a carbohydrate biosynthetic route. A gene for carbohydrate biosynthesis or glycosylation can be disrupted, thus leading to the production of a potentially new molecule lacking certain functionality of the sugar or the sugar itself. Alternatively, gene overexpression in native or heterologous host may cause the accumulation of novel compounds. Both gene disruption and expression can be combined to produce great structural diversity. An aglycone may be produced endogenously or added exogenously to a strain (so-called "biotransformation"), expressing necessary genes for production and attachment of an activated sugar, and its subsequent modification. We use the term "bioenzymatic approach" to cover a variety of methods, where cells are manipulated genetically to produce metabolic pathways for *in vivo* production of novel NPs. Both carbohydrate biosynthetic and GT genes can be manipulated to generate novel glycoforms.

Production of bioactive glycosylated NPs by microorganisms remains a main arena of bioenzymatic manipulations. Because genes for the biosynthesis of carbohydrate moieties of NPs of bacterial origin are usually clustered, it greatly facilitates their exploitation for production of novel NPs. First reports on the biosynthesis of novel glyco-NPs in plants have recently been published [3], and this research area is likely to grow taking into account the vast repertoire and importance of NPs produced by plants.

To date, the biosynthesis of many glycosylated NPs has been analyzed. The opportunity to modulate biological activities of NPs via glycan modification has spurred the interest in applying the bioenzymatic strategies to various families of NPs. Taking into account the biogenetic principle in classification of NPs, below we describe the scope of novel molecules available through *in vivo* manipulations of glycan moiety of parent compounds.

3.1 Aromatic Polyketides

Aromatic polyketides are structurally diverse, often polycyclic molecules that are derived from unreduced polyketone chains. This group of compounds is produced with the help of type II polyketide synthase (PKS), a complex of enzymes that catalyzes the iterative decarboxylative condensation of malonyl-CoA extender units with an acyl starter unit [70]. The carbon framework of aromatic polyketides is further decorated with different functionalities, and carbohydrates are often one of them. Their presence has profound effects on physico-chemical and biological properties of aromatic polyketides. For example, anthracycline aglycones are stable and unpolar, while polyglycosylated anthracyclines are quite polar and often soluble even in water [7]. Sugars are crucial for the ability of many antitumor aromatic polyketides to recognize and bind DNA and, therefore, their modification may lead to better drugs and valuable insights into cancer biology.

3.1.1 Anthracyclines and Tetracenomycins

Anthracyclines are typical aromatic polyketides, probably the most important ones in terms of their clinical utility. They display potent anticancer activities mediated through a variety of mechanisms, with DNA intercalation and DNA topoisomerase I or II inhibition being the most notable ones [71]. Many anthracyclines are glycosylated, for example medically useful doxorubicin **33**, aclacinomycin **34**, and nogalamycin **35** contain different glycan moieties. Manipulation of the carbohydrate portion of naturally occurring anthracyclines may improve their biological properties, as was elegantly demonstrated for **33** [72]. Particularly, TDP-L-daunosamine **36** biosynthetic pathway was changed in the **33** producer *S. peucetius* to give stereoisomer TDP-L-4-epidaunosamine **37**. This was achieved through disruption of native ketoreductase gene *dnmV* in *S. peucetius* and expression of ketoreductase gene *avrE* (from avermectin biosynthetic gene cluster of *S. avermitilis*) in the *dnmV* mutant. Altered activated sugar was attached to aglycon (ε-rhodomycinone **38**), leading to 4-epidoxorubicin **39**. The latter is more effective against certain leukemias than **33**. Compound **39** is not novel in a strict sense, since it was known before as a result of chemical synthesis. Nevertheless, its combinatorial biosynthesis provides clear benefits for industrial production and signals about utility of genetics-driven methods for anthracycline modification.

Several hybrid glycosylated anthracyclines were produced via mixing the genes for **33–35** biosyntheses. Expression of genes involved in glycosylating **34** in *S. peucetius* mutant caused the accumulation of L-rhamnosyl-ε-rhodomycinone **40a** [73]. Aclacinomycin-deficient *S. galilaeus* mutants were used as hosts for expression of *S. nogalater* genes for the biosynthesis of **35**. The 4′-ketoreductase gene *snogC* led to production of aklavinone-4′-epi-2-deoxyfucose **40b** by glycosylation-deficient *S. galilaeus* H039, while introduction of PKS genes involved in the biosynthesis of **35** into the aclacinomycin-minus strain H028 prompted the accumulation of novel glycosylated anthracyclines [74]. Chemical mutagenesis of *S. galileaus* also led to a set of glycosylation-deficient mutants that accumulated aclacinomycins with either shorter or altered glycosidic chains. For instance, the production of aklavinone containing a tri-2-deoxyfucosyl chain was observed in strain H075 blocked in the early steps of formation of dTDP-L-rhodinose, a precursor of terminal sugar in **34** triglycoside chain [75]. In the wild type strain, 2-deoxyfucose is added by GT AknK as a second sugar during **34** assembly [76]. Apparently, when imbalance in deoxysugar biosynthesis occurs, AknK (or other GTs encoded outside of **34** gene cluster) is able to catalyze the formation of novel glyco-NPs. As will be shown below, this kind of promiscuous enzymatic activity is rather common for certain GTs involved in assembly of polyglycosylated NPs. It points to the possibility that kinetic properties of GTs and protein–protein interactions may play an important role in the order of sugar attachment. Overall, these results demonstrate flexibility of the aforementioned genes and strains in directing the biosynthesis of novel glyco-NPs; more extensive genetic "mix and match" experiments may yield greater structural diversity centered on the medically important scaffold of **33**.

Recently, carbohydrate moieties of several other anthracyclines were diversified through bioenzymatic approaches. The *O*-methyltransferase gene *oleY* was overexpressed in a L-rhamnose-containing steffimycin **41a** producer *Streptomyces steffisburgensis* and shown to convert steffimycin into 3′-*O*-methylsteffimycin **41b** [77]. In *Streptomyces albus*, coexpression of a major portion of the steffimycin (*stf*) gene cluster with various plasmids directing the biosyntheses of different 2,6-deoxysugars has led to generation of 12 new derivatives. Of all the new steffimycins, D-boivinosyl-8-demethoxysteffimycin **42** and D-digitoxosyl-8-demethoxy-10-deoxysteffimycin **43** showed improved growth inhibition properties against several human tumor cell lines [77]. New glycosylated derivatives of aranciamycin **4**, another related to steffimycin anthracycline, were obtained after expression of genes for aranciamycin aglycone and its glycosylation (AraGT) in *Streptomyces fradiae* A0 and *S. diastatochromogenes*, capable of making several dTDP-deoxysugars (D-amicetose, L-axenose **44**, L-rhodinose, D-olivose) [78]. Except for L-olivose, all other deoxysugars were successfully attached with AraGT to the aglycone precursor. Aranciamycins B **45**a and D **45b** carrying trideoxysugars L-rhodinose and D-amicetose, respectively, showed significant tumor inhibition activities. Curiously, it was revealed that nonglycosylated aranciamycin derivative possesses the highest activity so far described for the aranciamycin family compounds [79]. Cosmomycins (e.g., cosmomycin D **46**) are another interesting group of anthracyclines with two glycoside tails attached at C7 and C10 position. The latter is not frequently glycosylated, and studies on GTs able to transfer sugars to C10 may lead to novel molecules. To this end, disruption of two GT genes within cosmomycin gene cluster of *Streptomyces olindensis* yielded five new cosmomycins carrying shorter glycoside chains. It turned out that GTs CosG and CosK are able to transfer deoxysugars to both carbon positions of the aglycone, and that CosK may accept either 2-deoxy-L-fucose or L-rhodinose [80].

Approaches used to diversify steffimycin and aranciamycin have been successfully applied to elloramycin **47**, an anthracycline-like rhamnose-containing compound that falls into tetracenomycin group of antibiotics [70, 81–83] (Fig. 6).

3.1.2 Aureolic Acids

Members of this group feature tricyclic carbon framework decorated with two oligosaccharide chains. Mithramycin **48** and chromomycin A3 **49** are two aureolic acids that have found their use in cancer chemotherapy. These compounds, interacting with Mg^{2+}, bind GC-rich DNA in a nonintercalative way, and sugar chains are indispensable in stabilizing metal–antibiotic complexes. Carbohydrates are

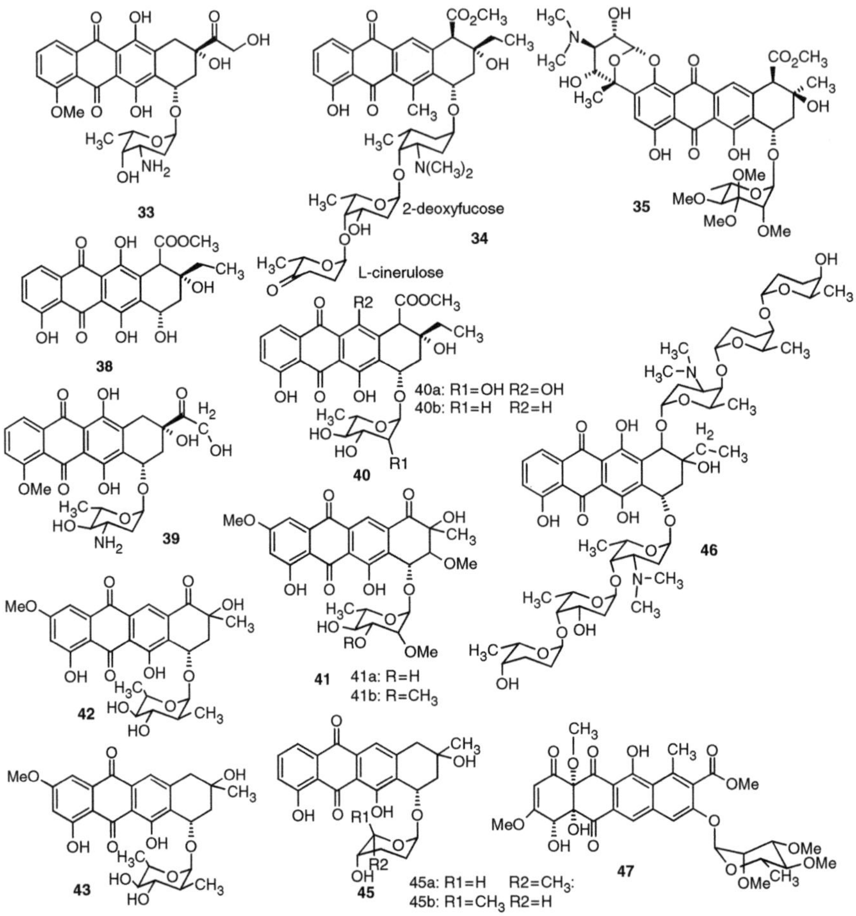

Fig. 6 Structures of selected anthracyclines

important for antitumor activity of **48** and **49**, since their nonglycosylated derivatives are inactive [84].

Disruption of GT genes was one of the sources of new glycosylated mithramycins and chromomycins carrying truncated glycoside chains. Also, demethyl- and deacetyl derivatives of chromomycins were generated through gene deletions. *S. argillaceus* strains deficient in one or all GT genes were used as hosts for expression of the C-GT gene *urdGT2*. UrdGT2 transfers D-olivose moiety during urdamycin A biosynthesis in *S. fradiae* Tu2717 (see following section). In *S. argillaceus* mutants producing nonglycosylated mithramycin precursor, UrdGT2 (alone or in combination with O-GT LanGT1) elicited the production of novel C-glycosylated compounds that contain olivose or mycarose residues [84]. UrdGT2 appears to be a very flexible enzyme, capable of transferring unnatural sugar (mycarose) to the position of mithramycin aglycon that is not glycosylated by mithramycin GTs in wild type *S. argillaceus* cells. Finally,

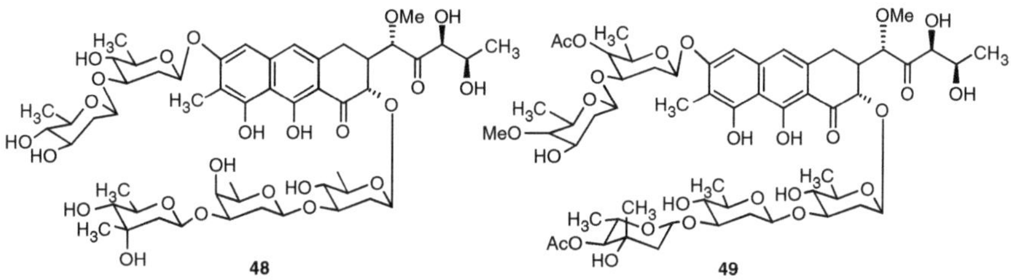

48 49

Fig. 7 Structures of selected aureolic acids

expression of "sugar" plasmids in *S. argillaceus* resulted in hybrid mithramycins that carry new deoxysugars in glycan chains (D-digitoxose, D-boivinose, D-amicetose, and others). Several of these compounds displayed better anticancer properties [85] (Fig. 7).

3.1.3 Angucyclines and Related Compounds

A uniquely shaped tetracyclic framework, where one ring is angularly condensed to three others, is the most salient feature of angucyclines [70]. Many angucyclines are glycosylated and at least for several angucycline families sugars are important for antibacterial and antitumor activity [86]. The knockouts of genes for biosynthesis of glycan part of angucyclines led to novel compounds, for example several jadomycin B **50** intermediates were obtained in this way [87, 88]. Bioenzymatic synthesis of new compounds based on landomycin A **51**, gilvocarvin V **52a**, and urdamycin A **53a** scaffolds will be detailed here due to extensive knowledge on these families and their notable bioactivities.

GT gene replacements in producer of landomycin A **51** *S. cyanogenus* S136 have yielded novel underglycosylated landomycins, some of which also lacked certain hydroxyl groups on polyketide scaffold [89]. Overexpression of the GT gene *lanGT3* in its native strain, *S. cyanogenus*, represents another striking example of how manipulation of the expression level of a GT may lead to novel compounds. Under typical conditions, *lanGT3* controls the attachment of fourth sugar (D-olivose) during the assembly of hexasaccharide chain of **51** [90]. Tetrasaccharide intermediate to **51** was never detected in culture broths of *S. cyanogenus*, probably because of a very rapid conversion of this intermediate to **51**. Introduction of *lanGT3* under transcriptional control of the strong promoter *ermEp* into *S. cyanogenus* caused accumulation of considerable amounts of tetrasaccharide landomycins [91]. Similarly, expression of rhodinosyltransferase gene *lanGT4* in *S. cyanogenus* led to accumulation of two novel landomycins carrying L-rhodinose residues in "unnatural" positions [92]. Landomycins having the same aglycon and saccharide chains of all possible lengths (including aglycon) were subjected to analysis of their activity against several tumor cell lines. Although **51** (the longest sugar chain) was generally the most active compound, the antitumor activities of landomycins do not increase linearly with the increase in sugar chain length. Particularly, tri- and hexasaccharides are equally potent inhibitors of lung cancer cells [91].

Gilvocarvin V **52a** falls into the aryl C-glycoside group of NPs, which is clearly different from angucyclines. However, biosynthesis of **52a** proceeds through an angucyclic intermediate [70], and therefore this compound is reviewed here. The five-membered furanose ring is a unique feature of gilvocarvins, which is believed to contribute to their strong antitumor activity. Disruption of putative ketoreductase gene *gilU* within **52a** gene cluster and subsequent expression of the *gilU*-minus cluster in *S. lividans* TK24 has led to isolation of several new compounds carrying 4′-OH fucofuranose residue instead of fucofuranose. One of these compounds, 4′-hydroxy gilvocarvin V **52b**, was shown to display improved anticancer properties [93]. Disruption of GT genes involved in **53** biosynthesis yielded many new compounds with shortened glycan chains [94, 95].

Production of several novel angucyclines referred to as urdamycins (Fig. 8) has been achieved through combinatorial expression of landomycin GT genes *lanGT1*

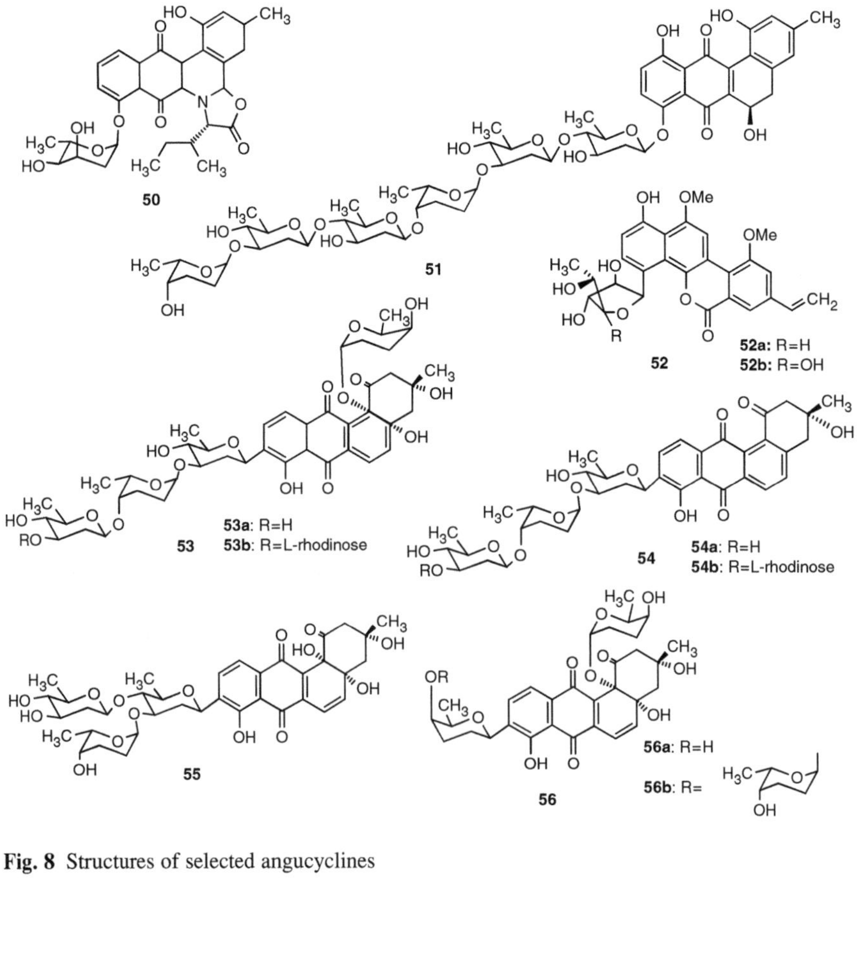

Fig. 8 Structures of selected angucyclines

or *lanGT1* plus *lanGT4* in *S. fradiae* mutant A-x making urdamycinone B **54a** [96]. Overexpression of *lanGT4* in wild type **53** producer resulted in production two new urdamycins U **53b** and V **54b** [97] that carry linear tetracyclic chains with L-rhodinose as the terminal sugar. Manipulations of **53a** pathway provided elegant evidence that shuffling of GT genes may result in a catalyst with activity unknown for parent genes. Particularly, identification and PCR-based mutagenesis of a short stretches of *urdGT1b* and *urdGT1c* GT genes allowed the preparation of a chimeric gene that, upon overexpression in appropriate *S. fradiae* strain, led to production of urdamycin P **55** having branched glycoside moiety [98]. This work showed the possibility of altering and combining both donor- and acceptor substrate specificities of different GTs being shuffled, adding new dimension to combinatorial biosynthesis of glyco-NPs. Besides the manipulation of **53a** GTs, disruption of sugar biosynthetic genes was another source of novel urdamycins. For example, knockout of NDP-4-keto-2,6-dideoxyglucose ketoreductase gene *urdR* yielded urdamycins M **56a** and R **56b** carrying D- and L-configured NDP-bound rhodinoses. Impaired production of NDP-D-olivose and promiscuity of UrdGT2 in *urdR*-deficient *S. fradiae* strain account for the production of these unusual urdamycin-like compounds [99]. There is little information on biological activities of novel urdamycins described herein. However, it's been shown that urdamycin J displays increased antitumor activity as compared to **53a** [95], implying that underglycosylated urdamycins resulted from GT gene disruptions may be an interesting subject for bioassays.

3.2 *Macrolides*

Macrolides are reduced polyketides that feature macrocyclic lactone ring. They are produced by type I (modular) PKSs. Many macrolides exhibit strong antibacterial activity and are successfully used in clinic and in animal nutrition. Also, examples of insecticidal, antitumor, and antiviral activities of macrolides are known. *S. venezuelae* producing pikromycin **57** and methymycin **58** (Fig. 9) will serve here as a model system to show all the variety of bioenzymatic approaches towards novel glycosylated macrolides. **57/58** are chosen because (1) desosamine, the only sugar that they contain, is crucial for their antibacterial activity, (2) genes for desosamine biosynthesis and attachment in *S. venezuelae* are well studied and shown to encode promiscuous enzymes, and (3) there was decade-long attention of both industry and academia to development of improved macrolides on their basis via biocombinatorial manipulations. Examples of novel "sweetened" macrolides derived from parent molecules other than **57/58** will also be given.

The gene cluster of *S. venezuelae* involved in the biosynthesis of **57** is notable for its inherent ability to produce two sets of related macrolides. Compounds **57/58** are representatives of these two sets that derive from 12- and 14-membered

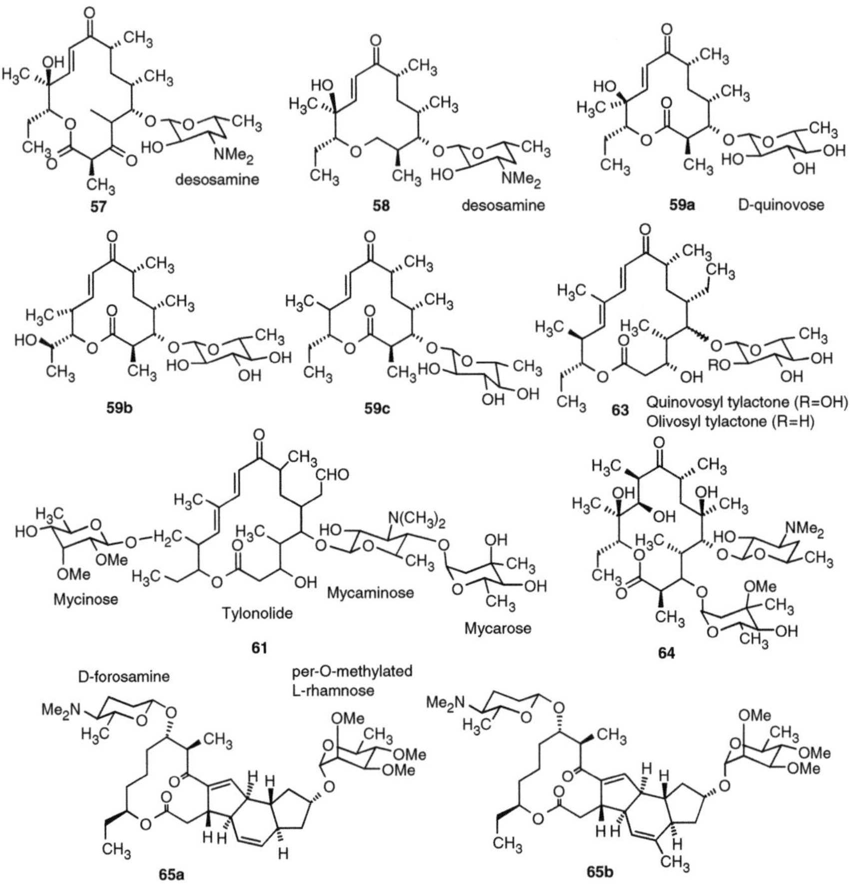

Fig. 9 Structures of selected macrolides

macrolactones, respectively. All of these antibiotics contain the single aminosugar desosamine **59**, which also exists in few other macrolides. Disruption of genes for desosamine biosynthesis within *S. venezuelae* chromosome led to several **57** analogs carrying desosamine intermediates attached to the macrolactone. For instance, inactivation of the dehydratase gene *desI* caused accumulation of D-quinovose-containing macrolides **59a/59b/59c** [100] (Fig. 9). Expression of the 3,5-epimerase gene *strL* from the streptomycin biosynthetic cluster in a *desI*-deficient strain has led to accumulation of novel macrolides carrying α-L-rhamnose [101]. Likewise, the ketoisomerase genes *tyl1a* (from the tylosin **61** gene cluster) and *fdtA* (from the S-layer polysaccharide biosynthetic pathway of *Aneurinibacillus thermoaerophilus*) endowed the *desI* mutant with the ability to produce macrolides bearing mycaminose or 4-*epi*-D-mycaminose, respectively [102, 103]. Deletion of the entire gene cluster for desosamine biosynthesis and attachment yielded *S. venezuelae* strain YJ003 that produces 12- and 14-membered aglycones. Plasmids carrying all the genes necessary for biosynthesis of noviose and olivose and their attachment to aglycones were expressed in YJ003 resulting in novel olivosyl- and noviosyl-containing derivatives [104, 105]. In the same way, a set of new macrolide derivatives was obtained that featured 4-amino-4,6-dideoxy-L-glucose **62** [106]. Another host, referred to as YJ069, lacking all genes for the biosynthesis of aglycones and accumulating the early desosamine precursor (TDP-4-keto-6-deoxy-D-glucose), was used to produce novel macrolides. For that purpose, a replicative plasmid containing genes for the biosynthesis and attachment of certain TDP-hexose was introduced into YJ069 and various aglycones were fed to the resulting strain. Utilizing this approach, researchers were able to produce quinovose- and olivose-decorated tylactones **63** [107]. It should be noted that in this and several other cases mentioned above the glycosylation efficiency was rather low (less than 10% of aglycone was converted into glycoform), despite the use of GT genes from either the pikromycin or the tylosin pathways.

The presented data show that GT gene from the desosamine cluster of *S. venezuelae*, as well as desosamine biosynthetic genes themselves, possess broad substrate specificity, which might be exploited to diversify various macrolides further. Besides the manipulations of these genes in the native host, they proved to be useful in several heterologous expression experiments. Particularly, genes for desosamine formation from *S. narbonensis* were used to redirect partly the tylosin biosynthesis towards the accumulation of substantial amounts of desosamine-containing tylactone (tylosin was also coproduced) [108]. In another experiment, genes of *S. venezuelae* involved in TDP-D-desosamine biosynthesis and attachment were integrated into genome of *S. lividans* strain K4-114, a host optimized for heterologous antibiotic production [109]. The resulting strain K39-22 was used in two ways. First, different natural and unnatural aglycones were fed to K39-22 and shown to be converted to the expected glycosylated compounds. Second, plasmids encoding macrolide type I PKS were expressed in K39-22 and also led to accumulation of novel desosamine-containing macrolides. In both aforementioned approaches the conversion rate did not exceed 20%, and a limiting production of the TDP-D-desosamine is thought to be the reason for that. The majority of 14 novel compounds exhibited antibacterial activity [110].

Erythromycin A **64** and spinosyns A and D **65a/65b** are important macrolides produced by *Saccharopolyspora erythraea* and *Sacch. spinosa*, respectively. Compound **65a** contains per-O-methylated L-rhamnose and D-forosamine attached O-glycosidically to the aglycone. Gene knockouts in respective strains as well as expression of various "sugar cassettes" (along with appropriate GT gene(s)) in *Sacch. erythraea* mutants led to production of many derivatives of **64** and **65a** and tylosin [111–114].

3.3 Polyene Macrolides

The presence of three to eight conjugated double bonds distinguishes polyenes from other classes of reduced polyketides. Polyenes interact with the sterol molecules in fungal cell membrane (usually ergosterol) and therefore are widely used as antifungal agents. Also, they are active against enveloped viruses, parasites, and prions [115]. Carbohydrates are often found as components of polyene macrolides. The deoxyaminosugar mycosamine decorates most of known polyenes. Recently, unusual derivatives of nystatin A_1 **66** that carry deoxysugars mycarose or digitoxose in addition to mycosamine were isolated from the *Streptomyces noursei* ATCC 11455. Interestingly, the mycarose containing derivative shows increased antifungal activity compared to **66** [116]. Genes responsible for the attachment of these deoxysugars to polyenes are not yet identified.

While all classes of glyco-NPs discussed above contain sugars derived from TDP-glucose, mycosamine is likely to stem from GDP-mannose. Mycosamine is essential for structural stability and antifungal activity of polyenes [117]. Chemical modification of mycosamine (especially its aminogroup) is a validated approach towards improved polyenes [118], and bioenzymatic approaches might be successful as well. The majority of polyene-producing streptomycetes are rather challenging subject of genetic engineering due to unstudied peculiarities of homologous recombination and lack of efficient "vehicles" for foreign DNA introduction [119]. Nevertheless, disruption of GDP-ketosugar aminotransferase gene from the FR008 biosynthetic cluster was achieved and caused the accumulation of different aglycones and small amounts of ketosugar-containing FR008 analog [120]. Since this derivative shows good activity, it will be interesting to apply heterologous expression approaches (which work so well for macrolides) to polyene producers, so that larger number of novel derivatives could be generated and tested. Given the fact that polyene GTs use GDP-sugars and that they form a phylogenetically separate branch of NP GTs [90], it is not clear at the moment what it will take to produce polyenes containing rationally designed sugar moieties.

3.4 Aminocoumarins

The 3-amino-4,7-dihydroxy-coumarin moiety is a core unit of this family of compounds. The amino group of coumarin scaffold is further decorated with derivatives of pyrrole or benzoic acid moieties. Also, branched deoxysugar 5-*C*-methyl-L-rhamnose (noviose) is installed onto one of the hydroxyls of the coumarin. Clorobiocin **67** and novobiocin **68** can be considered prototypical aminocoumarins. These molecules are potent inhibitors of bacterial gyrase and topoisomerase IV and **68** is used to treat human infections. Several works also showed that analogs of **68**, in combination with other drugs, can improve the chemotherapy of certain tumors [121].

Deletion of carbohydrate biosynthetic genes within cosmids directing the biosynthesis of **67** and **68** and their subsequent expression in *S. coelicolor* was a fruitful way to generate novel aminocoumarins [122–124]. These and other genetically engineered aminocoumarins were subjected to biological assays (both *in vitro* and *in vivo*) and **67** was shown to be the most active compound [125] (Fig. 10).

3.5 Glycopeptides

According to the general definition, these molecules are oligopeptides decorated with carbohydrates. The term glycopeptides is usually associated with vancomycin **3** family antibiotics; however, other families of NPs exist that are, in fact, glycopeptides due to chemical and biosynthetic features. Those are bleomycins (bleomycin B2, **69**), ramoplanin **70**, mannopeptimycins **71**, and salmochelin **72** [126]. Glycopeptides are produced by nonribosomal peptide synthetases (NRPSs) or, in case of bleomycins, by mixed NRPS-PKS complexes. Because of this, glycopeptides may contain proteinogenic as well as nonproteinogenic aminoacids which are further modified via acylation, hydroxylation, cyclization, etc. [127]. Glycopeptides are considered "last resort" antibiotics and used to treat human infections [128]. Bleomycin is a component of several anticancer chemotherapies [129]. Carbohydrates contribute to biological activity of glycopeptides and, therefore, their manipulations may lead to new insights into biology of these fascinating molecules.

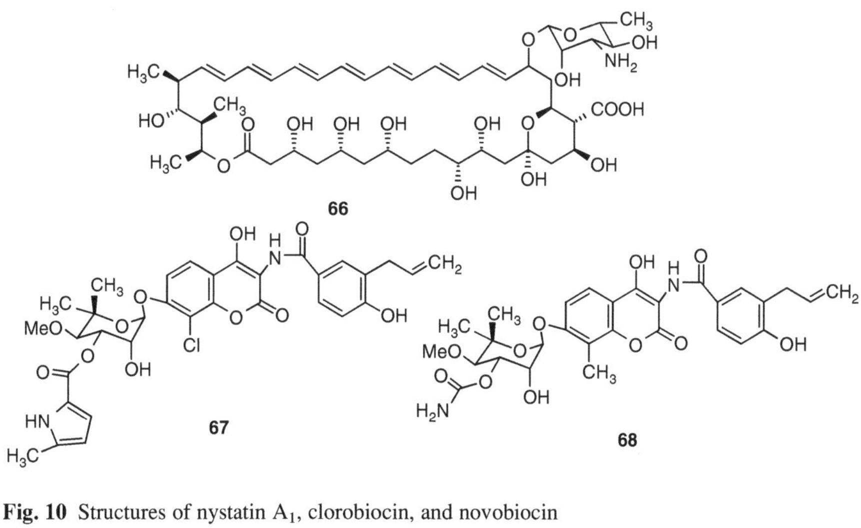

Fig. 10 Structures of nystatin A_1, clorobiocin, and novobiocin

In early work, cloning of GT genes from the **3** producer *Amycolatopsis orientalis* subsp. permitted the identification of putative glucosyltransferase genes *gtfB* and *gtfE*. Their heterologous expression in producers of teicoplanin-like heptapeptides A41030 and A47934 resulted in accumulation of respective glucosylated derivatives [126]. This result showed that bioenzymatic manipulation of glycopeptides is feasible and could lead to novel molecules. GT gene disruption was also successfully employed in *Nonomuraea* sp. to produce novel underglycosylated glycopeptide derivative [130].

Interesting derivatives of the bleomycin family were generated through manipulation of genes involved in the biosynthesis of carbohydrate portions of tallysomycins (tallysomycin B **73**) and bleomycin. Disruption of the carbamoyltransferase gene *blmD* resulted in the production of a derivative lacking the carbamoyl group at the sugar. DNA cleavage activity of this derivative was decreased 10 times as compared to bleomycin [131]. Inactivation of the GT gene *tlmK* from the gene cluster of tallysomycin has led to accumulation of five new compounds lacking 4-amino-4,6-dideoxy-L-talose [132] (Fig. 11).

3.6 *Indolocarbazoles*

A typical member of this family contains an indolo[2,3-α]pyrrolo[3,4-*c*]carbazole framework decorated with sugars. Indolocarbazoles are potent antitumor agents, whose activity is mediated via various mechanisms (inhibition of topoisomerases and protein kinases, DNA intercalation). Glycosylated indolocarbazoles are more active than the corresponding aglycones [133]. Rebeccamycin **74** and staurosporine **75**, the best studied members of the family, will be reviewed here to show how novel indolocarbazoles with altered sugar portions can be generated *in vivo*. Compound **74** and related compounds contain sugar attached to one of the indole nitrogens, while in **75** sugar is linked to both nitrogens. This rare doubly attached carbohydrate moiety arises from the activity of a GT that controls the first glycosidic bond formation, while a P450 oxygenase is responsible for the second C–N bond [133]. An efficient *S. albus*-based heterologous expression system was established for indolocarbazole production which allowed isolation of many novel derivatives [134–136]. All novel compounds generated to date exhibited antitumor activity similar to that of parent compounds; however, some of them are more selective inhibitors of certain protein kinases [136].

3.7 *Orthosomycins*

Members of this group consist of a long saccharide chain linked to dichloroisoeverninic acid. Orthosomycins display strong antibacterial activities. Poor solubility

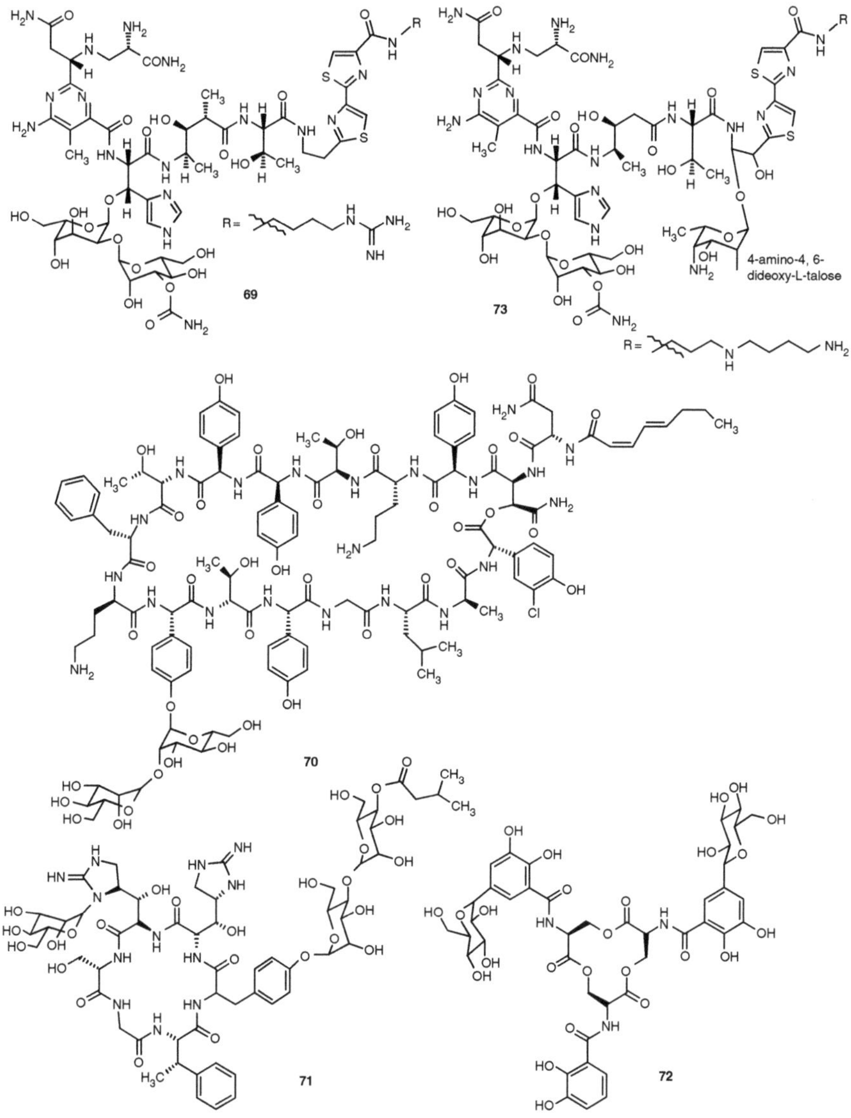

Fig. 11 Structure of selected glycopeptides

in water is one of their drawbacks, and modification of their carbohydrate portion may be one of the ways to overcome this problem. For example, disruption of one or several sugar methyltransferase genes involved in avilamycin A **76** biosynthesis led to isolation of novel derivatives lacking corresponding methyl groups, all of which showed improved polarity [137]. It has to be noted that undermethylated analogs of **76** were less stable than the parent compound; this was especially true for a derivative lacking three methyl groups. Similarly, “truncated” derivatives of **76**, lacking acetyl group on the terminal eurekanate residue or eurekanate itself, were generated through deletion of genes *aviB1–aviO2* or the GT gene *aviGT4*, respectively [138, 139]. The importance of carbohydrate structure for **76** activity was once again highlighted during manipulations of the gene *aviX12* encoding an unusual SAM radical epimerization enzyme. Here, the *aviX12*-deficient *S. viridochromogenes* mutant produced avilamycin N1 that contains glucose in glycoside tail instead of mannose in **76**. Antibacterial activity of this derivative was severely impaired as compared to **76** [140].

3.8 Phosphoglycolipids

This family is exemplified by moenomycin A **77**, a major representative of moenomycin complex produced by *S. ghanaensis* (ATCC14672). Compound **77** is one of the most potent antibiotics known to date; it is active against many vancomycin- and methicillin-resistant pathogens [141]. Moenomycin A suffers from poor pharmacokinetics, prompting the search for improved analogs that are suitable for treating human diseases.

Several derivatives of **77** were recently generated using GT gene disruption and heterologous expression techniques [142–144]. Of notable interest is the compound neryl-moenomycin A (n-MmA) **78**, which is simpler than the long-thought minimal pharmacophore and yet displays antibiotic activity [143, 144].

3.9 Plant Glycosylated Terpenoids

Terpenoids constitute the most diverse class of secondary metabolites of plants both in terms of structure and function. They are engaged in plant communication, resistance to disease, pollination, and pest control. In industry, they are used as fragrances, medicines, and food additives. All of them are produced through condensation of basic isoprene units, isopentenyl diphosphate, and its isomer dimethyl allyl diphosphate. Cyclization of nascent carbon chains is followed by their modification via oxidation, reduction, and transfer of various groups, including sugars. An interesting example of terpenoid glycosylation was observed when strawberry nerolidol synthase gene *FaNESI* was expressed in plastids of *Arabidopsis* and potato. The transgenic plants produced not only expected metabolite *E*-8-hydroxylinalool **79**, but also its glucosylated derivatives. Plastids of these plants appear to contain as-yet-unidentified flexible GT [145, 146] (Fig. 12).

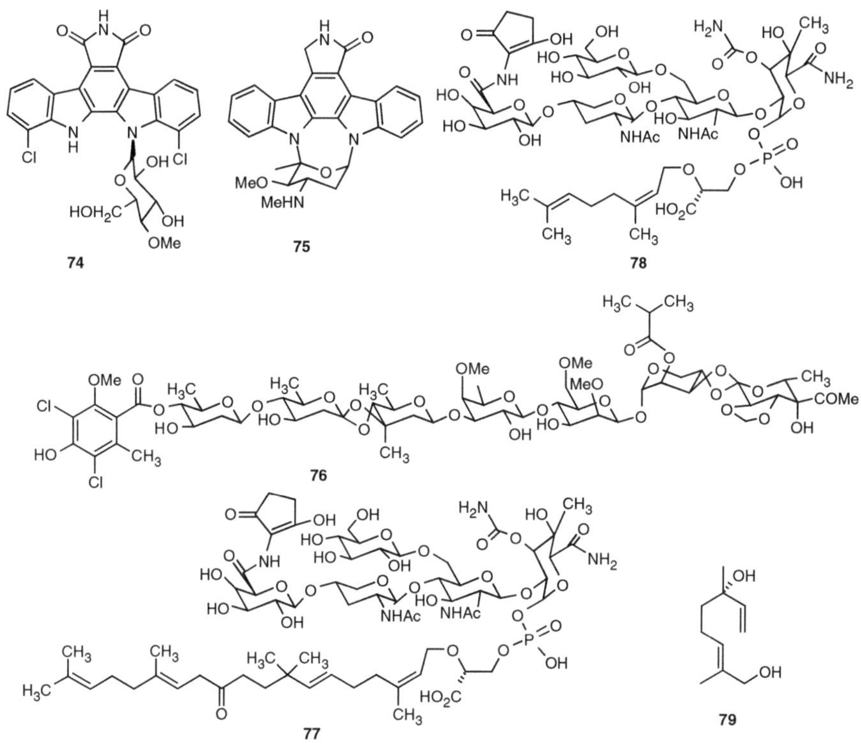

74 75 78 76 77 79

Fig. 12 Structures of selected natural products

4 Chemoenzymatic Synthesis of Novel "Glyco"-Natural Products

4.1 Chemoenzymatic Glycodiversification

Successful application of chemoenzymatic methods requires obtaining purified GTs or other carbohydrate-active enzymes, in soluble and active form, as well as suitable aglycone scaffolds and NDP-sugars [20, 147]. With the availability of NDP-sugars generated by the enzymatic strategies, numerous novel carbohydrate-containing natural products were generated chemoenzymatically.

4.2 Glycosylation of Natural Products In Vitro

4.2.1 Aromatic Polyketides

Anthracyclines were also intensively altered chemoenzymatically [148, 149]. The trisaccharide chain of aclacinomycin A **34** [150] consists of three deoxysugars,

L-rhodosamine, 2-deoxy-L-fucose, and L-cinerulose. This glycoside chain is assembled by two GTs, AknS and AknK. The first transfers TDP-L-rhodosamine to the aglycone to obtain rhodosaminyl-aklavinone. Then the second enzyme, AknK, attaches TDP-2-deoxy-L-fucose followed by TDP-L-rhodinose which is then oxidized to yield **34**. AknK can also transfer 2-deoxysugar to the axial 4′-OH of anthracycline monosaccharides, such as daunomycin **80**, adriamycin **81**, and idarubicin **82**, to generate anthracycline disaccharides [151]. Another protein, AknT, which was proposed to be a saturable activator for the AknS catalytic subunit, increases the activity of AknS by 40-fold in the 2-component AknS/AknT complex [152]. With this complex, Lu and colleagues produced *in vitro* novel anthracycline monosaccharides [147] (Fig. 13).

4.2.2 Macrolides

As mentioned above, macrolides are composed of a macrocyclic polyketide scaffold and appended deoxysugars which alter their activity, specificity, and resistance mechanisms [153–156]. Davis and coworkers tested the substrate tolerance of three family 1 macrolide GTs, namely MGT from *S. lividans* [157], OleD and OleI from the oleandomycin-producing bacterium *S. antibioticus* [158, 159]. Surprisingly, all of these GTs can utilize hydrophobic aglycones, including oleandomycin, flavonols, coumarins, and 3,4-dichloroaniline, as acceptors. *In vitro* combination of these GTs with polyketide aglycones and NDP-sugars generated novel polyketide antibiotics [160].

Sorangicins, other macrolides, have the same active binding site on the bacterial RNA polymerase as rifampicin, an important antitubercular antibiotic [161]. In comparison to rifampicin, sorangicins were shown to have greater tolerance to conformational changes in the RNA polymerase [162]. SorF is a GT in the sorangicin gene cluster from *Sorangium cellulosum* So ce12 [163]. It catalyzes the glucosylation of sorangicin A with UDP-α-D-glucose. *In vitro* substrate specificity test of SorF showed striking flexibility toward UDP-α-D-glucose, UDP-α-D-galactose, UDP-α-D-xylose, dTDP-α-D-glucose, dTDP-α-D-6-deoxy-4-keto-D-glucose and

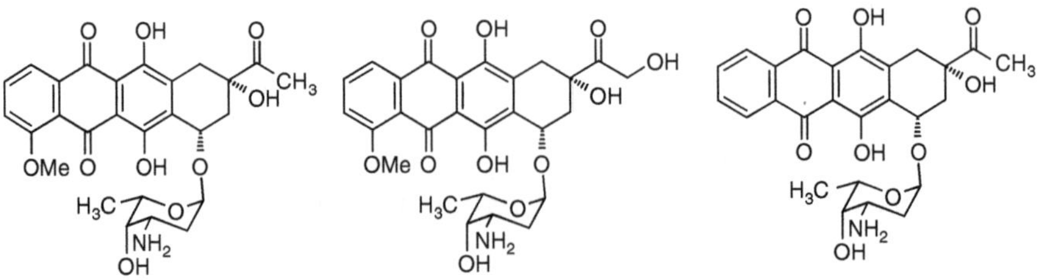

80 81 82

Fig. 13 Structures of daunomycin, adriamycin, and idarubicin

dTDP-β-L-rhamnose. Novel sorangicin derivatives **83** were produced *in vitro* from sorangicin and these NDP-sugars using SorF [164].

Avermectins, 16-membered macrocyclic lactones produced by *S. avermitilis*, act on the γ-aminobutyric acid (GABA)-related chloride ion channels unique to nematodes, insects, and arachnids, with little or no toxicity to mammals [165]. A GT, AveBI, is proposed to catalyze the attachment of both TDP-L-oleandroses in a stepwise, tandem manner, to the avermectin aglycone [166]. Using five aglycones produced by acid-mediated hydrolysis of avermectin B1a (**84-1**, **84-2**, **84-3**, **84-4**, **84-5**) and ivermectin as acceptors, and 22 NDP-sugars (generated chemically or chemoenzymatically) as donors, Thorson and coworkers produced 50 different glycosylated avermectins **84** via an AveBI-catalyzed glycolrandomization process [167] (Fig. 14).

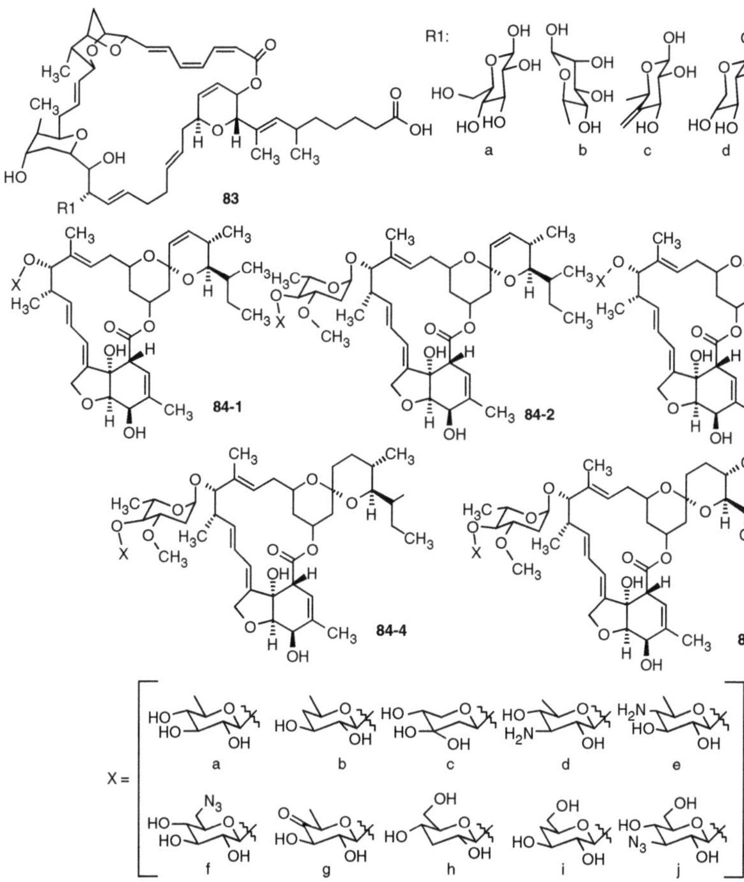

Fig. 14 Structures of sorangicins and avermectins

4.2.3 Aminocoumarins

As mentioned above, aminocoumarin antibiotics, such as novobiocin **68**, clorobiocin **67**, and coumermycin A_1 **85**, are very potent against Gram-positive bacteria, including methicillin-resistant *Staphylococcus* strains [168]. NovM is the noviosyl transferase responsible for attaching TDP-noviose to novobiocic acid [169]. However, although over 40 natural and "unnatural" nucleotide sugars were employed in the IVG experiment with NovM, only three NDP-sugars were accepted as substrates, revealing that NovM had stringent substrate specificity [170].

4.2.4 Glycopeptides

Reductive *N*-alkylation of one glycopeptide antibiotic, LY264826, on the disaccharide amino function created a new compound, oritavancin **86**, which is approximately 500 times more active than vancomycin **3** against vancomycin-resistant enterococci [171]. Hence, modifications on the sugar substituents would be a promising method to create novel active glycopeptides. Vancomycin has an L-vancosaminyl-1,2-D-glucosyl disaccharide attached to the 4-hydroxyphenylglycine residue. GTs GtfE and GtfD are responsible for the transfer of glucose and vancosamine, respectively [172]. It has been shown that GtfE was highly flexible toward both the aglycone and NDP-sugars [173, 174]. Using the chemically synthesized UDP-glucose analogs and chemoenzymatically synthesized TDP-deoxyglucose and TDP-aminoglucose as substrates, Losey and coworkers found that GtfE catalyzed the formation of 2-, 3-, 4-, and 6-deoxy-glucosyl and 2-, 3-, 4-, and 6-amino-glucosyl substituted vancomycin and teicoplanin **87** derivatives. Then the heptapeptide scaffolds with 3-, 4-, and 6-deoxyglucose and 3-, 4-, and 6-aminoglucose attached could be further elongated by GtfD with 4-epi-vancosamine and 4-amino-6-deoxy-glucose, respectively [175]. Subsequent modification of these amino groups by chemical acylation has been shown to be an efficient approach to increase the activity against vancomycin-resistant enterococci, as exemplified by oritavancin **85** [176].

4.2.5 Enedeyines

Among the four calicheamicin **88** GTs, three of them (CalG1, CalG3, CalG4) were characterized to catalyze reversible reactions. Both CalG1 and CalG3 showed flexibility toward a number of TDP-sugars. Remarkably, CalG3 can use all the five commercially available NDP-glucoses, as donor substrates. The reversible reactions catalyzed by these GTs were exploited to generate novel glycosylated calicheamicins. Using eight calicheamicin derivatives and ten TDP-sugars as substrates, CalG1 catalyzed several sugar exchange reactions and yielded more than 70 compounds [177, 178]. In the presence of TDP, CalG4 transferred the aminopentose moiety from four calicheamicin derivatives to an exogenous aglycone acceptor. Similarly, another ten unique calicheamicin variants were produced by CalG3 which catalyzed sugar exchange reactions with 22 TDP-sugars and a calicheamicin analog.

4.2.6 Flavonoids

Flavonoids are a class of plant secondary metabolites involved in the interactions of plant cells with their environments [179, 180]. They have many clinical effects, such as antitumor, antiinflammatory, and antimicrobial activities [181]. VvGT1 is a UDP-glucose: flavonoid 3-O-GT isolated from the red grape (*Vitis vinifera* L., cv. Shiraz). Glucosylation converts the unstable precursor, cyanidin, into a stable anthocyanin, cyanidin 3-O-glucoside **89**. It was shown that VvGT1 can utilize a wide range of UDP-sugars: UDP-5S-Glc, UDP-Xyl, UDP-Man, UDP-Gal, and UDP-GlcNAc, as well as GDP-Glc and dTDP-Xyl. Using UDP-Glc as donor, VvGT1 was found to transfer it to a wide variety of flavonoids and other compounds [182] (Fig. 15).

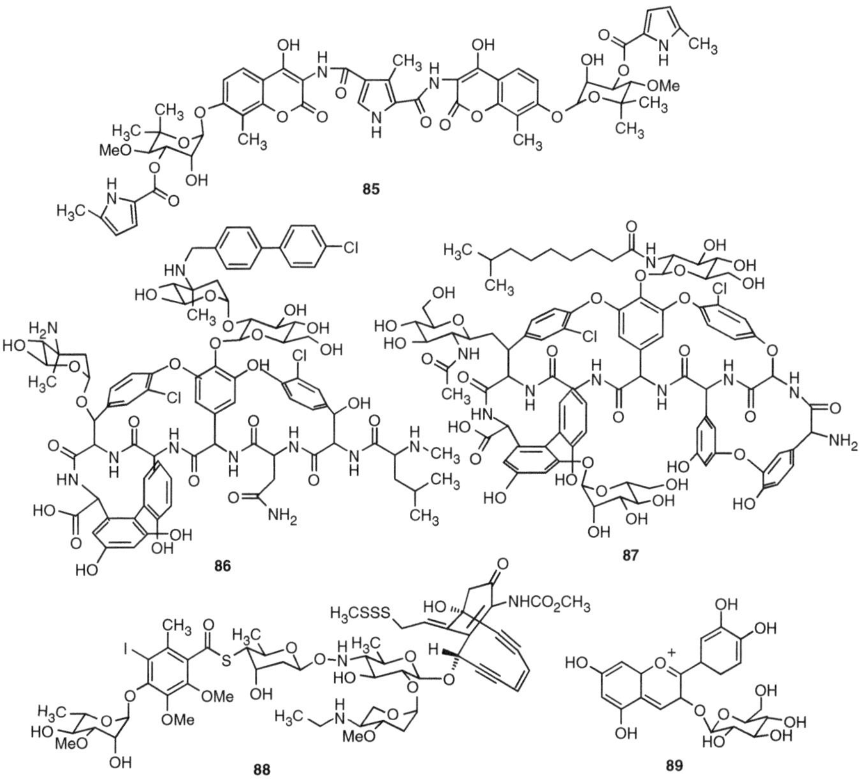

Fig. 15 Structures of selected compounds

4.2.7 Glycosylation of Natural Products Using Natural and Mutant Forms of OleD

While chemoenzymatic methods have successfully created many novel "glyco"-natural products, this route was still restricted by enzyme specificity and availability of promiscuous GTs [183]. OleD is the oleadomycin GT from *Streptomyces antibioticus* [184]. Using a high-throughput fluorescence-based GT screening system, Thorson and coworkers discovered an enhanced triple mutant (A242V/S132F/P67T, referred as ASP) of OleD that exhibited remarkable improvement in proficiency and substrate promiscuity [185]. The acceptor spectrum of Asp includes a diverse range of "drug-like" structures, such as anthraquinones, indolocarbozoles, polyenes, cardenolides, steroids, macrolides, β-lactams, and enediynes. Notably, using simple aromatic compounds like phenol, thiophenol, and aniline as acceptors, ASP also catalyzed formation of O-, S-, and N-glucosides, as well as iteratively glycosylated thiophenol. ASP is the first GT capable of catalyzing O-, S-, and N-glycosidic bond formation. Moreover, glycosidic bond formation was also detected from reactions of ASP with oximes, hydrazines, hydrazides, N-hydroxyamides, and O-substituted oxyamines [183].

4.2.8 Production of Glycosidase-Resistant Oligosaccharides Using Thiosugars

As essential structures in glycoproteins and glycolipids, oligosaccharides play many important roles in cellular regulation [186], protein folding [187], and immune modulation [188]. It has become clear that glycosylation is essential to many of the signaling pathways that turn a normal cell into a cancer one. Compounds that inhibit specific glycosylation reactions may potentially block the pathway in carcinogenesis. Carbohydrates have been recognized as novel cancer prevention drugs [189]. However, a main disadvantage of natural carbohydrate drugs, especially O-glycosidically-linked oligosaccharides, is their metabolic instability in biological systems [190].

5S-Glycosides, the ring sulfur analogs of native glycosides, are resistant to glycosidases and are able to confer metabolic stability to oligosaccharide-based drugs [191]. As they are difficult to synthesize by chemical strategies, enzymatic synthesis of such products by GTs is more practical. It had been shown that UDP-5-thiogalactose (UDP-5SGal) [192] and UDP-*N*-acetyl-5-thiogalactosamine (UDP-5-SGalNAc) [193] are substrates for galactosyltransferases, giving 5SGalβ(1,4)GlcNAc and 5SGalNacβ(1,4)GlcNAc. Using the mannosyltransferase ManT and the fucosyltransferase FucT, Tsuruta and coworkers attached the chemically synthesized GDP-5-thio-D-mannose (5SMan) and GDP-5-thio-L-fucose (5SFuc) to the glycosyl acceptors, generated a 5-thiomannose-containing disaccharide **90** and a 5-thiofucose-containing trisaccharide **91** (Fig. 16).

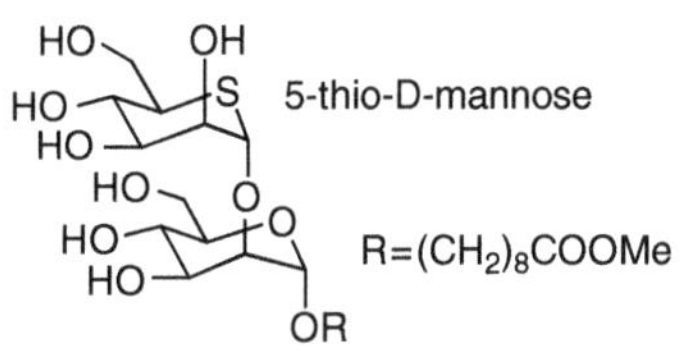

90

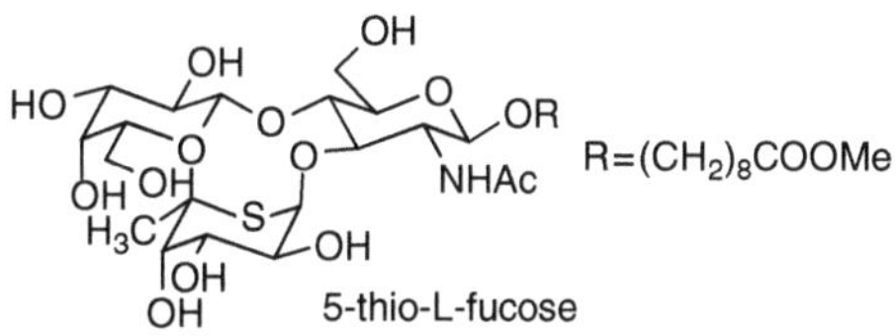

91

Fig. 16 Structures of 5S-sugar containing oligosaccharides

4.3 Chemoenzymatic Modification of Carbohydrate Moieties of Natural Products

Carbohydrate postglycosylation modifications, including methylations, acylations, and attachment of complex chromophores, are also used for glycodiversification of NPs in chemoenzymatic approaches.

4.3.1 Acylation

The critical difference between teicoplanin **87** and vancomycin **3**, a representative glycopeptide, is the presence of the acyl chain which has been implicated in its mechanism of antimicrobial activity [194, 195]. An acyltransferase (tAtf) was found to be responsible for transferring this acyl chain to the glucosamine moiety to form the teicoplanin lipoglycopeptide scaffold. Sosio and colleagues characterized aAtf and tAtf, a protein involved in the biosynthesis of the glycopeptides A-40,926 [196]. It was found that both enzymes have a preference for long-chain acyl CoAs (C_6–C_{14}). The best substrate for aAtf was the C_{12} acyl chain of lauroyl-CoA, while tAtf preferred decanoyl-CoA [197]. With respect to their specificity toward the acceptors, it was shown that the Atfs could use both vancomycin and teicoplanin scaffolds, with glucosyl, 2-aminoglucosyl, and 6-aminoglucosyl acceptors. Using tAtf, GtfD, and decanoyl-CoA, Kruger and coworkers generated two novel lipoglycopeptides chemoenzymatically. Further information on this interesting topic can be found in [198–201].

4.3.2 Methylation

Sugar methylation is a key tailoring step in the biosynthesis of many natural products [69]. Several sugar methyltransferases (MTase(s)) can act on a range of different substrates using different cofactors. The RebM O-methyltransferase of the rebeccamycin **74** cluster accepts an array of alternative substrates, displaying promiscuity toward both α-and β-glycosidic analogs [202]. RebM was shown, unlike other typical multimeric sugar O-methyltransferases (e.g., OleY [203], TylE and TylF [204, 205], MycF [206], CouP [207], and NovP [208]), to function as monomer. In addition it has a broad pH range and cannot be enhanced by divalent metals. Most importantly, RebM was the first secondary metabolite-associated MTase which can tolerate not only a broad range of acceptor substrates but also nonnatural cofactor analogs of AdoMet [209]. By combining *in vivo N*-glucosylation (RebG) and *in vitro* methylation (RebM) with a range of synthetic indolocarbazole surrogates, novel indolocarbazole were generated [202].

4.3.3 Attachment of Complex Chromophores

In **68**, the 3′-hydroxyl position of the 4-methoxy-L-noviosyl ring is carbamoylated, whereas in **67** and coumermycin A_1 **85**, it is acylated by a 5-methyl-2-pyrrolylcarbonyl moiety [210]. The acyltransferases Clo/CouN7 transfer this group from the carrier proteins Clo/CouN1 to the 3′-hydroxyl of the L-noviosyl scaffold [211]. Using a set of either chemically or chemoenzymatically prepared substrates, Fridman and coworkers tested the promiscuity of the protein pairs CouN1/CouN7, and produced 21 novel novobiocin analogs. This result shows that the biosynthetic pair of enzymes CouN1/ CouN7 is promiscuous with respect to the pyrrolylcarbonyl groups. One of the derivatives was chosen as a model compound and was produced in milligram quantities. It was demonstrated that the addition of 5-methylthiophene to the novobiocin scaffold restored antibacterial activity that was lost upon removal of the carbamoyl group from the natural compound.

5 Concluding Remarks

Both chemo- and bioenzymatic methods have already yielded considerable carbohydrate diversity around different skeleta. Glycosylation of novel positions and alteration of sugar biosynthetic pathways were realized for different classes of NPs. Several issues have to be solved to tap more fully into the potential of the described glycodiversification techniques.

First, the yield of many novel glycoforms is very low, showing the limits of substrate promiscuity of respective enzymes. Identification of a bottleneck step in the biosynthetic route and replacement of the "suboptimal" catalyst with a better one may ameliorate such problems, as was demonstrated in the case of aminocoumarin glycodiversification. Screening of producers of similar glyco-NPs for suitable enzyme or protein engineering of existing ones may be a source of improved catalyst. In some cases, there would be a need to narrow down the substrate specificity of an enzyme, when the *in vivo* production of a desired compound is accompanied by accumulation of side products. In this case, protein engineering could also be helpful, as was nicely demonstrated for GT ElmGT [212]. Besides those issues, bioenzymatic approaches may suffer from genetic instability of recombinant biosynthetic pathways, especially when they reside on autonomous genetic elements. When the production of a given glyco-NP is to go beyond the laboratory bench, it is worth investing the construction of a producer carrying all the necessary biosynthetic elements integrated into its genome. Also, bioenzymatic techniques may underperform because of poor expression of genes cloned from different organisms that differ in codon usage and details of transcription initiation. Promoter and codon optimization through PCR-based mutagenesis or gene synthesis are used to overcome such difficulties. Codon optimization is also a method of choice when *E. coli*-based expression of heterologous proteins for chemoenzymatic approach is not satisfactory.

Second, current glycodiversification efforts aim at "stitching" known building blocks, and generation of genuinely new compounds is rare. Syntheses of thiosugar-containing oligosaccharides [193, 213], glycosylated colchicines [214], and 1,2-dihydroxyanthraquinone [215] are among a few examples of use of unnatural sugars or glycosylation of naturally nonglycosylated compounds. Therefore, creation of enzymes and pathways for entirely new glycoconjugates remains a grand challenge for protein and metabolic engineering. Error-prone PCR or saturation mutagenesis in conjunction with high-throughput directed-evolution approaches should speed up the development of novel carbohydrate active enzymes. GTs, central players in biosynthesis of glyco-NPs, are a focus of such approaches. Several model studies on high-throughput screening of GTs are available. Several screens of GTs were published that are based on changes in fluorescence of resulting glycoconjugates upon glycosylation [185, 213, 216]. These approaches have proven useful for generation of GTs with novel specificity or enhanced promiscuity [217], but may yield catalysts biased towards recognition of a specific (fluorescent) substrate, be it donor or acceptor. A potentially more useful screen has been outlined that is based on detection of proton release during glycotransfer with the help of pH indicator bromothymol blue [218]. This approach was validated on blood group GTs GTA and GTB. Natural substrates can be used here, avoiding the danger of "biased" enzymes. It remains to be studied whether this method is applicable to NP GTs.

An interesting colony-based screen for glycosylated antibiotic biosynthesis in *E. coli* has been developed. It takes advantage of the fact that glycosylation of macrolide aglycone leads to antibiotically active erythromycin, and *E. coli* clones competent in glycosylation can be detected after overlaying the plate with erythromycin-sensitive *Bacillus subtilis* strain [219]. This approach has been used to improve the 6-deoxyerythronolide B glycosylation efficiency; apparently, it can be applied to those compounds with which biosynthesis and glycosylation can be reconstituted and phenotypically assayed in *E. coli*. Notwithstanding these limitations, this screen offers the simplicity of genetic manipulations of *E. coli*, the ability to rely on cellular machinery for production of substrates and catalysts, and the opportunity to apply evolutionary strategy on a pathway-wide level. Some of the aforementioned approaches could, in principle, be used to screen more than 10^6 cells/reactions within a reasonable timeframe. Even higher throughput (10^7–10^{12}) can be attained with the use of phage display technologies successfully employed to obtain improved nucleic acid polymerases [220]. GTs, bisubstrate enzymes catalyzing immensely diverse chemistry, are a much more challenging target for phage display. It is not trivial to display big proteins, such as GTs, on a phage surface and attachment of one of the GT substrates can be problematic. Nevertheless, using MurG, an essential GT involved in Lipid II **2** formation, the first solutions to the above problems have been described [221].

All the aforementioned techniques await wider implementation in order to assess their advantages and shortcomings, as well as to define the fields for their optimal use. The vast chemical diversity of glyco-NPs is made possible by just few basic GT created folds, which begs for directed-evolution-based GT engineering strategies. These, in combination with exploitation of carbohydrate-tailoring genes and enzymes, traditional screening for novel NPs and chemoenzymatic synthesis of unnatural sugars will enrich a toolkit for biology-led generation of glyco-NPs.

Acknowledgments Work in the laboratory of V. F. was generously supported by grants from Ministry of Education and Science of Ukraine and INTAS.

References

1. Magnet S, Blanchard JS (2005) Molecular insights into aminoglycoside action and resistance. Chem Rev 105:477–497
2. Silver LL (2005) Does the cell wall of bacteria remain a viable source of targets for novel antibiotics. Biochem Pharmacol 71:996–1005
3. Dudareva N, Pichersky E (2008) Metabolic engineering of plant volatives. Curr Opin Biotechnol 19:181–189
4. Handelsman J (2004) Metagenomics: application of genomics to uncultured microorganisms. Microbiol Mol Biol Rev 68:669–685
5. Keller N, Turner G, Bennett JW (2005) Fungal secondary metabolism – from biochemistry to genomics. Nat Rev Microbiol 3:937–948
6. Kohanski MA, Dwyer DJ, Wierzbowski J et al (2008) Mistranslation of membrane proteins and two-component system activation trigger antibiotic mediated cell death. Cell 135:679–690
7. Laatsch H, Fotso S (2008) Naturally occurring anthracyclines. Top Curr Chem. doi:10.1007/128_2008_5
8. Patrick WM, Quandt EM, Swartzlander DB et al (2007) Multicopy suppression underpins metabolic evolvability. Mol Biol Evol 24:2716–2722
9. Wagner B, Sieber SA, Baumann M et al (2006) Solvent engineering substantially enhances the chemoenzymatic production of surfactin. ChemBioChem 7:595–597
10. Thibodeaux CJ, Melançon CEI, H-w L (2008) Natural-product sugar biosynthesis and enzymatic glycodiversification. Angew Chem Int Ed Engl 47:9814–9859
11. Thorson JS, Hosted TJ, Jiang J et al (2001) Nature's carbohydrate chemists: the enzymatic glycosylation of bioactive bacterial metabolites. Curr Org Chem 5:139–167
12. Barton WA, Biggins JB, Jiang J et al (2002) Expanding pyrimidine diphosphosugar libraries via structure-based nucleotidylyltransferase engineering. Proc Natl Acad Sci USA 99:13397–13402
13. Timmons SC, Jakeman DL (2007) Stereoselective chemical synthesis of sugar nucleotides via direct displacement of acylated glycosyl bromides. Org Lett 9:1227
14. Marlow AL, Kiessling LL (2001) Improved chemical synthesis of UDP-galactofuranose. Org Lett 3:2517

15. Wittmann V, Wong C-H (1997) ^{1}H-Tetrazole as catalyst in phosphomorpholidate coupling reactions: efficient synthesis of GDP-fucose, GDP-mannose, and UDP-galactose. J Org Chem 62:2144
16. Jiang J, Biggins JB, Thorson JS (2000) A general enzymatic method for the synthesis of natural and "unnatural" UDP- and TDP-nucleotide sugars. J Am Chem Soc 122:6803–6804
17. Salas JA, Méndez C (2005) Biosynthesis pathways for deoxysugars in antibiotic-producing actinomycetes: isolation, characterization and generation of novel glycosylated derivatives. J Mol Microbiol Biotechnol 9:77–85
18. Yang J, Hoffmeister D, Liu L et al (2004) Natural product glycorandomization. Bioorg Med Chem 12:1577–1584
19. Yang J, Fu X, Liao J et al (2005) Structure-based engineering of *E. coli* galactokinase as a first step toward *in vivo* glycorandomization. Chem Biol 12:657–664
20. Blanchard S, Thorson JS (2006) Enzymatic tools for engineering natural product glycosylation. Curr Opin Chem Biol 10:263–271
21. Thibodeaux CJ, Melancon CE, Liu HW (2007) Unusual sugar biosynthesis and natural product glycodiversification. Nature 446:1008–1016
22. Farinas ET, Bulter T, Arnold FH (2001) Directed enzyme evolution. Curr Opin Biotechnol 12:545–551
23. Tao H, Cornish VW (2002) Milestones in directed enzyme evolution. Curr Opin Chem Biol 6:858–864
24. Jiang J, John BB, Jon ST (2001) Expanding the pyrimidine diphosphosugar repertoire: the chemoenzymatic synthesis of amino- and acetamidoglucopyranosyl derivatives13. Angew Chem Int Ed Engl 40:1502–1505
25. Lavine JE, Cantlay E, Roberts C et al (1982) Purification and properties of galactokinase from *Tetrahymena thermophila*. Biochim Biophys Acta 717:76–85
26. Dey PM (1983) Galactokinase of *Vicia faba* seeds. Eur J Biochem 136:155–159
27. Thomas P, Bessell EM, Westwood JH (1974) The use of deoxyfluoro-D-galactopyranoses in a study of yeast galactokinase specificity. Biochem J 139:661–664
28. Yang J, Fu X, Jia Q et al (2003) Studies on the substrate specificity of *Escherichia coli* galactokinase. Org Lett 5:2223–2226
29. Debouck C, Riccio A, Schumperli D et al (1985) Structure of the galactokinase gene of *Escherichia coli*, the last gene of the *gal* operon. Nucleic Acids Res 13:1841–1853
30. Hoffmeister D, Yang J, Liu L et al (2003) Creation of the first anomeric d/l-sugar kinase by means of directed evolution. Proc Natl Acad Sci USA 100:13184–13189
31. Thoden JB, Holden HM (2003) Molecular structure of galactokinase. J Biol chem 278:33305–33311
32. Yang J, Lesley L, Thorson JS (2004) Structure-based enhancement of the first anomericglucokinase. ChemBioChem 5:992–996
33. Hoffmeister D, Thorson JS (2004) Mechanistic implications of *Escherichia coli* galactokinase structure-based engineering. ChemBioChem 5:989–992
34. Kudo F, Kawabe K, Kuriki H et al (2005) A new family of glucose-1-phosphate/glucosamine-1-phosphate nucleotidylyltransferase in the biosynthetic pathways for antibiotics. J Am Chem Soc 127:1711–1718
35. Murrell JM, Liu W, Shen B (2004) Biochemical characterization of the SgcA1 α-D-glucopyranosyl-1-phosphate thymidylyltransferase from the enediyne antitumor antibiotic C-1027 biosynthetic pathway and overexpression of sgcA1 in *Streptomyces globisporus* to improve C-1027 production. J Nat Prod 67:206–213
36. Lennart L, Rudolf K, Peter RR et al (1993) Purification, characterization and HPLC assay of *Salmonella* glucose-1-phosphate thymidylyltransferase from the cloned *rfbA* gene. Eur J Biochem 211:763–770
37. Blankenfeldt W, Asuncion M, Lam JS et al (2000) The structural basis of the catalytic mechanism and regulation of glucose-1-phosphate thymidylyltransferase (RmlA). EMBO J 19:6652–6663

38. Jiang J, Christoph A, Thorson JS (2003) Application of the nucleotidylyltransferase E_p toward the chemoenzymatic synthesis of dTDP-desosamine analogues. ChemBioChem 4:443–446
39. Thorson JS, William AB, Dirk H et al (2004) Structure-based enzyme engineering and its impact on *in vitro* glycorandomization. ChemBioChem 5:16–25
40. Barton WA, Lesniak J, Biggins JB et al (2001) Structure, mechanism and engineering of a nucleotidylyltransferase as a first step toward glycorandomization. Nat Struct Mol Biol 8:545–551
41. Zuccotti S, Zanardi D, Rosano C et al (2001) Kinetic and crystallographic analyses support a sequential-ordered bi-bi catalytic mechanism for *Escherichia coli* glucose-1-phosphate thymidylyltransferase. J Mol Biol 313:831–843
42. Barton WA, Biggins JB, Jiang J et al (2002) Expanding pyrimidine diphosphosugar libraries via structure-based nucleotidylyltransferase engineering. Proc Natl Acad Sci USA 99:13397–13402
43. Elling L (1995) Effect of metal ions on sucrose synthase from rice grains–a study on enzyme inhibition and enzyme topography. Glycobiology 5:201–206
44. Zervosen A, Römer U, Elling L (1998) Application of recombinant sucrose synthase-large scale synthesis of ADP-glucose. J Mol Catal B Enzym 5:25–28
45. Römer U, Nadja N, Köckenberger W et al (2001) Characterization of recombinant sucrose synthase 1 from potato for the synthesis of sucrose analogues. Adv Synth Catal 343:655–661
46. Römer U, Schrader H, Günther N et al (2004) Expression, purification and characterization of recombinant sucrose synthase 1 from *Solanum tuberosum* L. for carbohydrate engineering. J Biotechnol 107:135–149
47. Zervosen A, Elling L, Kula MR (1994) Continuous enzymatic synthesis of 2′-deoxythymidine-5′-α-(D-glucopyranosyl) diphosphate. Angew Chem Int Ed Engl 33:571–572
48. Lavie A, Schlichting I, Vetter IR et al (1997) The bottleneck in AZT activation. Nat Med 3:922–924
49. Zervosen A, Stein A, Adrian H et al (1996) Combined enzymatic synthesis of nucleotide (deoxy) sugars from sucrose and nucleoside monophosphates. Tetrahedron 52:2395–2404
50. Johnson DA, H-w L (1998) Mechanisms and pathways from recent deoxysugar biosynthesis research. Curr Opin Chem Biol 2:642–649
51. He X, Agnihotri G, H-w L (2000) Novel enzymatic mechanisms in carbohydrate metabolism. Chem Rev 100:4615–4662
52. Liu H, Thorson JS (1994) Pathways and mechanisms in the biogenesis of novel deoxysugars by bacteria. Annu Rev Microbiol 48:223–256
53. Sterner Rh, Liebl W (2001) Thermophilic adaptation of proteins. Crit Rev Biochem Mol Biol 36:39–106
54. Zhang Z, Tsujimura M, J-i A et al (2005) Identification of an extremely thermostable enzyme with dual sugar-1-phosphate nucleotidylyltransferase activities from an acidothermophilic archaeon, *Sulfolobus tokodaii* strain 7. J Biol Chem 280:9698–9705
55. Bae J, Kim K-H, Kim D et al (2005) A practical enzymatic synthesis of UDP sugars and NDP glucoses. ChemBioChem 6:1963–1966
56. Mizanur RM, Pohl NLB (2009) Phosphomannose isomerase/GDP-mannose pyrophosphorylase from *Pyrococcus furiosus*: a thermostable biocatalyst for the synthesis of guanidinediphosphate-activated and mannose-containing sugar nucleotides. Org Biomol Chem 7:2135–2139
57. Järvinen N, Mäki M, Räbinä J et al (2001) Cloning and expression of *Helicobacter pylori* GDP-l-fucose synthesizing enzymes (GMD and GMER) in *Saccharomyces cerevisiae*. Eur J Biochem 268:6458–6464
58. Conklin PL, Norris SR, Wheeler GL et al (1999) Genetic evidence for the role of GDP-mannose in plant ascorbic acid (vitamin C) biosynthesis. Proc Natl Acad Sci USA 96:4198–4203

59. Nao S, Yoshio N, Yasuo Y et al (2002) Guanosine diphosphate-4-keto-6-deoxy-D-mannose reductase in the pathway for the synthesis of GDP-6-deoxy-D-talose in *Actinobacillus actinomycetemcomitans*. Eur J Biochem 269:5963–5971
60. Albermann C, Piepersberg W (2001) Expression and identification of the RfbE protein from *Vibrio cholerae* O1 and its use for the enzymatic synthesis of GDP-D-perosamine. Glycobiology 11:655–661
61. Mäki M, Järvinen N, Räbinä J et al (2002) Functional expression of *Pseudomonas aeruginosa* GDP-4-keto-6-deoxy-D-mannose reductase which synthesizes GDP-rhamnose. Eur J Biochem 269:593–601
62. Mizanur RM, Zea CJ, Pohl NL (2004) Unusually broad substrate tolerance of a heat-stable archaeal sugar nucleotidyltransferase for the synthesis of sugar nucleotides. J Am Chem Soc 126:15993–15998
63. Endo T, Koizumi S (2000) Large-scale production of oligosaccharides using engineered bacteria. Curr Opin Struct Biol 10:536–541
64. Koeller KM, Wong C-H (2000) Synthesis of complex carbohydrates and glycoconjugates: enzyme-based and programmable one-pot strategies. Chem Rev 100:4465–4494
65. Rodríguez L, Aguirrezabalaga I, Allende N et al (2002) Engineering deoxysugar biosynthetic pathways from antibiotic-producing microorganisms: a tool to produce novel glycosylated bioactive compounds. Chem Biol 9:721–729
66. Lombó F, Gibson M, Greenwell L et al (2004) Engineering biosynthetic pathways for deoxysugars: branched-chain sugar pathways and derivatives from the antitumor tetracenomycin. Chem Biol 11:1709–1718
67. Méndez C, Salas JA (2002) Engineering glycosylation in bioactive compounds by combinatoiral biosynthesis. In: Wohlleben W (ed) Biocombinatorial approaches for drug finding, volume 51. Springer, Heidelberg
68. Lombo F, Olano C, Salas JA et al (2009) Sugar biosynthesis and modification. In: Hopwood DA (ed) Methods in enzymology, vol 458. Academic Press, New York
69. Salas JA, Méndez C (2007) Engineering the glycosylation of natural products in actinomycetes. Trends Microbiol 15:219–232
70. Hertweck C, Luzhetskyy A, Rebets Y, Bechthold A (2007) Type II polyketide synthases: gaining a deeper insight into enzymatic teamwork. Nat Prod Rep 24:162–190
71. Minotti G, Menna P, Salvatorelli E, Cairo G, Gianni L (2004) Anthracyclines: molecular advances and pharmacologic developments in antitumor activity and cardiotoxicity. Pharmacol Rev 56:185–229
72. Madduri K, Kennedy J, Rivola G et al (1998) Production of the antitumor drug epirubicin (4'-epidoxorubicin) and its precursor by a genetically engineered strain of *Streptomyces peucetius*. Nat Biotechnol 16:69–74
73. Raty K, Kunnari T, Hakala J et al (2000) A gene cluster from *Streptomyces galilaeus* involved in glycosylation of aclarubicin. Mol Gen Genet 264:164–172
74. Torkell S, Kunnari T, Palmu K et al (2001) The entire nogalamycin biosynthetic gene cluster of *Streptomyces nogalater*: characterization of a 20-kb DNA region and generation of hybrid structures. Mol Genet Genomics 266:276–288
75. Raty K, Hautala A, Torkkell S et al (2002) Characterization of mutations in aclacinomycin A-non-producing *Streptomyces galilaeus* strains with altered glycosylation patterns. Microbiology 148:3375–3384
76. Lu W, Leimkuhler C, Oberthur M et al (2004) AknK is an L-2-deoxyfucosyltransferase in the biosynthesis of the anthracycline aclacinomycin A. Biochemistry 43:4548–4558
77. Olano C, Abdelfattah MS, Gullon S et al (2008) Glycosylated derivatives of steffimycin: insights into the role of the sugar moieties for the biological activity. ChemBioChem 9:624–633
78. Luzhetskyy A, Mayer A, Hoffmann J et al (2007) Cloning and heterologous expression of the aranciamycin biosynthetic gene cluster revealed a new flexible glycosyltransferase. ChemBioChem 8:599–602

79. Luzhetskyy A, Hoffmann J, Pelzer S et al (2008) Aranciamycin analogs generated by combinatorial biosynthesis show improved antitumor activity. Appl Microbiol Biotechnol 80:15–19
80. Garrido LM, Lombo F, Baig I et al (2006) Insights in the glycosylation steps during biosynthesis of the antitumor anthracycline cosmomycin: characterization of two glycycosyltransferase genes. Appl Microbiol Biotechnol 73:122–131
81. Blanco G, Patallo EP, Brana AF et al (2001) Identification of a sugar flexible glycosyltransferase from *Streptomyces olivaceus*, the producer of the antitumor polyketide elloramycin. Chem Biol 8:253–263
82. Perez M, Lombo F, Zhu L et al (2005) Combining sugar biosynthesis genes for the generation of L- and D-amicetose and formation of two novel antitumor tetracenomycins. Chem Commun 12:1604–1606
83. Fischer C, Rodriguez L, Patallo EP et al (2002) Digitoxosyltetracenomycin C and glucosyltetracenomycin C, two novel elloramycin analogues obtained by exploring the sugar donor substrate specificity of glycosyltransferase ElmGT. J Nat Prod 65:1685–1689
84. Lombo F, Menendez N, Salas JA et al (2006) The aureolic acid family of antitumor compounds: structure, mode of action, biosynthesis and novel derivatives. Appl Microbiol Biotechnol 73:1–14
85. Perez M, Baig I, Brana AF et al (2008) Generation of new derivatives of the antitumor antibiotic mithramycin by altering the glycosylation pattern through combinatorial biosynthesis. ChemBioChem 9:2295–2304
86. Rix U, Fischer C, Remsing LL et al (2002) Modification of post-PKS tailoring steps through combinatorial biosynthesis. Nat Prod Rep 19:542–580
87. Wang L, White RL, Vining LC (2002) Biosynthesis of the dideoxysugar component in jadomycin B: genes in the *jad* cluster of *Streptomyces venezuelae* ISP5230 for L-digitoxose assembly and transfer to the angucycline aglycone. Microbiology 148:1091–1103
88. Erb A, Luzhetskyy A, Bechthold A et al (2009) Cloning and sequencing of the biosynthetic gene cluster for saquayamycin Z and galtamycin B and the elucidation of the assembly of their saccharide chains. ChemBioChem 10:1392–1401
89. Luzhetskyy A, Zhu L, Gibson M et al (2005) Generation of novel landomycins M and O through targeted gene disruption. ChemBioChem 6:675–678
90. Luzhetskyy A, Vente A, Bechthold A (2005) Glycosyltransferases involved in the biosynthesis of biologically active natural products that contain oligosaccharides. Mol Biosyst 1:117–126
91. Zhu L, Luzhetskyy A, Luzhetska M et al (2007) Generation of new landomycins with altered saccharide patterns through over-expression of the glycosyltransferase gene *lanGT3* in the biosynthetic gene cluster of landomycin A in *Streptomyces cyanogenus* S-136. ChemBioChem 8:83–88
92. Erb A, Krauth C, Luzhetskyy A et al (2009) Differences in substrate specificity of glycosyltransferases involved in landomycins A and E biosynthesis. Appl Microbiol Biotechnol 83:1067–1076
93. Liu T, Kharel MK, Zhu L et al (2009) Inactivation of the ketoreductase *gilU* gene of the gilvocarvin biosynthetic gene cluster yields new analogues with partly improved biological activity. ChemBioChem 10:278–286
94. Trefzer A, Hoffmeister D, Kunzel E et al (2000) Function of glycosyltransferase genes involved in urdamycin A biosynthesis. Chem Biol 7:133–142
95. Kunzel E, Faust B, Oelkers C et al (1999) Inactivation of the *urdGT2* gene, which encodes a glycosyltransferase responsible for the C-glycotransfer of activated D-olivose, leads to formation of the novel urdamycins I, J and K. J Am Chem Soc 121:11058–11062
96. Trefzer A, Fischer C, Stockert S et al (2001) Elucidation of the function of two glycosyltransferase genes (*lanGT1* and *lanGT4*) involved in landomycin biosynthesis and generation of new oligosaccharide antibiotics. Chem Biol 8:1239–1252
97. Hoffmeister D, Weber M, Drager G et al (2004) Rational saccharide extension by using the natural product glycosyltransferase LanGT4. ChemBioChem 5:369–371

98. Hoffmeister D, Wilkinson B, Foster G et al (2002) Engineered urdamycin glycosyltransferases are broadened and altered in substrate specificity. Chem Biol 9:287–295
99. Hoffmeister D, Drager G, Ichinose K et al (2003) The *C*-glycosyltransferase UrdGT2 is unselective toward D- and L-configured nucleotide-bound rhodinose. J Am Chem Soc 125:4678–4679
100. Borisova SA, Zhao L, Sherman DH et al (1999) Biosynthesis of desosamine: construction of a new macrolide carrying a genetically designed sugar moiety. Org Lett 1:133–136
101. Yamase H, Zhao L, Liu H-W (2000) Engineering a hybrid sugar biosynthetic pathway: production of L-rhamnose and its implication on dihydrostreptose biosynthesis. J Am Chem Soc 122:12397–12398
102. Melancon CE III, Yu WL, Liu HW (2005) TDP-mycaminose biosynthetic pathway revised and conversion of desosamine pathway to mycaminose pathway with one gene. J Am Chem Soc 127:12240–12241
103. Melancon CE III, Liu HW (2007) Engineered biosynthesis of macrolide derivatives bearing the non-natural deoxysugars 4-epi-D-mycaminose and 3-N-methylamino-3-deoxy-D-fucose. J Am Chem Soc 129:4896–4899
104. Hong JSJ, Park SH, Choi CY et al (2004) New olivosyl derivatives of methymycin/pikromycin from an engineered strain of *Streptomyces venezuelae*. FEMS Microbiol Lett 238:391–399
105. Pageni BB, Oh TJ, Lee HC et al (2008) Metabolic engineering of noviose: heterologous expression of *novWUS* and generation of a new hybrid antibiotic, noviosylated 10-deoxymethynolide/narbonolide, from *Streptomyces venezuelae* YJ003-OTBP1. Biotechnol Lett 30:1609–1615
106. Pageni BB, Oh TJ, Liou K et al (2008) Genetically engineered biosynthesis of macrolide derivatives including 4-amino-4, 6-dideoxy-L-glucose from *Streptomyces venezuelae* YJ003-OTBP3. J Microbiol Biotechnol 18:88–94
107. Jung WS, Han AR, Hong JSJ et al (2007) Bioconversion of 12-, 14-, and 16-membered ring aglycones to glycosylated macrolides in an engineered strain of *Streptomyces venezuelae*. Appl Microbiol Biotechnol 76:1373–1381
108. Butler AR, Bate N, Kiehl DE et al (2002) Genetic engineering of aminodeoxyhexose biosynthesis in *Streptomyces fradiae*. Nat Biotechnol 20:713–716
109. Ziermann R, Betlach MC (1999) Recombinant polyketide synthesis in *Streptomyces*: engineering of improved host strains. Biotechniques 26:106–110
110. Tang L, McDaniel R (2001) Construction of desosamine containing polyketide libraries using a glycosyltransferase with broad substrate specificity. Chem Biol 8:547–555
111. Gaisser S, Lill R, Wirtz G et al (2001) New erythromycin derivatives from *Saccharopolyspora erythraea* using sugar O-methyltransferase from the spinosyn biosynthetic gene cluster. Mol Microbiol 41:1223–1231
112. Gaisser S, Martin CJ, Wilkinson B et al (2002) Engineered biosynthesis of novel spinosyns bearing altered deoxyhexose substituents. Chem Commun 21:618–619
113. Gaissser S, Carletti I, Schell U et al (2009) Glycosylation engineering of spinosyn analogues containing an L-olivose moiety. Org Biomol Chem 7:1705–1708
114. Schell U, Haydock SF, Kaja AL et al (2008) Engineered biosynthesis of hybrid macrolide polyketides containing D-angolosamine and D-mycaminose moieties. Org Biomol Chem 6:3315–3327
115. Martin JF, Aparicio JF (2009) Enzymology of the polyenes pimaricin and candicidin biosynthesis. Methods Enzymol 459:215–242
116. Bruheim P, Borgos SEF, Tsan P et al (2004) Chemical diversity of polyene macrolides produced by *Streptomyces noursei* ATCC11455 and recombinant strain ERD44 with genetically altered polyketide synthase nysC. Antimicrob Agents Chemother 48:4120–4129
117. Nedal A, Sletta H, Brautaset T et al (2007) Analysis of the mycosamine biosynthesis and attachment genes in the nystatin biosynthetic gene cluster of *Streptomyces noursei* ATCC11455. Appl Environ Microbiol 73:7400–7407

118. Preobrazhenskaya MN, Olsufyeva EN, Solovieva SE et al (2009) Chemical modification and biological evaluation of new semisynthetic derivatives of 28, 29-didehydronystatin A1 (S44HP), a genetically engineered antifungal polyene macrolide antibiotic. J Med Chem 52:189–196
119. Zotchev SB, Caffrey P (2009) Genetic analysis of nystatin and amphotericin biosynthesis. Methods Enzymol 459:243–258
120. Chen S, Huang X, Zhou X et al (2003) Organizational and mutational analysis of a complete FR-008/candicidin gene cluster encoding a structurally related polyene complex. Chem Biol 10:1065–1076
121. Rappa G, Shyam K, Lorico A et al (2000) Structure-activity studies of novobiocin analogs as modulators of the cytotoxicity of etoposide (VP-16). Oncol Res 12:113–119
122. Heide L, Gust B, Anderle C et al (2008) Combinatorial biosynthesis, metabolic engineering and mutasynthesis for the generation of new aminocoumarin antibiotics. Curr Top Med Chem 8:667–679
123. Freitag A, Rapp H, Heide L et al (2005) Metabolic engineering of aminocoumarins: inactivation of the methyltransferase gene *cloP* and generation of new clorobiocin derivatives in a heterologous host. ChemBioChem 6:1411–1418
124. Freitag A, Li S-M, Heide L (2006) Biosynthesis of the unusual 5, 5-gem-dimethyl-deoxysugar noviose: investigation of the C-methyltransferase gene *cloU*. Microbiology 152:2433–2442
125. Flatman RH, Eustaquio A, Li S-M et al (2006) Structure-activity relationships of aminocoumarin-type gyrase and topoisomerase IV inhibitors obtained by combinatorial biosynthesis. Antimicrob Agents Chemother 50:1136–1142
126. Wolter F, Schoof S, Sussmuth R (2007) Synopsis of structural, biosynthetic, and chemical aspects of glycopeptide antibiotics. Top Curr Chem 267:143–185
127. Fischbach M, Walsh CT (2006) Assembly-line enzymology for polyketide and nonribosomal peptide antibiotics: logic, machinery, and mechanisms. Chem Rev 106:3468–3496
128. Kahne D, Leimkuhler C, Lu W et al (2005) Glycopeptide and lipoglycopeptide antibiotics. Chem Rev 105:405–428
129. Galm U, Hager MH, Van Lanen SG et al (2005) Antitumor antibiotics: bleomycin, enediynes, and mitomycin. Chem Rev 105:739–758
130. Sosio M, Stinchi S, Beltrametti F et al (2003) The gene cluster for the biosynthesis of the glycopeptide antibiotic A40926 by *Nonomuraea* species. Chem Biol 10:541–549
131. Galm U, Wang L, Wendt-Pienkowski E et al (2008) *In vitro* manipulation of the bleomycin biosynthetic gene cluster in *Streptomyces verticillus* ATCC15003 revealing new insights into its biosynthetic pathway. J Biol Chem 283:28236–28245
132. Wang L, Tao M, Wendt-Pienkowski E et al (2009) Functional characterization of *tlmK* unveiling unstable carbinolamide intermediates in the tallysomycin biosynthetic pathway. J Biol Chem 284:8256–8264
133. Sanchez C, Mendez C, Salas JA (2006) Indolocarbazole natural products: occurrence, biosynthesis, and biological activity. Nat Prod Rep 23:1007–1045
134. Sanchez C, Zhu L, Brana AF et al (2005) Combinatorial biosynthesis of antitumor indolocarbazole compounds. Proc Natl Acad Sci USA 102:461–466
135. Salas AP, Zhu L, Sanchez C et al (2005) Deciphering the late steps in the biosynthesis of the anti-tumor indolocarbazole staurosporine: sugar donor substrate flexibility of the StaG glycosyltransferase. Mol Microbiol 58:17–27
136. Sanchez C, Salas AP, Brana AF et al (2009) Generation of potent and selective kinase inhibitors by combinatorial biosynthesis of glycosylated indolocarbazoles. Chem Commun 4118–4120
137. Weitnauer G, Hauser G, Hofmann C et al (2004) Novel avilamycin derivatives with improved polarity generated by targeted gene disruption. Chem Biol 11:1403–1411
138. Hofmann C, Boll R, Heitmann B et al (2005) Genes encoding enzymes responsible for biosynthesis of L-lyxose and attachment of eurekanate during avilamycin biosynthesis. Chem Biol 12:1137–1143

139. Treede I, Hauser G, Muhlenweg A et al (2005) Genes involved in formation and attachment of a two-carbon chain as a component of eurekanate, a branched-chain sugar moiety of avilamycin A. Appl Environ Microbiol 71:400–406
140. Boll R, Hofmann C, Heitmann B et al (2006) The active conformation of avilamycin A is conferred by AviX12, a radical AdoMet enzyme. J Biol Chem 281:14756–14763
141. Ostash B, Walker S (2005) Bacterial transglycosylase inhibitors. Curr Opin Chem Biol 9:456–459
142. Ostash B, Saghatelian A, Walker S (2007) A streamlined metabolic pathway for the biosynthesis of moenomycin A. Chem Biol 14:257–267
143. Yuan Y, Fuse S, Ostash B et al (2008) Structural analysis of the contacts anchoring moenomycin to peptidoglycan glycosyltransferases and implication for antibiotic design. ACS Chem Biol 3:429–436
144. Ostash B, Doud E, Lin C et al (2009) Complete characterization of the seventeen step moenomycin biosynthetic pathway. Biochemistry 48:8830–8841
145. Aharoni A, Giri AP, Deuerlein S et al (2003) Terpenoid metabolism in wild-type and transgenic *Arabidopsis* plants. Plant Cell 15:2866–2884
146. Aharoni A, Jongsma MA, Kim TY et al (2006) Metabolic engineering of terpenoid biosynthesis in plants. Phytochem Rev 5:49–58
147. Lu W, Leimkuhler C, Gatto GJ et al (2005) AknT is an activating protein for the glycosyltransferase AknS in L-aminodeoxysugar transfer to the aglycone of aclacinomycin A. Chem Biol 12:527–534
148. Fujii I, Ebizuka Y (1997) Anthracycline biosynthesis in Streptomyces galilaeus. Chem Rev 97:2511–2524
149. Temperini C, Messori L, Orioli P et al (2003) The crystal structure of the complex between a disaccharide anthracycline and the DNA hexamer d(CGATCG) reveals two different binding sites involving two DNA duplexes. Nucleic Acids Res 31:1464–1469
150. Larsen AK, Escargueil AE, Skladanowski A (2003) Catalytic topoisomerase II inhibitors in cancer therapy. Pharmacol Ther 99:167–181
151. Lu W, Leimkuhler C, Oberthur M et al (2004) AknK is an L-2-deoxyfucosyltransferase in the biosynthesis of the anthracycline aclacinomycin A. Biochemistry 43:4548–4558
152. Lu W, Leimkuhler C, Gatto GJ et al (2005) AknT is an activating protein for the glycosyltransferase AknS in L-aminodeoxysugar transfer to the aglycone of aclacinomycin A. Chem Biol 12:527–534
153. Weissman KJ, Leadlay PF (2005) Combinatorial biosynthesis of reduced polyketides. Nat Rev Microbiol 3:925–936
154. Ogasawara Y, Katayama K, Minami A et al (2004) Cloning, sequencing, and functional analysis of the biosynthetic gene cluster of macrolactam antibiotic vicenistatin in *Streptomyces halstedii*. Chem Biol 11:79–86
155. Quirós LM, Aguirrezabalaga I, Olano C et al (1998) Two glycosyltransferases and a glycosidase are involved in oleandomycin modification during its biosynthesis by *Streptomyces antibioticus*. Mol Microbiol 28:1177–1185
156. Douthwaite S (2001) Structure-activity relationships of ketolides vs. macrolides. Clin Microbiol Infect 7:11–17
157. Jenkins G, Cundliffe E (1991) Cloning and characterization of two genes from *Streptomyces lividans* that confer inducible resistance to lincomycin and macrolide antibiotics. Gene 108:55–62
158. Quiros LM, Salas JA (1995) Biosynthesis of the macrolide oleandomycin by *Streptomyces antibioticus*. J Biol Chem 270:18234–18239
159. Quiros LM, Carbajo RJ, Brana AF et al (2000) Glycosylation of macrolide antibiotics. Purification and kinetic studies of a macrolide glycosyltransferase from Streptomyces antibioticus. J Biol Chem 275:11713–11720
160. Yang M, Proctor MR, Bolam DN et al (2005) Probing the breadth of macrolide glycosyltransferases: *in vitro* remodeling of a polyketide antibiotic creates active bacterial uptake and enhances potency. J Am Chem Soc 127:9336–9337

161. Xu M, Zhou YN, Goldstein BP et al (2005) Cross-resistance of *Escherichia coli* RNA polymerases conferring rifampin resistance to different antibiotics. J Bacteriol 187: 2783–2792
162. Campbell EA, Pavlova O, Zenkin N et al (2005) Structural, functional, and genetic analysis of sorangicin inhibition of bacterial RNA polymerase. EMBO J 24:674–682
163. Irschik H, Jansen R, Gerth K et al (1987) The sorangicins, novel and powerful inhibitors of eubacterial RNA ploymerase isolated from myxobacteria. J Antibiot 40:7–13
164. Maren K, Carsten R, Herbert I et al (2007) SorF: a glycosyltransferase with promiscuous donor substrate specificity *in vitro*. ChemBioChem 8:813–819
165. Geary TG (2005) Ivermectin 20 years on: maturation of a wonder drug. Trends Parasitol 21:530–532
166. Luzhetskyy A, Fedoryshyn M, Dürr C et al (2005) Iteratively acting glycosyltransferases involved in the hexasaccharide biosynthesis of landomycin A. Chem Biol 12:725–729
167. Zhang C, Albermann C, Fu X et al (2006) The *in vitro* characterization of the iterative avermectin glycosyltransferase AveBI reveals reaction reversibility and sugar nucleotide flexibility. J Am Chem Soc 128:16420–16421
168. Francis TFT, Onkar MPS, Tadeusz S et al (1997) The high-resolution crystal structure of a 24-kDa gyrase B fragment from *E. coli* complexed with one of the most potent coumarin inhibitors, clorobiocin. Proteins Struct Funct Genet 28:41–52
169. Steffensky M, Li S-M, Heide L (2000) Cloning, overexpression, and purification of novobiocic acid synthetase from *Streptomyces spheroides* NCIMB 11891. J Biol Chem 275:21754–21760
170. Albermann C, Soriano A, Jiang J et al (2003) Substrate specificity of NovM: implications for novobiocin biosynthesis and glycorandomization. Org Lett 5:933–936
171. Cooper RDG, Snyder NJ, Zweifel MJ et al (1996) Reductive alkylation of glycopeptide antibiotics: synthesis and antibacterial activity. J Antibiot 49:575–581
172. Losey HC, Peczuh MW, Chen Z et al (2001) Tandem action of glycosyltransferases in the maturation of vancomycin and teicoplanin aglycones: novel glycopeptides. Biochemistry 40:4745–4755
173. Solenberg PJ, Matsushima P, Stack DR et al (1997) Production of hybrid glycopeptide antibiotics *in vitro* and in *Streptomyces toyocaensis*. Chem Biol 4:195–202
174. Fu X, Albermann C, Jiang J et al (2003) Antibiotic optimization via *in vitro* glycorandomization. Nat Biotechnol 21:1467–1469
175. Losey HC, Jiang J, Biggins JB et al (2002) Incorporation of glucose analogs by GtfE and GtfD from the vancomycin biosynthetic pathway to generate variant glycopeptides. Chem Biol 9:1305–1314
176. Norris EA, Thalia IN (2003) Mechanism of action of oritavancin and related glycopeptide antibiotics. FEMS Microbiol Rev 26:511–532
177. Zhang C, Griffith BR, Fu Q et al (2006) Exploiting the reversibility of natural product glycosyltransferase-catalyzed reactions. Science 313:1291–1294
178. Zhang C, Bitto E, Goff RD et al (2008) Biochemical and structural insights of the early glycosylation steps in calicheamicin biosynthesis. Chem Biol 15:842–853
179. Kren V, Martínková L (2001) Glycosides in medicine: "the role of glycosidic residue in biological activity". Curr Med Chem 8:1313–1338
180. D'Auria JC, Gershenzon J (2005) The secondary metabolism of *Arabidopsis thaliana*: growing like a weed. Curr Opin Plant Biol 8:308–316
181. Halliwell B, Rafter J, Jenner A (2005) Health promotion by flavonoids, tocopherols, tocotrienols, and other phenols: direct or indirect effects? antioxidant or not? Am J Clin Nutr 81:268S–276S
182. Offen W, Martinez-Fleites C, Yang M et al (2006) Structure of a flavonoid glucosyltransferase reveals the basis for plant natural product modification. EMBO J 25:1396–1405
183. Gantt RW, Goff RD, Williams GJ et al (2008) Probing the aglycon promiscuity of an engineered glycosyltransferase13. Angew Chem Int Ed Engl 47:8889–8892

184. Hernández C, Olanoa C, Méndeza C et al (1993) Characterization of a *Streptomyces antibioticus* gene cluster encoding a glycosyltransferase involved in oleandomycin inactivation. Gene 134:139–140
185. Williams GJ, Thorson JS (2008) A high-throughput fluorescence-based glycosyltransferase screen and its application in directed evolution. Nat Protoc 3:357–362
186. Zachara NE, Hart GW (2002) The emerging significance of O-GlcNAc in cellular regulation. Chem Rev 102:431–438
187. Helenius A, Aebi M (2001) Intracellular functions of N-linked glycans. Science 291: 2364–2369
188. Daniels MA, Hogquist KA, Jameson SC (2002) Sweet 'n' sour: the impact of differential glycosylation on T cell responses. Nat Immunol 3:903–910
189. Alper J (2003) Glycobiology: turning sweet on cancer. Science 301:159–160
190. Franz AH, Gross PH, Samoshin VV (2009) ChemInform abstract: syntheses of small cluster oligosaccharide mimetics. ChemInform. doi:10.1002/chin.200915274
191. Mehta S, Andrews JS, Svensson B et al (2002) Synthesis and enzymic activity of novel glycosidase inhibitors containing sulfur and selenium. J Am Chem Soc 117:9783–9790
192. Yuasa H, Hindsgaul O, Palcic MM (2002) Chemical-enzymic synthesis of 5′-thio-N-acetyllactosamine: the first disaccharide with sulfur in the ring of the non-reducing sugar. J Am Chem Soc 114:5891–5892
193. Tsuruta O, Shinohara G, Yuasa H et al (1997) UDP-N-acetyl-5-thio-galactosamine is a substrate of lactose synthase. Bioorg Med Chem Lett 7:2523–2526
194. Li TL, Huang F, Haydock SF et al (2004) Biosynthetic gene cluster of the glycopeptide antibiotic teicoplanin: characterization of two glycosyltransferases and the key acyltransferase. Chem Biol 11:107–119
195. Dong SD, Oberthur M, Losey HC et al (2002) The structural basis for induction of VanB resistance. J Am Chem Soc 124:9064–9065
196. Sosio M, Stinchi S, Beltrametti F et al (2003) The gene cluster for the biosynthesis of the glycopeptide antibiotic A40926 by *Nonomuraea* species. Chem Biol 10:541–549
197. Kruger RG, Lu W, Oberthür M et al (2005) Tailoring of glycopeptide scaffolds by the acyltransferases from the teicoplanin and A-40, 926 biosynthetic operons. Chem Biol 12:131–140
198. Ge M, Chen Z, Onishi HR et al (1999) Vancomycin derivatives that inhibit peptidoglycan biosynthesis without binding D-Ala-D-Ala. Science 284:507–511
199. Kerns R, Dong SD, Fukuzawa S et al (2000) The role of hydrophobic substituents in the biological activity of glycopeptide antibiotics. J Am Chem Soc 122:12608–12609
200. Malabarba A, Ciabatti R (2001) Glycopeptide derivatives. Curr Med Chem 8:1759–1773
201. Malabarba A, Nicas TI, Thompson RC (1997) Structural modifications of glycopeptide antibiotics. Med Res Rev 17:69–137
202. Zhang C, Albermann C, Fu X et al (2006) RebG- and RebM-catalyzed indolocarbazole diversification. ChemBioChem 7:795–804
203. Rodriguez L, Rodriguez D, Olano C et al (2001) Functional analysis of OleY L-oleandrosyl 3-O-methyltransferase of the oleandomycin biosynthetic pathway in *Streptomyces antibioticus*. J Bacteriol 183:5358–5363
204. Bauer NJ, Kreuzman AJ, Dotzlaf JE et al (1988) Purification, characterization, and kinetic mechanism of S-adenosyl-L- methionine: macrocin O-methyltransferase from *Streptomyces fradiae*. J Biol Chem 263:15619–15625
205. Kreuzman AJ, Turner JR, Yeh WK (1988) Two distinctive O-methyltransferases catalyzing penultimate and terminal reactions of macrolide antibiotic (tylosin) biosynthesis. Substrate specificity, enzyme inhibition, and kinetic mechanism. J Biol Chem 263: 15626–15633
206. Masaharu I, Hideaki S, Yoshio T et al (1994) A gene encoding mycinamicin III O-methyltransferase from *Micromonospora griseorubida*. Gene 141:121–124

207. Meyers CLF, Oberthür M, Heide L et al (2004) Assembly of dimeric variants of coumermycins by tandem action of the four biosynthetic enzymes CouL, CouM, CouP, and NovN. Biochemistry 43:15022–15036
208. Pi N, Meyers CLF, Pacholec M et al (2004) Mass spectrometric characterization of a three-enzyme tandem reaction for assembly and modification of the novobiocin skeleton. Proc Natl Acad Sci USA 101:10036–10041
209. Zhang C, Weller RL, Thorson JS et al (2006) Natural product diversification using a non-natural cofactor analogue of S-adenosyl-L-methionine. J Am Chem Soc 128:2760–2761
210. Balibar CJ, Garneau-Tsodikova S, Walsh CT (2007) Covalent CouN7 enzyme intermediate for acyl group shuttling in aminocoumarin biosynthesis. Chem Biol 14:679–690
211. Fridman M, Balibar CJ, Lupoli T et al (2007) Chemoenzymatic formation of novel aminocoumarin antibiotics by the enzymes CouN1 and CouN7. Biochemistry 46:8462–8471
212. Ramos A, Olano C, Brana AF et al (2009) Modulation of deoxysugar transfer by the elloramycin glycosyltransferase ElmGT through site-directed mutagenesis. J Bacteriol 191:2871–2875
213. Aharoni A, Thieme K, Chiu CPC et al (2006) High-throughput methodology for the directed evolution of glycosyltransferases. Nat Methods 3:609–614
214. Ahmed A et al (2006) Colchicine glycorandomization influences cytotoxicitiy and mechanism of action. J Am Chem Soc 128:14224–14225
215. Durr C, Hoffmeister D, Wohlert SE et al (2004) The glycosyltransferase UrdGT2 catalyzes both C- and O-glycosidic sugar transfers. Angew Chem Int Ed 43:2962–2965
216. Mayer C, Jakeman DL, Mah M et al (2001) Directed evolution of new glycosynthases from *Agrobacterium* β-glucosidase: a general screen to detect enzymes for oligosaccharide synthesis. Chem Biol 8:437–443
217. Williams GJ, Zhang C, Thorson JS (2007) Expanding the promiscuity of natural-product glycosyltransferase by directed evolution. Nat Chem Biol 3:657–662
218. Persson M, Palcic MM (2008) A high-throughput pH indiciator assay for screening glycosyltransferase saturation mutagenesis libraries. Anal Biochem 378:1–7
219. Lee HY, Khosla C (2007) Bioassay-guided evolution of glycosylated macrolide antibiotics in *Escherichia coli*. PLoS Biol 5:0243–0250
220. Xia G, Chen L, Sera T et al (2002) Directed evolution of novel polymerase activities: mutation of a DNA polymerase into an efficient RNA polymerase. Proc Natl Acad Sci USA 99:6597–6602
221. Love KR, Swoboda JG, Noren CJ et al (2006) Enabling glycosyltransferase evolution: a facile substrate-attachment strategy for phage-display enzyme evolution. ChemBioChem 7:753–756

Top Curr Chem (2010) 297: 149–203
DOI: 10.1007/128_2010_79

Published online: 10 August 2010

Total (Bio)Synthesis: Strategies of Nature and of Chemists

Alexandra A. Roberts, Katherine S. Ryan, Bradley S. Moore, and Tobias A.M. Gulder

Abstract The biosynthetic pathways to a number of natural products have been reconstituted *in vitro* using purified enzymes. Many of these molecules have also been synthesized by organic chemists. Here we compare the strategies used by nature and by chemists to reveal the underlying logic and success of each total synthetic approach for some exemplary molecules with diverse biosynthetic origins.

Keywords Biocatalysis · Biosynthesis · In vitro reconstitution · Natural products · Total synthesis

Contents

A.A. Roberts*, K.S. Ryan*, B.S. Moore (✉), and T.A.M. Gulder (✉)
Center for Marine Biotechnology and Biomedicine, Scripps Institution of Oceanography, and Skaggs School of Pharmacy and Pharmaceutical Sciences, University of California at San Diego, La Jolla, CA 92093, USA
e-mail: bsmoore@ucsd.edu; tgulder@uni-bonn.de
*Denotes equal contribution

1 Introduction

Synthetic chemists have long undertaken the challenge of generating natural products de novo [1]. From a synthetic point of view, natural products are of particular interest because of their complex molecular frameworks, which are often equipped with numerous stereocenters and highly reactive functional groups. Their structures raise interesting problems of regio- and stereoselectivity to be solved during their chemical preparation. The field of total synthesis has thus been a driving force in the development of novel synthetic reactions and strategies to solve these issues [1]. Synthetic routes to natural products are often characterized by their *convergent* approaches: numerous intermediate scaffolds can be en route to a single product. Examples of convergent approaches are syntheses of staurosporinone (**1**), which are discussed in this review, where over ten different synthetic routes converge to a single product [2–14].

Biosynthetic mechanisms to one molecule rarely show such a diversity of approaches; rather, multiple products can often be traced back to a core set of precursors. The complexity in biosynthetic routes derives instead from the ability of nature to develop *divergent* pathways from this core set of simple building blocks to give distinct molecules. These building blocks include amino acids, sugars, lipids, nucleotides, and other compounds, such as acetate, propionate, and malonate, shunted from primary metabolism, which are converted to an astonishing diversity of structures [15]. Clear examples are seen, for instance, in monoterpene biosynthesis, where a single precursor molecule is converted to a huge variety of known monoterpenes in different organisms, including (–)-(4*S*)-limonene, 3-carene, α-thujene, (–)-*endo*-fenchol, (–)-β-pinene, and 1,8-cineole [16].

Recently, some research groups have embarked on the total *bio*synthesis of secondary metabolites [17]. Like traditional total synthesis, where the goal is to construct a desired target molecule from easily available precursors, total biosynthesis endeavors to take simple, biologically relevant starting materials and convert them into a natural product. The difference from a chemical synthetic approach is that all reactions are carried out with purified enzymes, generally those derived from a single, microbially-derived biosynthetic operon. Beyond demonstration of a biologically based synthesis, such work is often carried out as part of studies attempting to elucidate biosynthetic logic and enzyme mechanisms, to produce new derivatives by testing enzyme substrate specificities, or to circumvent a challenging, possibly low-yielding, organic synthesis.

This review will compare the strategies taken by synthetic chemists to produce natural products with the strategies utilized by nature. What types of molecular scaffolds are made in each case in the process of assembling a natural product? Rarely will the paths be identical, but are there features that reveal a more universal logic in the construction of these molecules? Or, by contrast, are there fundamental distinctions between organic synthesis and biosynthesis? To answer these questions we have chosen – from a biosynthetic point of view – a diverse selection of secondary metabolites, ranging from relatively simple terpenes and shikimate-derived compounds to structurally more complex polyketide synthase (PKS) and nonribosomal peptide synthetase (NRPS) products. While these molecules indeed represent a diversity of chemical structures and biological roles, they also have much in common: they are relatively small, they are derived from discrete biosynthetic gene clusters, and they are accessible and of interest to synthetic chemists. Our comparison between biosynthetic/chemo-enzymatic and classical synthetic strategies in the preparation of these natural products will allow us to discuss the utility and current limitations of an intersection between total synthesis and "total" biosynthesis.

2 Sesquiterpenes: (+)-5-Epi-Aristolochene and (–)-Premnaspirodiene

Both (+)-5-epi-aristolochene (**2**) and (–)-premnaspirodiene (**3**) (Fig. 1) are representative of key intermediates in the biosynthesis of terpenes. Nature has used a divergent strategy to make a complex array of terpenes, which serve diverse roles in chemical biology, acting as volatiles, flavors, defense molecules, steroids, and vitamins, and include pharmaceutically relevant compounds such as the antimalarial drug artemisinin (**4**) (see Fig. 38) and the anticancer drug taxol [18]. All terpenes are derived from the condensation of isopentenyl diphosphate (IPP) and dimethylallyl diphosphate (DMAPP) to give a core set of precursors including geranyl diphosphate (**5**), farnesyl diphosphate (**6**), and geranylgeranyl diphosphate (**7**) (Fig. 2). These molecules are converted, respectively, to monoterpenes (C_{10}) such as (–)-limonene (**8**) and (–)-camphene (**9**), sesquiterpenes (C_{15}) like **2** and **3**, and diterpenes (C_{20}), e.g., taxadiene (**10**) and casbene (**11**), through transformations that include electrophilic attacks, hydride, methyl, and methylene shifts, water quenching of carbocations, and proton abstractions [16]. These resulting intermediates are then further modified by enzymes such as cytochrome P450s, which catalyze reactions including stereospecific hydroxylations. For instance, capsidiol (**12**), an antifungal phytoalexin, derives from **2** [19] as a result of such additional modifications [20].

The formation of the sesquiterpene (+)-5-epi-aristolochene (**2**) represents, from a biosynthetic point of view, the transformation requiring the fewest steps among those discussed in this review. A single enzyme, tobacco 5-epi-aristolochene synthase (TEAS), converts the biosynthetic precursor, farnesyl diphosphate (**6**), to

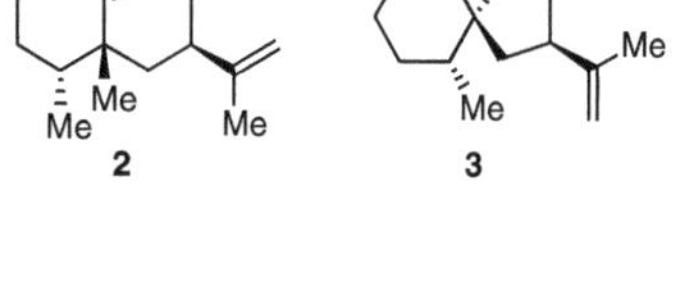

Fig. 1 Molecular structures of (+)-5-epi-aristolochene (**2**) and (–)-premnaspirodiene (**3**)

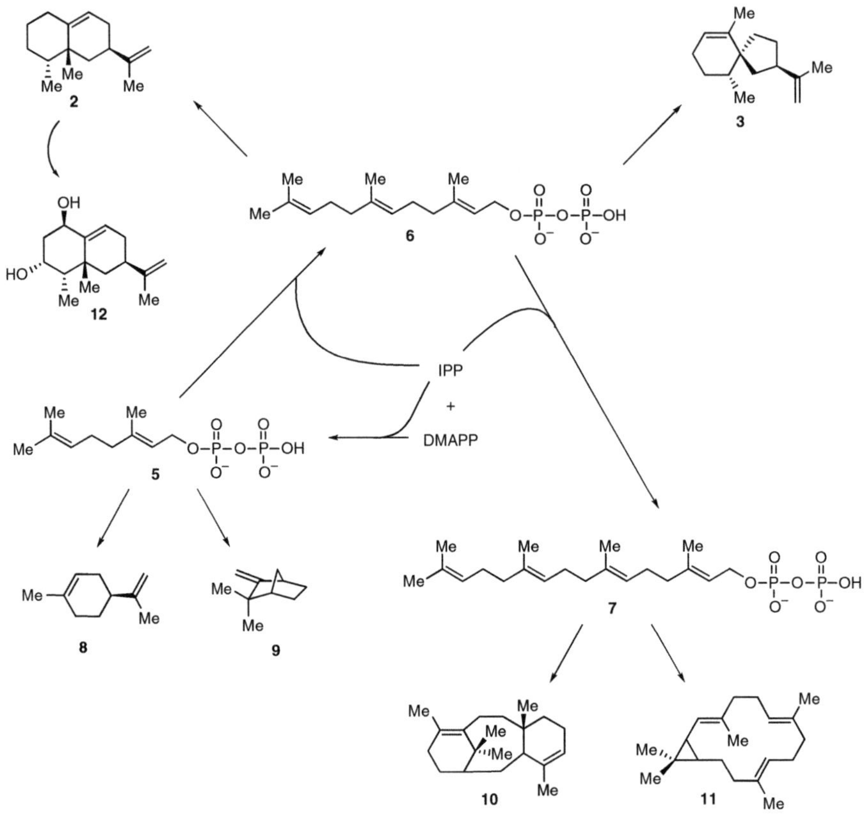

Fig. 2 Biosynthetic route to terpenes. Geranyl diphosphate (**5**); farnesyl diphosphate (**6**); geranylgeranyl diphosphate (**7**); (–)-limonene (**8**); (–)-camphene (**9**); taxadiene (**10**); casbene (**11**); capsidiol (**12**). IPP = isopentenyl diphosphate, DMAPP = dimethylallyl diphosphate

2 with a k_{cat}/K_M of 0.3 μM^{-1} min^{-1} [21, 22]. Nonetheless, this single transformation itself is complex: TEAS catalyzes two ring closures, a hydride and a methyl migration, and a proton abstraction to form a double bond (Fig. 3). Interest in TEAS was dramatically increased when it was discovered that a highly related enzyme, henbane premnaspirodiene synthase (HPS), isolated from *Hyoscyamus muticus*, with 75% identity at the amino acid level to TEAS, was able to catalyze a completely different outcome with the same precursor [23]. Both HPS and TEAS are thought to have similar reaction mechanisms, through intermediate **13**, where they diverge. While TEAS initiates a 1,2-methyl shift in **13** to give finally **2**, HPS instead triggers a 1,2-shift of the cycloalkyl substituent to form **3**. Overall, HPS catalyzes two ring closures, a methylene shift, and abstraction of a distinct proton to give (–)-premnaspirodiene (**3**), a spirovetivane with three stereocenters (Fig. 3).

How TEAS and HPS can give rise to such distinct terpene intermediates has been probed by crystallographic, biochemical, and site-directed mutagenesis approaches. The crystal structure of TEAS revealed the enzyme as a two-domain

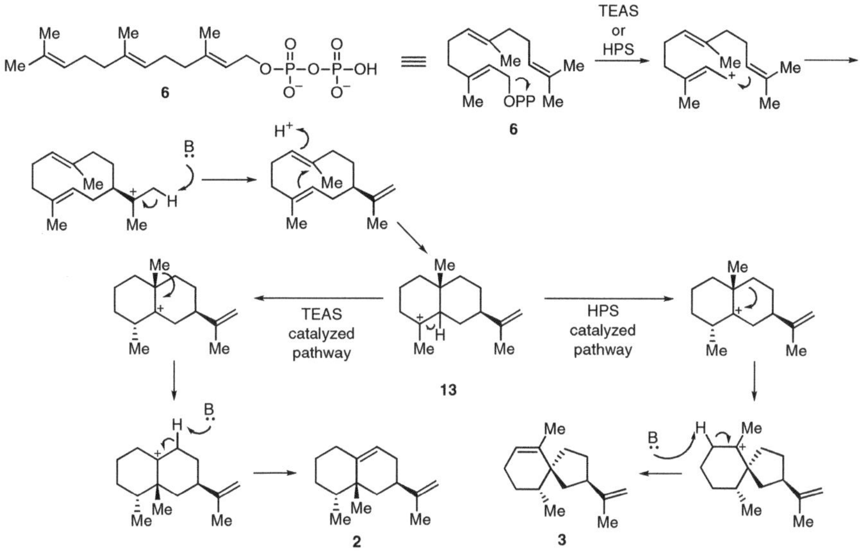

Fig. 3 Proposed mechanisms for the formation of (+)-5-epi-aristolochene (**2**) and (–)-premnaspirodiene (**3**) from farnesyl-diphosphate (**6**) by the action of TEAS and HPS, respectively

alpha-helical protein with a hydrophobic active site containing two Mg^{2+} ions for the coordination of the farnesyl diphosphate substrate **6**. A complex of the protein with analogs of **6** enabled identification of amino acid side chains likely to modulate key reaction steps [24]. The authors proposed a mechanism for (+)-5-epi-aristolochene (**2**) formation [24] consistent with earlier studies [23]. This crystallographic study was the groundwork for comparative analysis with HPS. Although the structure of HPS itself has not been determined, molecular modeling of HPS, followed by docking with energy-minimized putative substrates and contact mapping, identified nine amino acids, among those that were distinct between the two enzymes, likely to be responsible for determining the catalytic outcome. The activity of HPS could be installed in TEAS by replacement of these nine amino acids from HPS into the TEAS protein. Conversely, the activity of TEAS could be introduced by substitution of the nine TEAS residues into the HPS background [25]. A systematic evaluation of 418 of the 512 possible mutant combinations in the TEAS protein background led to a consideration of the catalytic landscape of the terpene cyclase activity, and a realization that single amino acid mutations did not necessarily cause predictable changes in enzyme activity [26].

Unlike the biosynthetic route to **2** and **3**, the synthetic approaches to these molecules are characterized, instead, by deconstruction of terpene natural products further along the biosynthetic pathway. Preparation of (+)-5-epi-aristolochene (**2**) has so far only been carried out using a semisynthetic strategy from the natural product **12** [27], which is available from pepper (*Capsicum annum*) fruits in high quantities [28]. The synthetic route started with *O*-acetylation of **12** to give

1,3-di-*O*-acetylcapsidiol **14**, followed by reduction to furnish 1-deoxycapsidiol (**15**) in 27% yield (Fig. 4). Alcohol **15** was derivatized to deliver a thiocarbonylimidazole adduct, which was finally converted to **2** in 54% combined yield. The synthesis is notable in that it occurs in a reverse order to that seen naturally; capsidiol (**12**) is thought to be biosynthetically derived from (+)-5-epi-aristolochene (**2**) [19, 29], and, semisynthetically, **2** is derived from **12** [27].

The semisynthesis of **3** utilized a ring contracting rearrangement reaction similar to the respective biosynthetic transformation (Fig. 5). The starting material santonin (**16**) [30], itself a natural product [31], was converted in a three step sequence into ester **17**, which in turn was equipped with a TMS group and reduced to give alcohol **18** in 37% overall yield. Epoxidation of the remaining double bond using oxone furnished **19**, the precursor of the rearrangement reaction, in 91% yield. In the following key synthetic step, the epoxide ring was opened using the Lewis acid BF_3, leading to the semistable carbocation **20**, which, after a 1,2-shift of the cycloalkyl residue to give **21**, and elimination of the TMS group, furnished 64% of the spirovetivane **22**. The preparation of **3** was concluded by mesylation of the primary alcohol, elimination of the mesylate leaving group to give **23**, and final removal of the secondary alcohol in 27% overall yield. (–)-Premnaspirodiene (**3**)

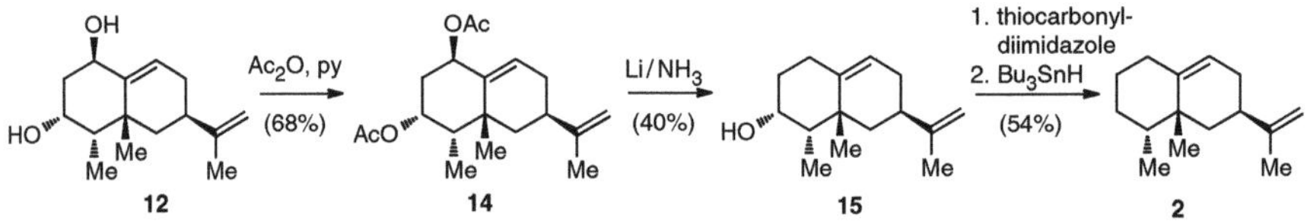

Fig. 4 Synthesis of (+)-5-epi-aristolochene (**2**) from the natural product capsidiol (**12**)

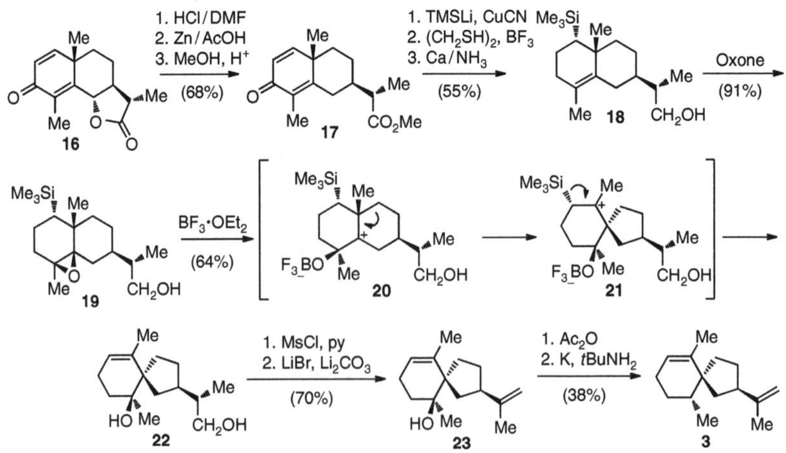

Fig. 5 Synthetic route to (–)-premnaspirodiene (**3**)

has also been synthesized as an intermediate en route to (–)-solavetivone from dihydrocarvone [32].

Although there are a limited number of syntheses to (+)-5-epi-aristolochene (**2**) and (–)-premnaspirodiene (**3**), it is clear that the biosynthetic approach, involving a single enzyme whose specific amino acid side chains determine the catalytic outcome, is more simple than the synthetic routes. Nonetheless, the synthesis to **3** is an elegant demonstration of the capacity of organic synthesis to generate the spirovetivane structure in a biosynthetically inspired manner. While this chemical route nicely utilized a biomimetic rearrangement reaction, it is far more complex with 12 steps separating santonin (**16**) from **3** (Fig. 5).

3 Salicylate Containing Siderophores: Yersiniabactin and Pyochelin

Iron is an essential requirement for life; however, in aerobic environments the majority of it is present in its insoluble ferric (Fe^{3+}) form. In order to sequester this metal from iron-limited habitats, some microorganisms synthesize siderophores, which are small molecules that have a high affinity for iron and other transition metals [33–35]. Siderophore production in *Yersinia* species was first proposed by Wake and coworkers, who observed that an iron-limited *Yersinia pestis* culture was able to promote the growth of another cocultured strain [36]. However, the siderophore yersiniabactin (**24**) was not isolated until almost two decades later by Haag and colleagues, who detected a catechol functional group and reported the molecular mass of the iron chelator to be 482 Da [37]. Subsequently, the structure of **24** was elucidated and was shown to be comprised of two thiazoline rings, a thiazolidine ring, a malonyl unit, and a phenol moiety derived from salicylate [38]. Crystallization studies of **24** with Fe^{3+} revealed that it forms a hexadentate ligand with iron via the phenolate, the alcoholate on the malonate-derived moiety, the terminal carboxylate, and the three cyclized cysteine nitrogens (see Fig. 7a) [39].

Various bacteria, predominantly pathogenic Enterobacteriaceae, have been shown to biosynthesize **24**, including *Y. pestis*, the causative agent for pneumonic and bubonic plague [40], the enteric pathogens *Yersinia pseudotuberculosis* [41] and *Yersinia enterocolitica* [42], as well as uropathogenic strains of *Escherichia coli* [43, 44] and the plant pathogen *Pseudomonas syringae* [45]. Yersiniabactin (**24**) is secreted by cells in low-iron environments. The siderophore's high affinity for Fe^{3+}, with a stability constant of $4 \times 10^{36}\ M^{-1}$ [46], allows it to sequester iron from lower affinity host heme-binding proteins, such as the transferrin glycoproteins which have stability constants of approximately $10^{20}\ M^{-1}$ [39]. As a result, **24** is a virulence factor essential for the pathogenicity of producing species [47–49]. Siderophore biosynthesis is, therefore, becoming an increasingly appealing therapeutic target in the development of novel antibiotics against infectious diseases, such as biosynthesis substrate mimics which inhibit compound production [50, 51].

Biosynthesis of **24** occurs via a nonribosomal peptide synthetase/polyketide synthase (NRPS/PKS) mechanism. The enzymatic machinery is encoded by the *ybt* gene cluster [52–54] which is located on a virulence conferring high-pathogenicity island (HPI), and which also includes the genes for regulation [55, 56], transport [57], and uptake [49, 58] of the siderophore and the ferric-siderophore complex. Although the HPI does not encode the necessary genes to be self-transmissible [59], it is still highly mobilizable, possibly via phage-mediated horizontal gene transfer [60] or by integration on a transmissible plasmid [61].

Assembly of **24** (Figs. 6 and 7a) proceeds in a linear, modular fashion across the NRPS/PKS interface [62]. First, the salicylate synthase YbtS converts chorismic acid (**26**) to salicylic acid (**27**) [63, 64], which is then activated as an adenylate by the salicyl-AMP ligase, YbtE. Salicyl-AMP is tethered as a thioester to the aryl

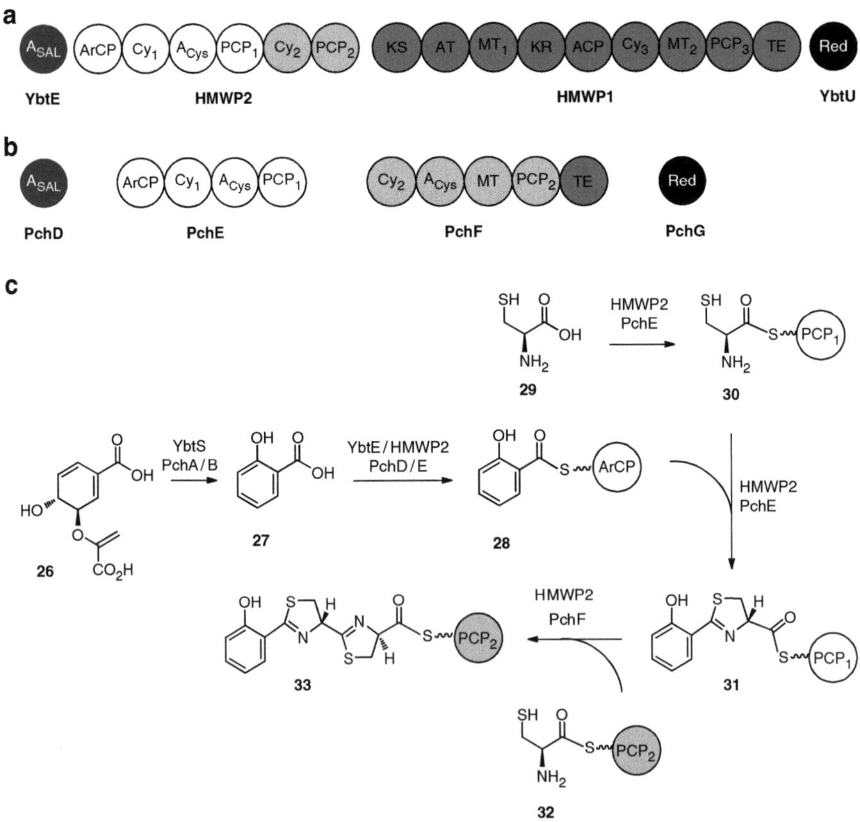

Fig. 6 Four enzymes (one salicylate-AMP ligase YbtE/PchD, two NRPS and/or NRPS/PKS enzymes HMWP1/PchE and HMWP2/PchF, and one reductase YbtU/PchG) are required for the *in vitro* biosynthesis of (**a**) yersiniabactin (**24**) and (**b**) pyochelin (**25**). (**c**) The initial stages of **24** and **25** biosynthesis proceed via a similar mechanism from chorismic acid **26** to the salicylate-bisthiazole intermediate **33**

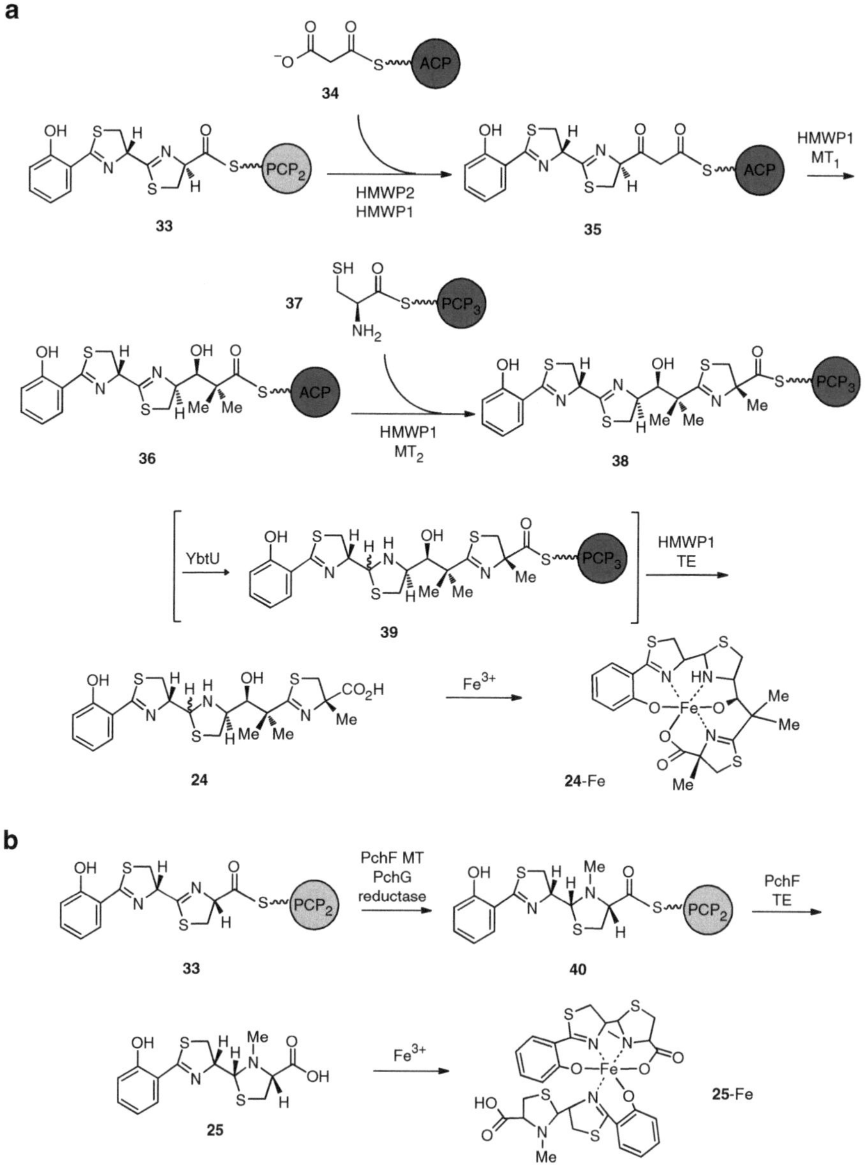

Fig. 7 Biosynthesis of (**a**) **24** and (**b**) **25** from intermediate **33**. One molecule of **24** can bind ferric iron while two molecules of **25** are required for Fe^{3+} chelation

carrier protein (ArCP) of the NRPS module, HMWP2, to give **28**. HMWP2 catalyzes the activation of L-cysteine (**29**) as a PCP-bound acyl adenylate **30**, which is subsequently epimerized [65], condensed with **28**, and cyclized to a thiazoline ring to yield **31**. A second L-cysteine molecule (**29**) is then activated as

32 and incorporated to the nascent peptide to yield the bisthiazoline intermediate **33**. Transfer of intermediates across the NRPS and PKS interface is facilitated by aryl-, peptidyl-, and acyl- carrier proteins (ArCP, PCP_1, PCP_2, ACP, and PCP_3) that are first activated into their holo forms by the addition of a phosphopantetheinyl arm, derived from coenzyme A (CoA), which is catalyzed by the phosphopantetheinyl transferase (PPTase) YbtD [66]. Following the incorporation of the two thiazoline rings, the substrate **33** is transferred from the terminal PCP_2 of HMWP2 to the PKS/NRPS hybrid enzyme, HMWP1, for elongation with an ACP-bound malonate unit **34** to yield **35**, which is subsequently double Cα-methylated and reduced to furnish **36**. The methyl groups are derived from *S*-adenosylmethionine (SAM) and are attached by a methyltransferase domain within the HMWP1 PKS machinery. HMWP1 also catalyzes the activation of the third L-cysteine (**29**) which is bound to PCP_3 as thioester **37** and is then incorporated, cyclized, and methylated to give **38**. Conclusion of the pathway occurs by transfer of the substrate, from the final PCP_3 domain to the terminal thioesterase (TE), for release of the mature yersiniabactin (**24**). The second thiazoline ring is reduced to thiazolidine **39** by the reductase, YbtU [62], presumably before release by the TE domain, although the timing of this reaction within **24** biosynthesis remains unknown. In addition to the four yersiniabactin (**24**) synthetase enzymes necessary for *in vitro* biosynthesis (YbtE, YbtU, HMWP1, and HMWP2), *in vivo* production also requires the type II thioesterase YbtT which is proposed to remove misprimed structures from the NRPS/PKS enzyme complex [67].

Siderophores produced via NRPS or NRPS/PKS mechanisms comprise the majority of currently reported one-pot *in vitro* reconstitutions, likely due to their elegantly simple and modular biosynthetic machinery, and also largely due to the efforts of the Walsh laboratory. Total biosynthesis of yersiniabactin (**24**) was carried out by this group with four substrates (salicylic acid (**27**), L-cysteine (**29**), SAM, and malonyl-CoA), three cofactors (NADPH, coenzyme A, and ATP) and five enzymes (YbtE, YbtU, HMWP2, HMWP1, and Sfp) [62]. The NRPS/PKS enzymes were phosphopantetheinylated in a single reaction by the PPTase Sfp from *Bacillus subtilis*. The 22 catalytic reactions resulting in the total biosynthesis of **24** proceeded with a relatively inefficient final turnover rate of 1.4 min^{-1}, which is attributable to the poor transfer of the substrate from HMWP2 (**33**) to the PKS/NRPS hybrid enzyme HMWP1 (**35**) [62], as well as to the large number of enzyme reactions required for *in vitro* reconstitution.

Total *in vitro* biosynthesis of a structurally similar siderophore, pyochelin (**25**), was also previously carried out by the same laboratory [68]. Pyochelin (**25**), which is produced by the human lung pathogens *Pseudomonas aeruginosa* and *Burkholderia cepacia* [69–71], possesses a phenol moiety followed by a thiazoline and thiazolidine ring [72], as in **24**. However, **25** does not contain the dimethyl malonate or the final thiazoline ring and, instead, possesses an *N*-methylation on the thiazolidine ring that is not present in **24**. Two molecules of **25** are required to bind Fe^{3+} [73, 74]. One molecule of **25** binds Fe^{3+} as a tetradentate ligand with the phenolate hydroxyl, the two nitrogens, and the terminal carboxylate. The other molecule can bind as a bidentate with either the hydroxyl and thiazoline nitrogen or the thiazolidine nitrogen and the carboxylate (Fig. 7b). Analogous to **24** reconstitution, the enzymes required for *in vitro* reconstitution of **25** included a salicyl AMP-ligase (PchD), two NRPS enzymes (PchE and PchF) for the incorporation and cyclization of two L-cysteines (**29**) to give **31** and **33** [65], and a reductase domain for conversion of the thiazoline ring to a thiazolidine ring (PchG) (Figs. 6 and 7b) [68, 75]. Subsequent SAM-mediated *N*-methylation of the thiazolidine ring, delivering **40**, and final release of **25** from the NRPS machinery is carried out by the methyltransferase and thioesterase encoded within the NRPS enzyme PchF (Fig. 7b). The overall rate of turnover in **25** *in vitro* biosynthesis was approximately 2 min^{-1}, which is consistent with the final efficiency of **24** reconstitution [62, 68]. However, the oxidized pyochelin-like intermediate in **24** biosynthesis (hydroxyphenyl-thiazolinyl-thiazolinyl-CO_2H) was reportedly formed with the higher catalytic efficiency of 6 min^{-1}, which indicates that the double module HMWP2 is more efficient at incorporating and cyclizing two cysteines than the two-enzyme counterpart (PchE and PchF) in the biosynthesis of **25**. Further, it seems that the incorporation of a dimethyl malonate and a third cyclized cysteine into **24** does not significantly reduce the *in vitro* catalytic efficiency when compared to that of **25**.

Only one strategy has been reported for the total chemical synthesis of **24** [76], which may be indicative of the complexity of this synthesis, as well as the relatively high fermentation yields (over 100 mg L^{-1}) and ease of purification of **24** from *Yersinia* spp. culture [37, 38]. Ino and Murabayashi [76] adapted their synthesis strategy from their previous total synthesis of the antimycoplasma antibiotic micacocidin (**41**) (see Fig. 9) [77]. Structurally, **41** is identical to **24** except for the absolute configuration at position C10, an additional pentyl chain at C3 of the phenol group, and an *N*-methylation on the thiazolidine ring. Total synthesis of **24** proceeded via condensation of two molecular building blocks, the aldehyde **42** (Fig. 8) and intermediate **43** from **41** synthesis (Fig. 9). The aldehyde **42** was synthesized by condensation of the Weinreb amide **44** with methoxy benzoyl chloride **45** to yield ester **46** (Fig. 8). Subsequent *N*-deprotection and treatment with $NaHCO_3$ resulted in intramolecular attack of the free amine at the ester to give the rearranged amide **47** in 100% yield from **44**. Following *O*-demethylation of the phenolic methoxy residue, **48** was constructed by converting the amide first into a thioamide **49** and finally into a thiazoline in 48% yield using Burgess reagent. After protection of the phenol group, reduction of the Weinreb amide **48** yielded aldehyde **42** (Fig. 8).

The remainder of **24** was synthesized as for micacocidin (**41**) [77] (Fig. 9). The synthesis started with conversion of L-cysteine (**29**) to the thiazolidine carboxylate **50** using the strategy of Kemp and Carey [78]. Notably, L-cysteine (**29**) is also utilized in the biosynthesis of **24** and **25** despite epimerization to the D-form during thiazoline ring formation [65], which highlights a similarity between the synthetic and biosynthetic logic. Following elongation to a β-keto-carboxylic acid **51** and reduction of the keto group to give alcohol **52**, **53** was formed by deprotonation of the secondary alcohol using NaH, leading to intramolecular carbamate formation, and subsequent hydrolysis of the methyl ester in 33% overall yield [77]. Condensation with 2-methyl-D-cysteine methyl ester delivered intermediate **54**, which was

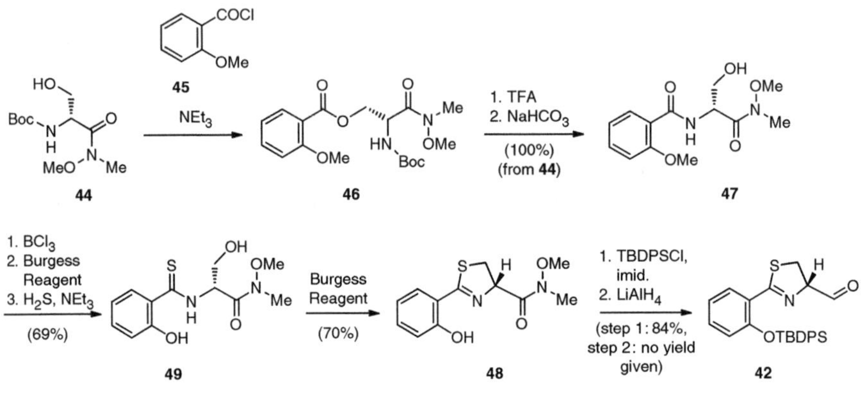

Fig. 8 Chemical route to aldehyde **42**, an intermediate in **24**, **25**, and **58** synthesis

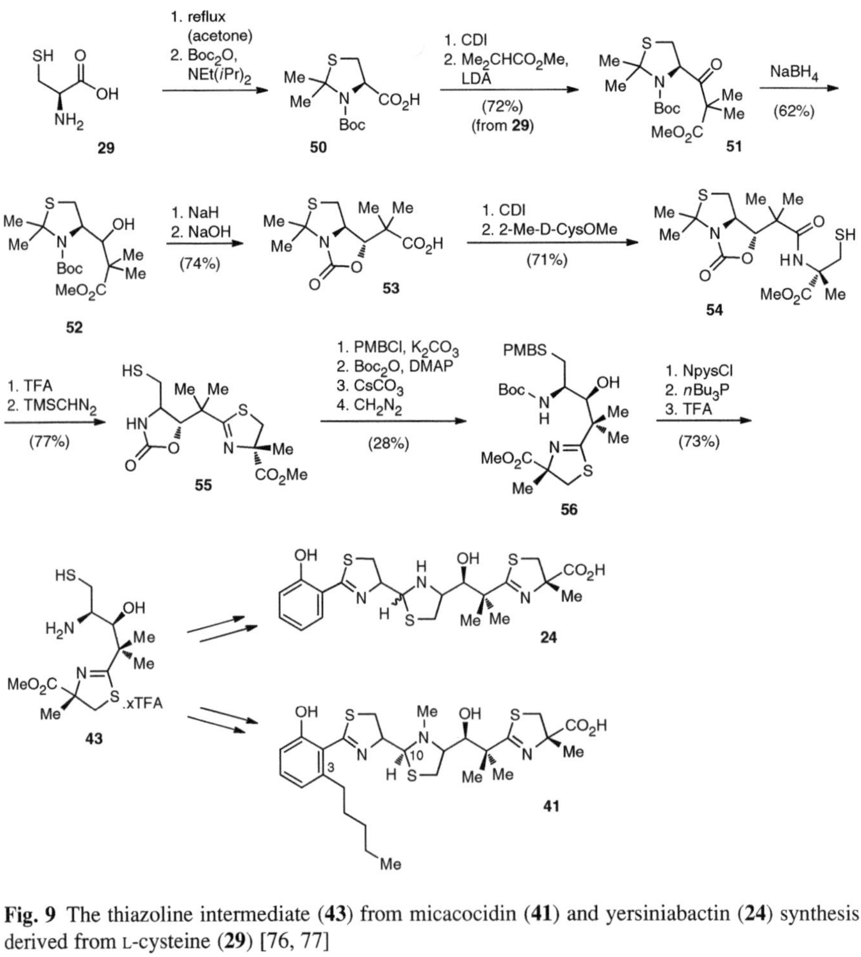

Fig. 9 The thiazoline intermediate (**43**) from micacocidin (**41**) and yersiniabactin (**24**) synthesis derived from L-cysteine (**29**) [76, 77]

transformed into **55** by cyclization of the thiazoline ring and removal of the acetonide (55% overall yield). Protection of the thiol and amino groups and cleavage of the oxazolidine ring afforded alcohol **56** in four steps and 28% yield. The synthesis was concluded by consecutive removal of the PMB and Boc protecting groups [76] to furnish 73% of **43** for subsequent condensation with **42** (Fig. 9). These units were linked by nucleophilic attack of the thiol and the amino group in **43** to the aldehyde function in **42** under basic conditions. Subsequent *O*-deprotection delivered ester **57** in 43% yield, which was saponified to yield the final compound, **24** (Fig. 10).

Interestingly, Ino and Murabayashi also synthesized the "truncated" yersiniabactin derivatives pyochelin I (**25**) and neopyochelin II (**58**) as a mixture during their analysis of the stereochemical preferences at the C9 and C10 chiral centers of **24**. For **25** synthesis (Fig. 11a), instead of condensation with the micacocidin intermediate **43**, aldehyde **42** was reacted with *N*-methyl-L-cysteine (**59**) and deprotected to yield a diastereomeric mixture of **25** and **60**, which was treated with zinc chloride to afford **25** and **58** in a 5:1 mixture in 46% overall yield [76]. This shared synthetic route, although not entirely biomimetic, does highlight the analogous pathway of **24** and **25** biosynthesis. A simpler three-step synthesis of **25** by Cox and colleagues did follow a completely biomimetic strategy, apart from the *N*-methylation of cysteine prior to cyclization of the thiazolidine ring. Synthesis occurred via condensation of salicylnitrile **61** with L-cysteine (**29**) to afford carboxylic acid **62**, followed by *t*-hexylborane reduction to give aldehyde **63** (Fig. 11b) [79]. Condensation of the aldehyde in **63** with *N*-methyl-L-cysteine (**59**) yielded the final compound **25** in 2% overall yield. Several groups have since optimized the synthetic access to aldehyde **63** in order to increase the overall yield to **25** [80–82].

The number of reactions required for total biosynthesis and chemical synthesis of yersiniabactin (**24**) is comparable, with 22 [62] and 30 [76, 77] reactions, respectively. However, two very different strategies are used by nature and by chemists in the production of this compound. The enzymatic pathway biosynthesizes the siderophore in a linear manner, while the chemical route convergently synthesizes two molecular halves and condenses them to the final product. The number of steps required for thiazolidine formation in the chemical synthesis of

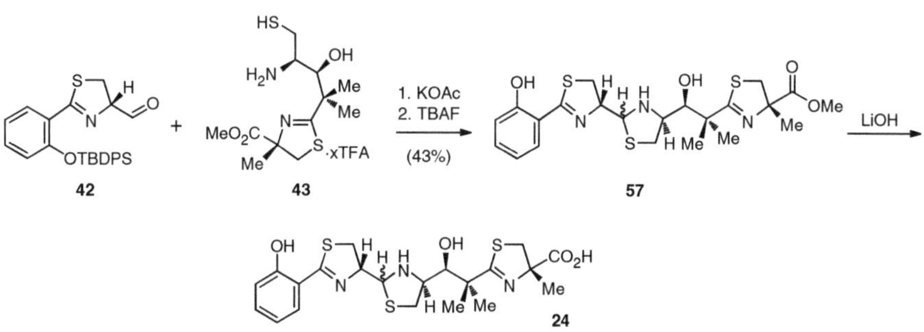

Fig. 10 Synthesis of **24** via condensation of aldehyde **42** with the micacocidin (**41**) synthesis intermediate **43**

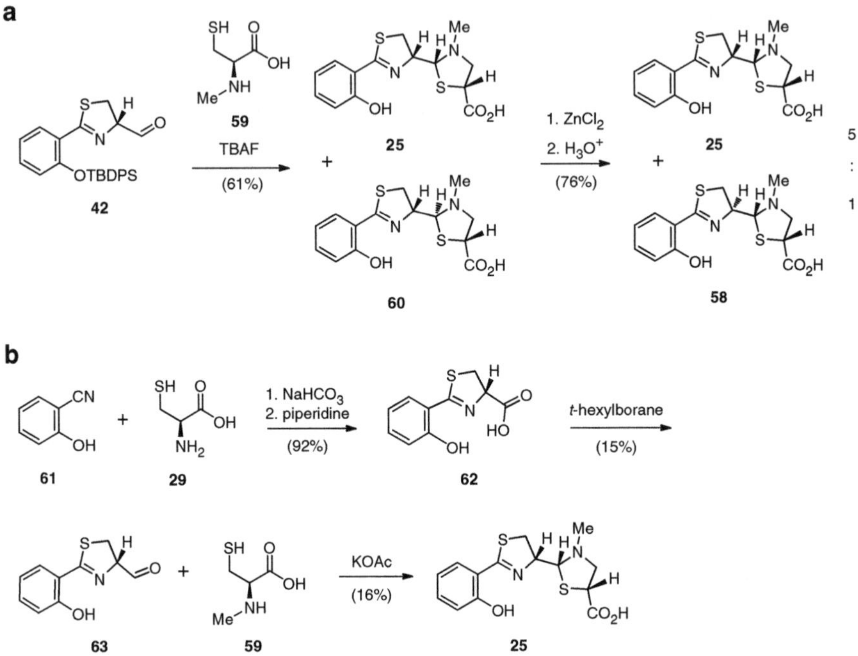

Fig. 11 Synthesis of **25** via (**a**) intermediate **42**, also utilized in **24** and **58** synthesis and (**b**) a biomimetic strategy from salicylnitrile **61** and L-cysteine **29**

24 by Ino and Murabayashi demonstrates the complexity of this approach over the elegant simplicity of the biosynthetic mechanism.

4 Dihydroxybenzoate Containing Siderophores: Enterobactin and Derivatives, Myxochelin A and Vibriobactin

The siderophore enterobactin (enterochelin) (**64**) is a cyclic lactone of three *N*-(2,3-dihydroxybenzoyl)-L-serine moieties produced by *E. coli* under iron stress. Enterobactin (**64**) was first isolated from iron-limited cultures of *Salmonella typhimurium* [83], *E. coli* [84], and *Aerobacter aerogenes* [84]. Structural analysis has confirmed that **64** chelates iron as a hexadentate ligand via the two hydroxyl groups on each catechol moiety (see Fig. 13) [85]. Of all the siderophores characterized to date, **64** has been shown to have the highest affinity for ferric iron, with a stability constant of 10^{52} M^{-1} [86, 87], which is remarkable, considering the affinity of EDTA for iron is 27 orders of magnitude lower. In mammals, serum albumin [88] and siderocalin [89, 90] bind the hydrophobic **64** which impedes siderophore-mediated transfer of iron to bacteria. Consequently, bacteria such as *E. coli* and

Salmonella enterica have evolved various linear and cyclic *C*-glucosylated analogs of **64**, known as salmochelins **67–70** (see Fig. 14) [91, 92], which are resistant to binding by these mammalian proteins [93–95]. A derivative of salmochelin, microcin E492 (**71**), has been isolated from *Klebsiella pneumonia* [96], and has been shown to be a pore-forming antibiotic that possesses an *N*-terminal, 84 amino acid peptide, which is attached to a linear *C*-glucosylated **64** derivative (**72**) [97–100]. The structural similarities of microcin E492 (**71**) with enterobactin (**64**) and salmochelins **67–70** may allow the uptake of this antibiotic into the inner membrane of other bacteria that possess siderophore-specific receptors, which could then effectively kill these bacterial competitors [93].

The initial stages of enterobactin (**64**) biosynthesis are common to many other 2,3-dihydroxybenzoic acid (DHB) containing siderophores (Fig. 12), including myxochelin (**65**) and vibriobactin (**66**). First, chorismic acid (**26**) is converted to isochorismic acid (**73**) by the isochorismate synthase EntC [101–103], then to 2,3-dihydro-2,3-dihydroxybenzoic acid (**74**) by the *N*-terminus of the isochorismate lyase (ISL) EntB [104, 105], and finally to the substrate **75** by the dehydrogenase EntA [106]. Further assembly occurs via an NRPS mechanism (Fig. 12) with the stepwise condensation of DHB (**75**) and an amino acid **76a–c** to form a DHB-amino acid monomer bound as a thioester to the PCP domain (**77a–c**). Specifically for enterobactin (**64**) biosynthesis (Figs. 12 and 13), three *N*-(2,3-dihydroxybenzoyl) serine units are synthesized in three iterative rounds by the NRPS enzymes, followed by cyclization of the three monomers into the final tricyclic lactone. The adenylation domain of EntE activates **75** as an acyl-adenylate **78**, which is then loaded onto the *C*-terminus aryl carrier protein domain (ArCP) of EntB to give **79** [104], while the adenylation domain of EntF activates L-serine (**76a**) as

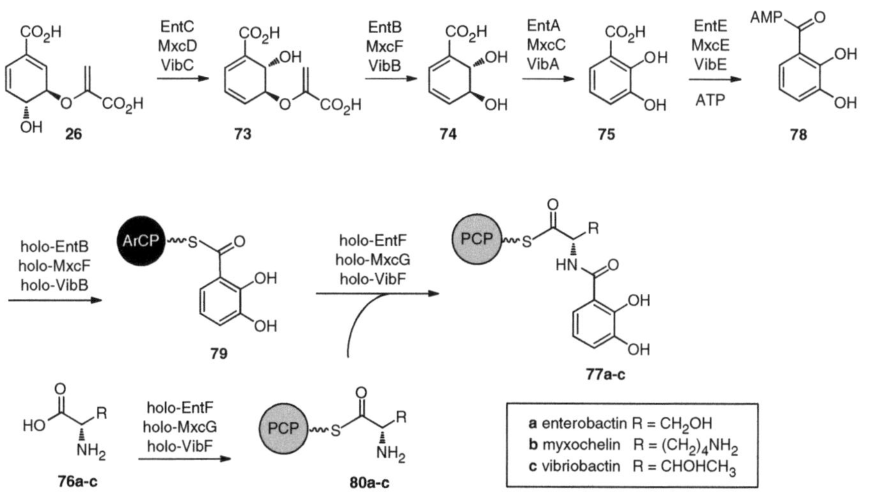

Fig. 12 The initial stages of **64**, **65**, and **66** biosynthesis proceed via the same mechanism, with the condensation of DHB (**75**) (from chorismic acid **26**) with a PCP-tethered amino acid **80** (serine (**a**), lysine (**b**), or threonine (**c**), respectively)

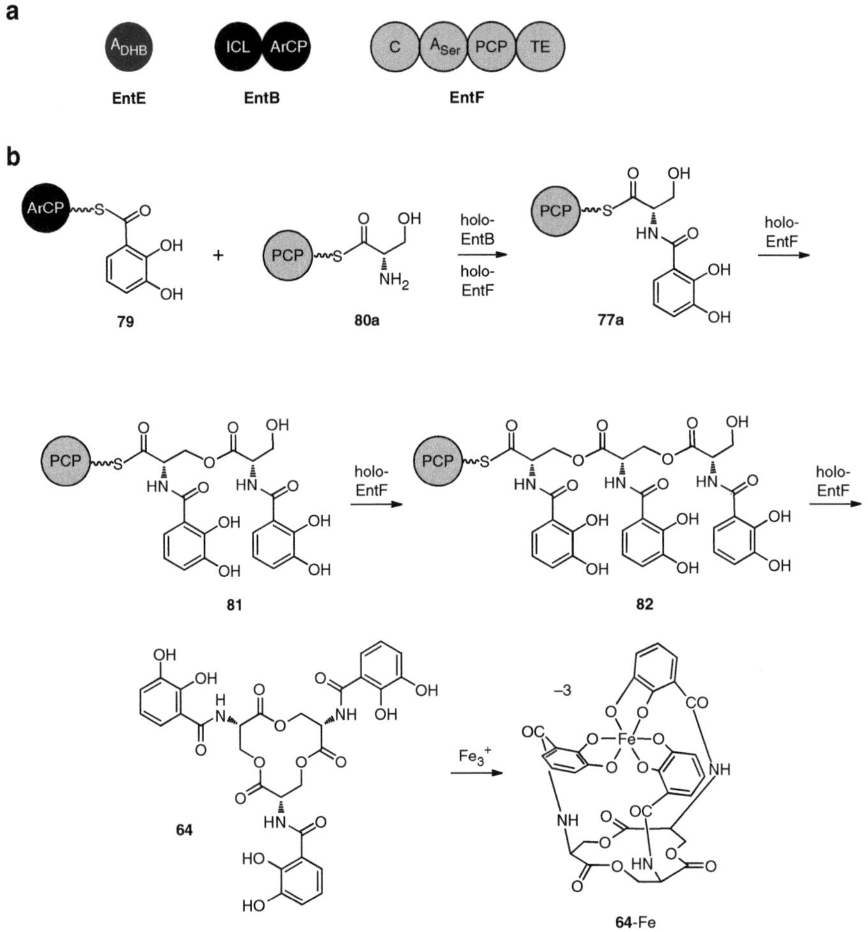

Fig. 13 (**a**) Three enzymes (EntE, EntB and EntF) are required for the *in vitro* reconstitution of **64**. (**b**) Biosynthesis of **64** by the stepwise condensation of three ArCP-bound DHB groups (**79**) and three PCP-bound L-serines (**80a**). Enterobactin (**64**) chelates ferric iron with high affinity via the six phenolate hydroxyl groups

L-serine-AMP. The condensation domain of EntF then catalyzes amide bond formation between each DHB-ArCP unit **79** and an L-serine-PCP molecule **80a** to form **77a** [107]. Subsequently, the terminal thioesterase of EntF catalyzes the stepwise ester bond formation between each of the three tethered *N*-(2,3-dihydroxybenzoyl)-L-serine moieties (**77a**, **81**, **82**) as they are biosynthesized, as well as the cyclization and release of the final 12-membered trilactone **64** [108] (Fig. 13).

Following this biosynthetic route to the core structure of **64**, diverse tailoring steps can lead to glycosylated derivatives. In salmochelin S4 (**67**) biosynthesis (Fig. 14), **64** is glucosylated at the C5 position of two of the catechol groups by the *C*-glycosyltransferase, IroB [109] to give first **83** and then **67**. A variety of

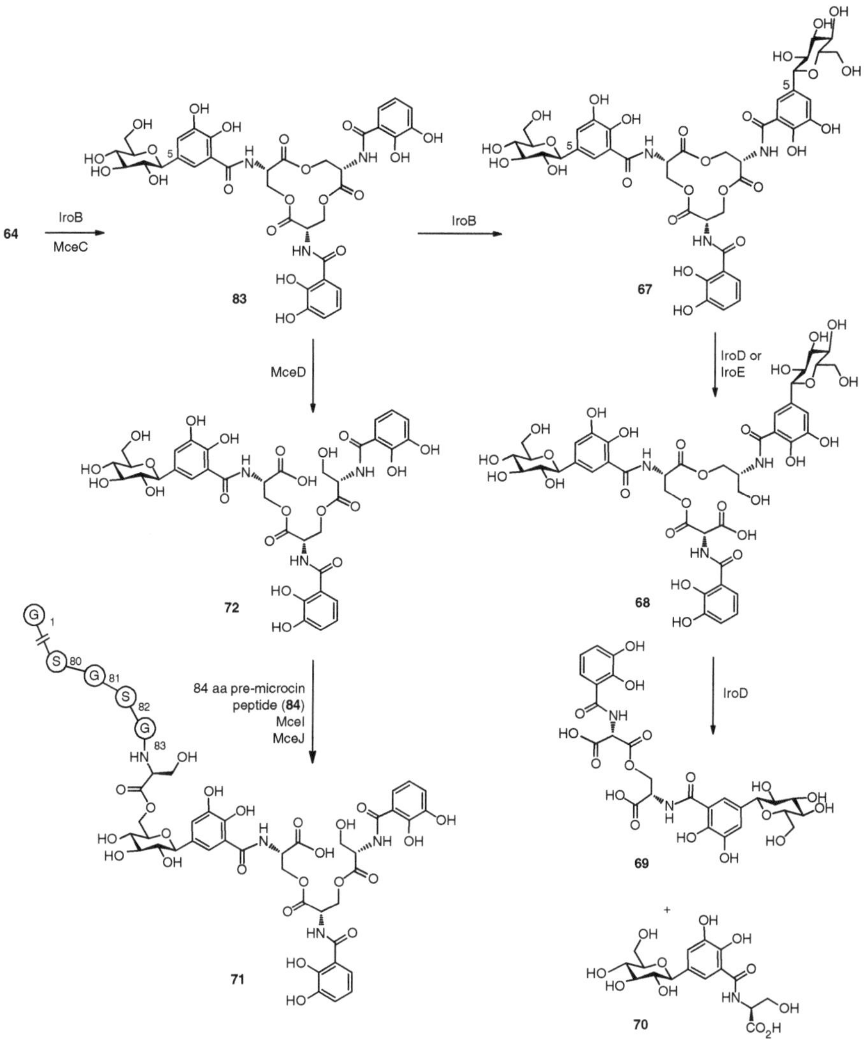

Fig. 14 Biosynthesis of microcin E492 (**71**), salmochelin S4 (**67**), and salmochelin SX (**70**) from the enterobacin (**64**) precursor

salmochelin analogs then arise from linearization of **67** into the DHB-serine trimer salmochelin S2 (**68**) by the esterases IroD or IroE; or cleavage into the respective monoglucosylated dimer salmochelin S1 (**69**) or glucosylated monomer salmochelin SX (**70**) via IroD, during iron release [110, 111]. For the biosynthesis of microcin E492 (**71**), the IroB homolog, MceC, and the trilactone hydrolase, MceD, function to glucosylate and cleave the precursor **64** to a linear monoglucosyl derivative **72** [99, 100]. The enzymes MceI and MceJ then form a heterodimer to attach the glucose group of the linear trimer to the *C*-terminal serine of an 84-amino acid long, ribosomally synthesized, peptide chain (**84**), which is cleaved from a precursor peptide encoded by the gene *mceA*, to give **71** (Fig. 14) [17, 99, 112].

In the *in vitro* reconstitution of **64** biosynthesis, the enzymes EntB and EntF were first activated to their holo forms via the cognate EntD PPTase before final purification [107]. Total biosynthesis then proceeded via a one-pot reaction with the three enzymes (EntE, holo-EntB, and holo-EntF), two substrates (DHB (**75**) and L-serine (**76a**)), and ATP as a cofactor [107]. The final catalytic efficiency of **64** total reconstitution was extremely high at 121 min^{-1}, which reflects the elegant simplicity of the biosynthetic strategy as well as the relatively small number of enzymes, substrates, and cofactors required for *in vitro* biosynthesis [107]. Incubation of **64** with the *C*-glycosyltransferase, IroB, *in vitro* resulted in sequential C5 glucosylation of the catechol groups, to form mono- (**83**), di- (**67**), and triglucosyl enterobactin derivatives [109]. In this way, salmochelin S4 (**67**) was reconstituted by IroB *in vitro* from the precursor **64** using the substrate UDP-glucose and the reducing agent TCEP. Although the majority of the microcin E492 (**71**) biosynthetic enzymes have been individually characterized, there are currently no reports on the total reconstitution of **71**, most probably because this would require the integration of both ribosomal and nonribosomal peptide syntheses *in vitro*.

There have been seven total syntheses of **64**, compared to the single synthesis of yersiniabactin (**24**). The greater focus on **64** synthesis may be due to its more potent activity, simpler synthetic strategy, and the comparably low yields that can be isolated from culture (0.6 mg L^{-1} from *S. typhimurium* [83], 15 mg L^{-1} from *E. coli* [84], and 22 mg L^{-1} from *A. aerogenes* [84]). Each of the syntheses of **64** follows one of two different strategies. The first two synthetic routes utilize the condensation of three serine derivatives followed by cyclization of the trilactone and subsequent addition of the DHB groups (**75**) [113, 114], which is in the opposite order to the biosynthetic mechanism. The synthesis described by Corey and Bhattacharyya (Fig. 15) [113] started with the generation of the *p*-bromophenacyl ester **85** from *N*-benzyloxycarbonyl-L-serine (**86**) in 98% yield. THP-protection of the alcohol function and removal of the acid protective group with Zn paved the way for generation of an activated thioester of **85**, which was in situ condensed with one equivalent of **85** to give **87** in 69% overall yield. Renewed deprotection of the acid and coupling of an in situ generated thioester with another equivalent of **85** delivered the cysteine trimer **88**, which upon deprotection of the alcohol and the acid function furnished compound **89** in 51% overall yield. Macrocyclization using the same thioester activation strategy as above gave the cyclic tripeptide **90** (40% yield), which was further transformed into the natural product **64** by hydrogenolytic cleavage of the Cbz groups and subsequent attachment of the 2,3-dihydroxybenzoyl moieties no yield given. The later synthesis by Rastetter [114] followed the same strategy but used different protection and deprotection methods. Both syntheses, however, resulted in low overall yields of approximately 1%. In a similar manner, Rogers utilized *N,N′*-dibenzyl-L-serine as the precursor for cyclization to the protected trilactone, which increased the yield of this step by five times over the previous two strategies [115].

The second synthesis approach, carried out by Shanzer and Libman, employed L-serine β-lactones **91**, derived from *N*-trityl L-serine (**92**) in 26% yield (Fig. 16a).

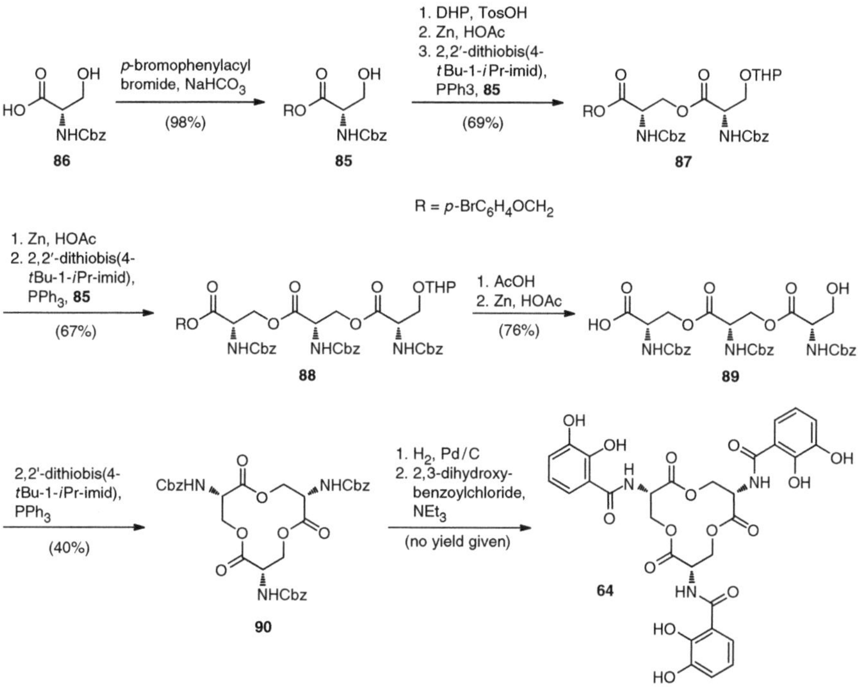

Fig. 15 The first chemical synthesis of **64**, which involves the condensation of three serine derivatives (**86**) with subsequent cyclization and addition of DHB groups

Stannoxane mediated cyclotrimerization directly gave **93** in 23% yield. Subsequent HCl-induced replacement of the trityl groups, and addition of the nitrophenyl esters of 2,3-bis(benzyloxy)benzoic acid to the free amines thus obtained, delivered **94** (no yield given), which was transformed quantitatively into the natural product **64** by final hydrogenolytic cleavage of the benzyl protective groups [116]. Although this route effectively reduced the number of steps required for the synthesis of **64**, the overall yield was only increased to approximately 6% [116]. In order to enhance the total yield of this organotin strategy, the low efficiency steps, namely the conversion of **92** to **91** and then to the tricyclic lactone, were targeted for optimization. These processes were replaced by converting an *N*-Boc-L-serine precursor to the corresponding β-lactone and subsequently to the trilactone [117]. This increased the overall yield to approximately tenfold of the original Shanzer and Libman strategy. A variation on the organotin template based strategy was used in order to bypass the relatively low efficiency β-lactonization of the *N*-tritylserine substrate (**92**), which increased the overall yield to approximately 50% [118]. Using L-serine methyl ester as a starting material, an *N*-trityl-L-serine methyl ester (**95**) was generated and used to produce the tritylated trilactone **93**. In an optimized strategy (Fig. 16b), **95** was reacted with 2,2-dibutyl-1,2,3-dioxastannolane as the template to afford **93**, which was then converted to the hexabenzylenterobactin derivative **94** by

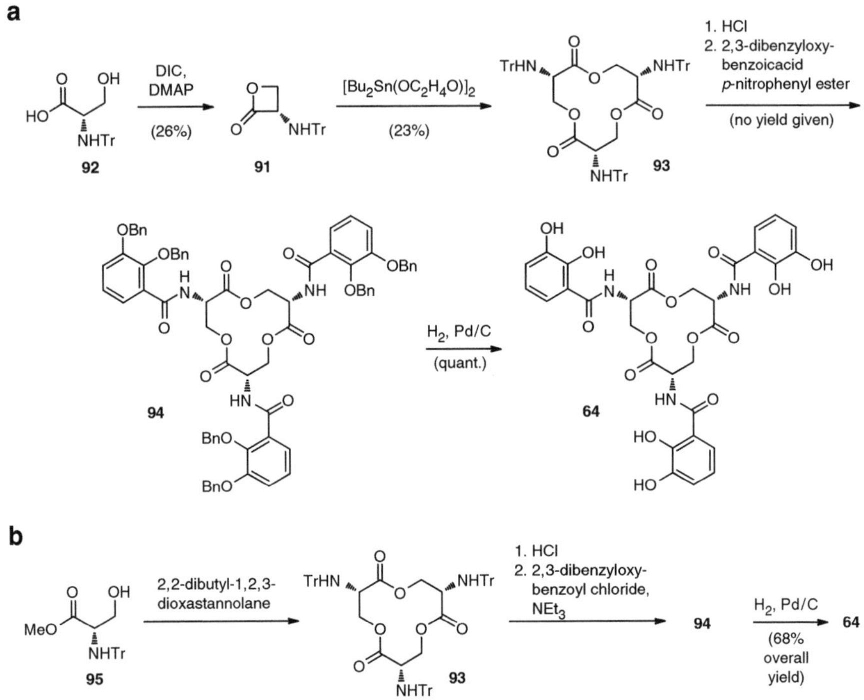

Fig. 16 Chemical syntheses of **64** via (**a**) the first and (**b**) the currently most efficient organotin template strategies

deprotection and treatment with 2,3-dibenzyloxybenzoyl chloride [119]. Following this, the benzyl protective groups were cleaved by Pd-C catalyzed hydrogenolysis, which yielded **64** in a highly efficient overall yield of 64%. The synthetic approaches to **64** have, thus far, not utilized the elegant, biomimetic route involving the condensation of DHB (**75**) to L-serine (**76a**) and subsequent cyclization of three monomers to form **64**. In fact, attempts at stepwise condensation of DHB-serine monomers have not been successful, due to undesired racemization of the monomers [120]. These unsuccessful attempts highlight an advantage of natural biosynthesis where the substrate specificity of enzymes can effectively dictate the biosynthetic route and simultaneously protect sensitive parts of the molecule, which suggests a fundamentally efficient logic to nature's biosynthetic rationale.

Recently, a total synthesis of salmochelin SX (**70**) has been reported which used cross-coupling of acetobromo-α-D-glucose **96** with arylzinc derivative **97** to furnish **98**, followed by full deacylation and perbenzylation to give **99**. Subsequent saponification of the methyl/benzyl ester, formation of the acid chloride, and addition of a protected L-serine unit yielded **100**. Final deprotection formed the glucosyl-DHB-serine **70** in 28% overall yield (Fig. 17) [121]. This strategy involved the stepwise addition of the aryl moiety and serine (**76a**) to the sugar portion, which is in the opposite order to the biosynthetic rationale.

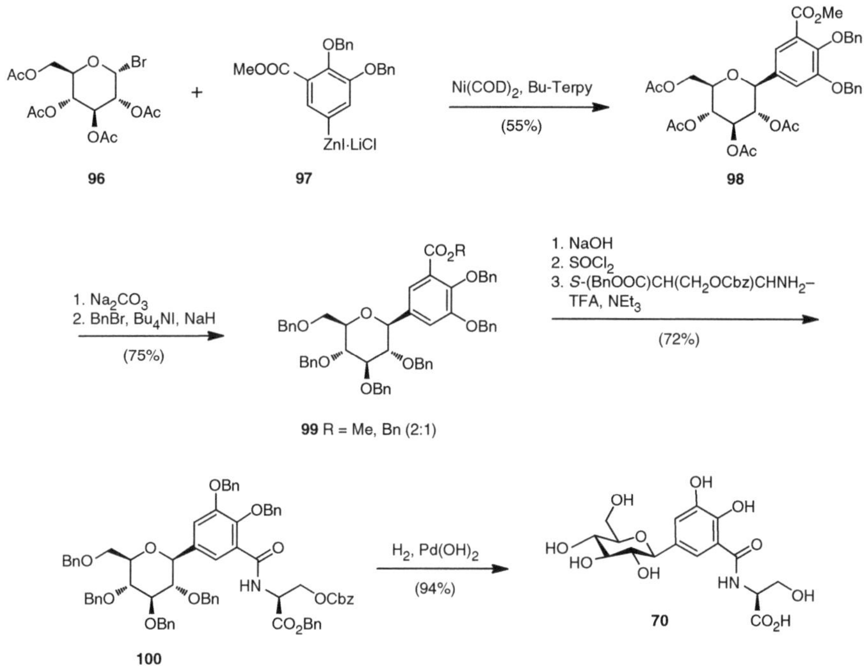

Fig. 17 Total synthesis of salmochelin SX (**70**)

Another DHB-containing siderophore that has been reconstituted *in vitro* is myxochelin A (**65**). This siderophore, which is produced by *Stigmatella aurantiaca* [122] and *Nonomuraea pusilla* [123], is comprised of two DHB residues (**75**) attached to the amino groups of lysine (**76b**), and has also been shown to possess antitumor activity [123]. The initial stages of **65** biosynthesis parallel those of **64**, with production of DHB (**75**) via the EntC, EntB, and EntA homologs, MxcD, MxcF, and MxcC (Fig. 12) [122]. In the same way, **75** is activated as DHB-AMP (**78**) by the EntE homolog MxcE and transferred to the ArCP domain of MxcF to form **79**. The NRPS module MxcG then activates lysine (**76b**) as an acyl adenylate bound to the PCP as **80b** and catalyzes the formation of amide bonds between this residue and two molecules of ArCP-bound DHB (**79**) (Fig. 18). This dual bisacylation activity, which could possibly occur sequentially via intermediate **77b** or simultaneously to give **101** directly, has previously only been reported in vibriobactin (**66**) biosynthesis [124]. Further, the unique mechanism of release of the final compound **65** from the NRPS via the action of a terminal reductase domain, which converts PCP-bound thioester **101** into aldehyde **102** or, possibly, directly into myxochelin A (**65**) [122], is found in only a handful of known biosynthetic systems. Reductive transamination, catalyzed by MxcL, can also further convert **102** to the amine myxochelin B **103**.

Total biosynthesis of **65** was performed using the enzymes MxcE, holo-MxcF, and holo-MxcG with the substrates DHB (**75**) and lysine (**76b**), and the cofactors

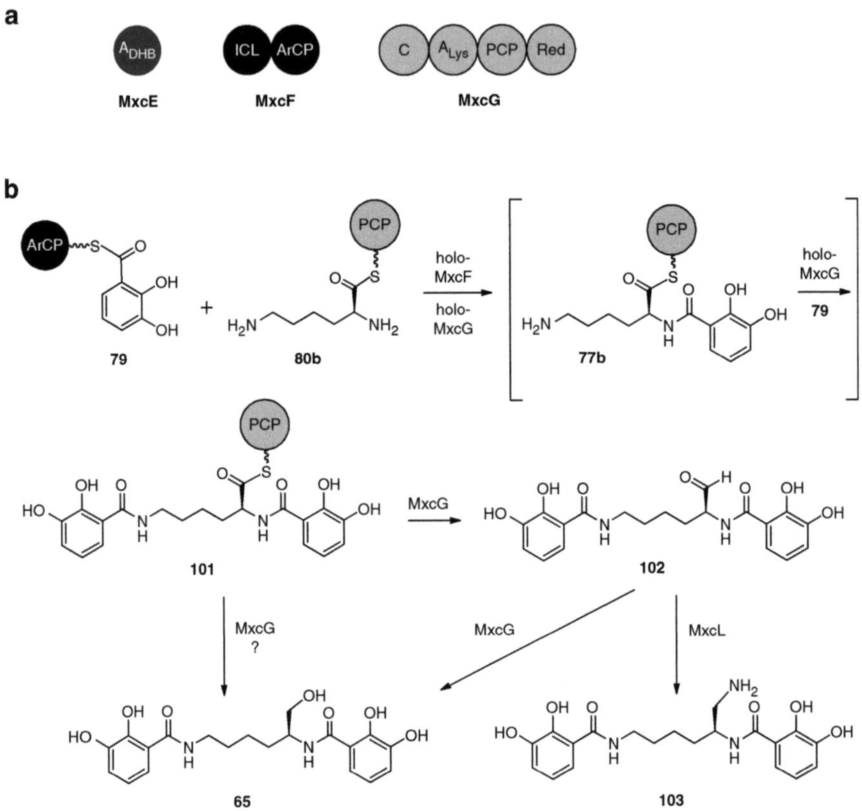

Fig. 18 (**a**) Total *in vitro* biosynthesis of **65** requires three enzymes, MxcE, MxcF, and MxcG. (**b**) Biosynthesis of **65** and **103** from two ArCP-bound DHB units (**79**) and one PCP-bound lysine (**80b**)

ATP and NADPH [125]. The apo-ArCP and -PCP domains of MxcF and MxcG were first activated by coexpression with the PPTase MtaA from myxothiazol biosynthesis [126]. Myxochelin A (**65**) was synthesized at a linear rate for the endurance of the assay (65 h), with over 800 nmol mL^{-1} of compound formed during that time [125].

The cholera-causing pathogen *Vibrio cholerae* is able to produce the siderophore vibriobactin (**66**) [127] via an NRPS mechanism [128, 129] similar to that in **64** and **65** biosynthesis. Once again, DHB (**75**) is biosynthesized using the pathway specific enzymes VibC, VibB, and VibA (EntC/MxcD, EntB/MxcF, and EntA/MxcC homologs) (Fig. 12). After acylation of DHB (**75**) by VibE (EntE/MxcE homolog), the acyl adenylate (**78**) is loaded onto the ArCP of VibB (EntB/MxcF homolog) to give **79** for condensation with the VibF-activated and -tethered aminoacyl-threonine **80c**. Following this, an amide bond is formed between these two building blocks yielding **77c**, and the threonine portion is cyclized to the corresponding oxazoline **104** by VibF (Fig. 19) [124]. A final DHB unit (**75**) is also loaded onto

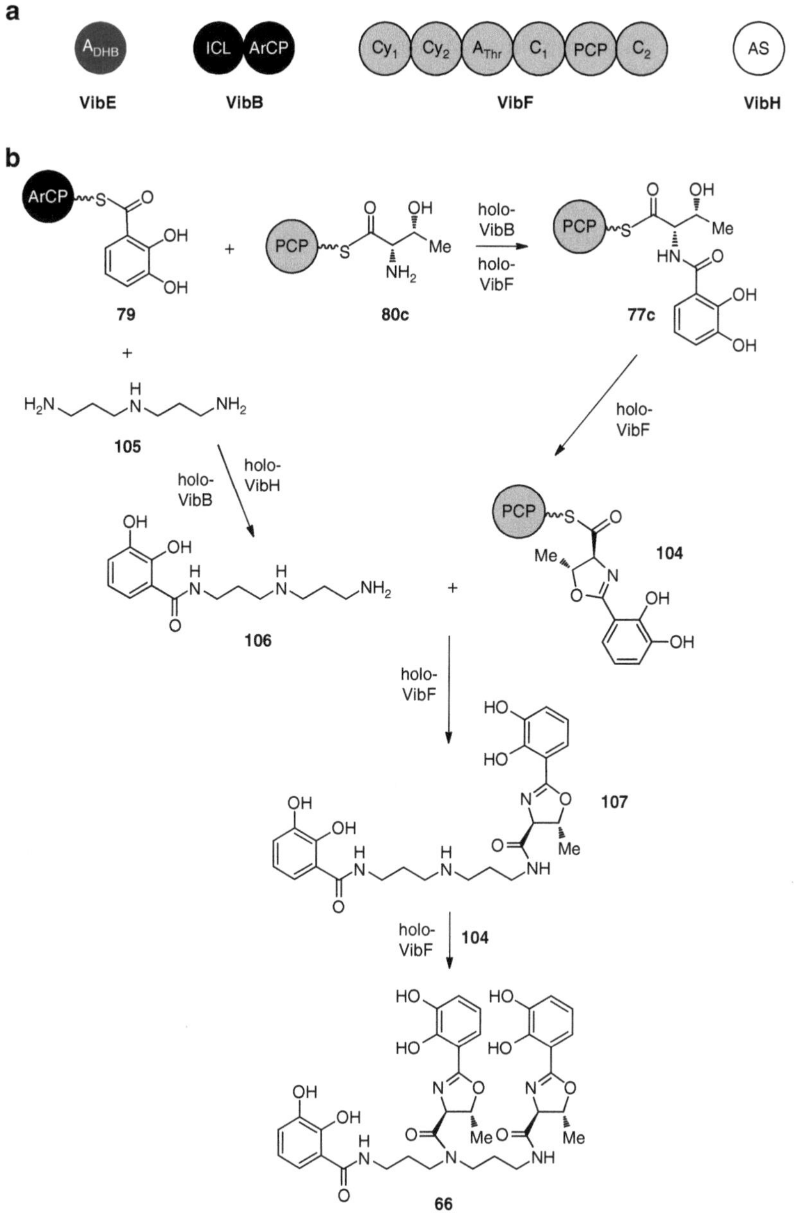

Fig. 19 (**a**) Total *in vitro* biosynthesis of **66** requires three enzymes, VibE, VibF, and VibH. (**b**) Biosynthesis of **66** from three ArCP-bound DHB units (**79**), two PCP-bound threonines (**80c**), and one norspermidine (**105**)

VibB as **79** for condensation with the primary amine of norspermidine **105** via the amide synthase VibH, resulting in unbound, soluble DHB-norspermidine **106** [128]. Two DHB-oxazoline monomers, still tethered as thioesters on VibF as **104**, are then sequentially transferred to the primary and secondary amines of the DHB-norspermidine molecule to yield **107** and **66**, respectively [124].

Reconstitution of the biosynthesis of **66** was carried out in a similar manner to the **64** and **65** total biosyntheses, using the enzymes VibB, VibH, VibE, VibF and Sfp, the substrates DHB (**75**), L-threonine (**76c**) and norspermidine (**105**), and the cofactors ATP and CoA [124]. Vibriobactin (**66**) was formed *in vitro* at a linear rate of 15 μM min^{-1} over 20 min with large scale total biosynthesis yielding 1.6 mg [124]. Interestingly, during the directed evolution of the VibB ArCP, the Walsh laboratory replaced EntB with mutant VibB enzymes in the *in vitro* biosynthesis of **64** in order to assay increased activity [130]. VibB and corresponding high activity mutants were able to complement EntB activity, albeit at a lower efficiency than the natural enzyme. However, this example demonstrates the effective strategy of nature to utilize one pathway in the production of several divergent compounds.

The total biosynthetic route to both **65** and **66** involves the condensation of one or two DHBs (**75**) with either lysine (**76b**), threonine (**76c**), or norspermidine (**105**), with further cyclization of the threonine residues to DHB-oxazolines and – in the case of **66** – condensation of the latter to the central backbone. Similarly, the synthetic strategy to **65** reflects this biosynthetic approach, with the reduction of a *t*-Boc-protected lysine **108** to the corresponding alcohol, removal of the protecting groups to give **109**, and subsequent installation of benzylated 2,3-DHB **110** (Fig. 20) [123]. Following this, the benzyl groups were removed to afford the final compound **65** in 23% overall yield. In contrast, **66** was first synthesized using an entirely different strategy. Although Bergeron and colleagues [131] began biomimetically, with the condensation of a norspermidine derivative **111** and 2,3-dimethoxybenzoyl chloride to yield a trisubstituted norspermidine **112** (Fig. 21), the synthesis diverges from the biosynthetic strategy from this point onwards. Compound **112** was *N*-deprotected to the monoacylated norspermidine **113**, which was then bisacylated with a thioester activated *N*-Boc-L-threonine

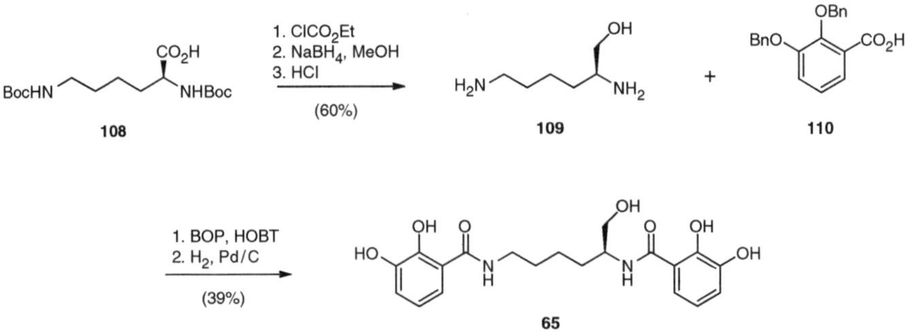

Fig. 20 Total synthesis of **65** from *t*-Boc protected lysine (**108**)

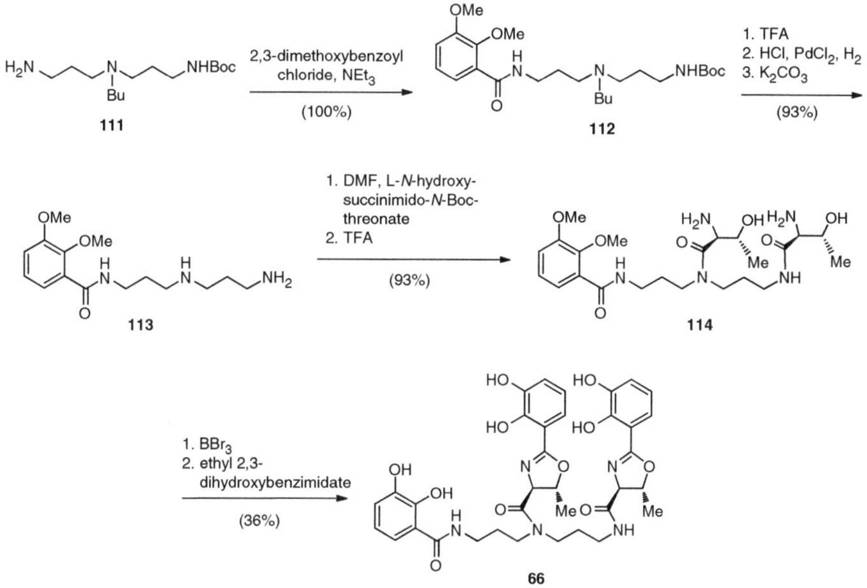

Fig. 21 Total synthesis of **66** via the original chemical synthesis route

derivative with subsequent removal of the Boc groups to give **114** (Fig. 21) [131]. After *O*-demethylation, the cyclization of the two L-threonine amides to oxazolines was carried out in ethyl 2,3-dihydroxybenzimidate, which also accomplished the addition of the two DHB moieties to give **66** in 31% overall yield. This strategy is in direct contrast to the biosynthetic route where dehydration and cyclization of the threonines is catalyzed after the condensation with DHB (**75**), but prior to the attachment of these residues to the norspermidine core. Using a more biomimetic rationale, Sakakura *et al.* reasoned that a higher yield to **66** could be achieved by an early construction of the DHB-oxazoline group (Fig. 22a) [132]. Thus, *N*-(*o,m*-dialkoxybenzoyl)-L-threonine (**115**), derived from DHB (**75**) in a five-step sequence via **116** in 72% yield, was cyclized to 2-(*o,m*-dialkoxyphenyl)oxazoline (**117**) in 94% yield using molybdenum catalysis. This catalyst effectively mimicked the activity of the cyclization domains (Cy) from **66** biosynthesis. The DHB-norspermidine derived molecule **118** was in turn synthesized from condensation of norspermidine (**105**) and **116** (Fig. 22b). Subsequently, the methyl ester **117** and the corresponding acid **119** were sequentially condensed to the primary and the secondary amines of **118** and **120**, respectively, exactly as occurs in biosynthesis. Finally, the *o*-xylylene groups were removed by hydrogenolysis to yield **66** with an overall yield of 64% from **105** [132].

A comparison of siderophores that have been both totally biosynthesized and chemically synthesized reveals that the more structurally simple compounds (pyochelin, myxochelin) can be synthesized following a biomimetic route, whereas more complex molecules (yersiniabactin, enterobactin, vibriobactin) often tend to

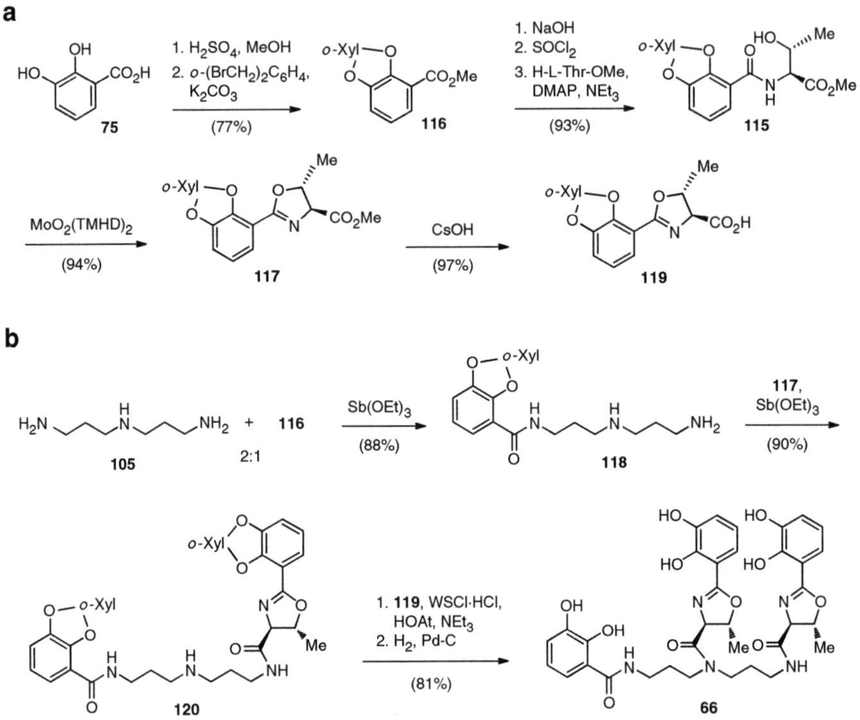

Fig. 22 Total synthesis of **66** via a biomimetic strategy

be synthesized via unique strategies. This trend highlights the specialized roles of *in vitro* biosynthesis and chemical synthesis in the production of highly functionalized natural products. However, the biomimetic approach to the structurally more complex **66** [132] demonstrates that an effective synthetic rationale can sometimes be ascertained from the biosynthetic mechanism.

5 Indole Alkaloids: Staurosporine Aglycone, K252c

The potent protein kinase inhibitor staurosporine (**121**) was first isolated in 1977 from the actinomycete strain AM-2282 [133], which was later reclassified as *Lentzea albida* [134]. Staurosporine (**121**) has received immense attention as a potential chemotherapeutic, possessing nanomolar inhibitory activity against protein kinase C (PKC) [135]. Its inhibitory activity against multiple protein kinases in cells has prompted an investigation of related molecules with less potent, but more PKC-specific, activity [12, 136]. A related molecule, UCN-01 or 7-hydroxystaurosporine (**122**), differing from staurosporine (**121**) only in the presence of a hydroxyl at C7 (Fig. 23) [137], is in Phase II clinical trials against renal cell carcinoma, melanoma, and lymphoma, as well as against small cell lung carcinoma, as a dual agent with topotecan [138], suggesting that molecules related to **121** are strong candidates for use as chemotherapeutics [139]. Staurosporine (**121**) has been pursued as a synthetic target, with the total synthesis reported by both the Danishefsky and Wood research groups [140–143].

An understanding of the biosynthetic pathway to **121** has also been hotly pursued. With the elucidation of the putative biosynthetic gene cluster for staurosporine in 2002 [144], as well as those of closely related bisindoles [145–149], numerous groups have sought to determine the biosynthetic logic of staurosporine production, to understand mechanisms of intriguing biosynthetic enzymes, and to generate modified natural products using combinatorial biosynthesis [144, 150–159]. Most of the early steps in staurosporine biosynthesis are now well explored. However, the enzymes StaG and StaN, involved in the unique glycosylation chemistry – catalyzing the coupling of the staurosporine aglycone (**1**) at the two indole nitrogens to an L-ristosamine sugar – while well-explored *in vivo* [160, 161], have not yet been shown to be active *in vitro*, perhaps due to the poor quality of expression of enzymes from recombinant systems.

In its own right, K252c (**1**), the staurosporine aglycone, also known as staurosporinone, is a molecule of high interest. Like staurosporine (**121**), it was first isolated as a natural product, from a strain that produces a variety of related molecules [162]. It also possesses strong activity against protein kinases. Although no single, one-pot, total biosynthesis has been demonstrated for **1**, each of the steps necessary for its production have been separately reported. Our discussion will, therefore, focus on K252c (**1**), given its synthesis both chemically and *in vitro*, using purified enzymes, and given its strong similarities to staurosporine (**121**), including structural, biosynthetic, and pharmaceutical features.

The biosynthesis of **1** follows a pattern seen for all bisindoles that have been biosynthetically investigated thus far [17, 163, 164]: two molecules of L-tryptophan (**123**) are oxidized and then dimerized to give an initial bisindole skeleton (Fig. 24). Historically, the biosynthesis of staurosporine (**121**) has been investigated in parallel with that of rebeccamycin (**124**), a related bisindole. Both molecules are thought to be biosynthesized through nearly identical routes to give the aglycone

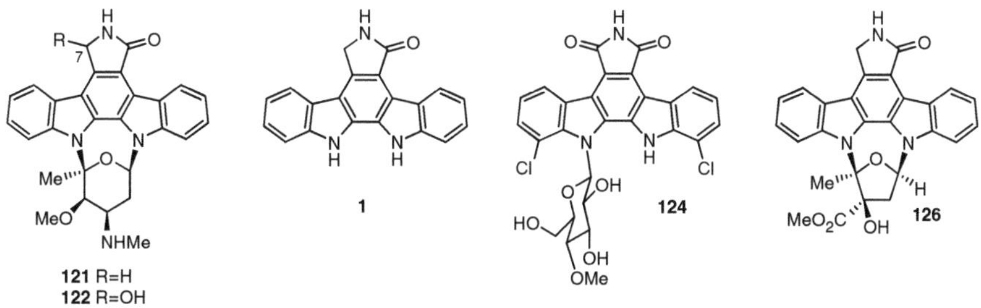

Fig. 23 Staurosporine (**121**) and related molecules

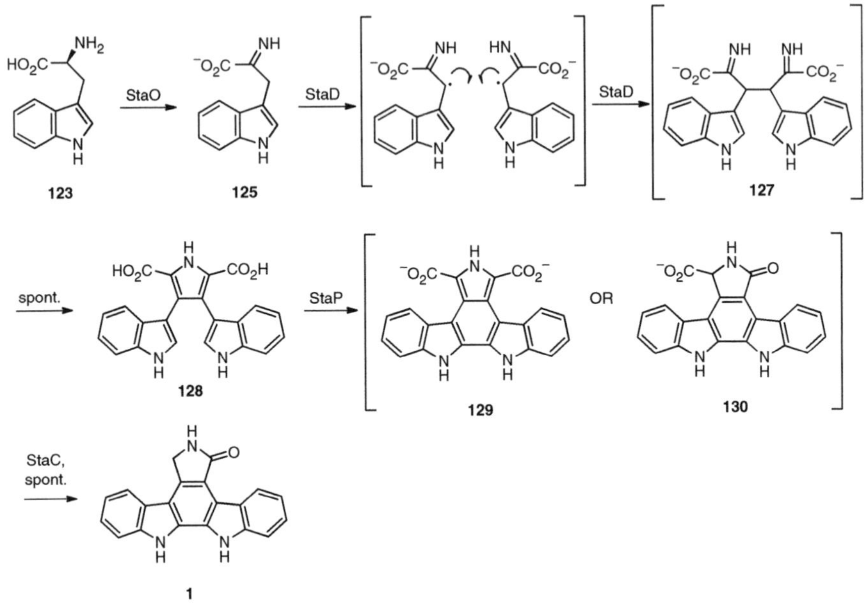

Fig. 24 Proposed biosynthetic route to **1** from L-tryptophan (**123**)

portions of the molecules, and the corresponding biosynthetic enzymes in each pathway are highly related. In the case of staurosporine (**121**), L-tryptophan (**123**) is first converted to indole-3-pyruvic acid imine (**125**) by the action of the flavin-dependent enzyme StaO. However, only RebO, the corresponding enzyme from the rebeccamycin (**124**) biosynthetic pathway, has been investigated *in vitro* [151, 152]. While RebO has a preference for 7-chloro-L-tryptophan (the rebeccamycin precursor) over L-tryptophan (**123**) (the staurosporine precursor) with a 57-fold greater k_{cat}/K_m for the chlorinated substrate [151], it is nonetheless capable of converting **123** and molecular oxygen to indole-3-pyruvic acid and hydrogen peroxide using flavin adenine dinucleotide (FAD) as a cofactor with a k_{cat}/K_m of 7.9 min^{-1} mM^{-1} [151]. Improved conversion of **123** may be possible with the use of StaO itself, related homologs such as InkO from the K252a (**126**) biosynthetic pathway [145], active in a heterologous expression system [165], or alternate StaO homologs identified from genomic sequencing of other staurosporine producers [166].

The next reaction in the biosynthetic pathway, the dimerization of two molecules of **125**, is thought to occur through radical bond formation to give rise to **127** (Fig. 24). This unusual reaction – dimerization of two unreactive carbon centers – is catalyzed by an equally unusual enzyme, StaD, a heme-containing enzyme with ~1,100 amino acids [158], which has relatively few sequence relatives in sequence databases. Each of the currently known StaD sequence relatives are thought to play equivalent roles in related biosynthetic pathways [145–149, 155, 159], and all characterized homologs contain heme iron. Work on the related enzyme RebD

(54% identity) has shown that the preferred substrate is indole-3-pyruvic acid imine (**125**) [152], a substrate preference that is likely to hold for StaD as well. Spontaneous chemistry is then thought to result in production of chromopyrrolic acid (**128**) [149, 152, 158].

The final set of biosynthetic reactions to **1** involves the four-electron oxidation of chromopyrrolic acid (**128**). This transformation can be carried out with **128**, the cytochrome P450 enzyme StaP, flavodoxin $NADP^+$ reductase, ferredoxin, NAD(P)H, and the putative flavin-dependent oxidoreductase StaC [153]. Crystal structures are currently available for StaP itself [157] and for RebC, the homolog of StaC from the rebeccamycin biosynthetic pathways [154]. These structural data coupled with chemical investigations [167] and modeling studies [156] suggest that StaP catalyzes the aryl-aryl coupling of **128** to give an intermediate that may be **129** or **130**. StaC is then likely to stabilize one of the intermediates in its active site – both suggested intermediates would decompose in solution [167], and the homolog RebC in crystalline form "traps" a tautomer of **130** in its active site [154] – and to react with the intermediate to generate the staurosporine aglycone (**1**) (Fig. 24).

Synthetic routes to **1** are numerous and varied (Table 1). Of all synthetic routes, none use the same pathway seen biosynthetically: L-tryptophan (**123**) converted to indole-3-pyruvic acid imine (**125**), dimerized to chromopyrrolic acid (**128**), and oxidized to give the staurosporine aglycone (**1**). However, some of the synthetic routes are decidedly biomimetic. For instance, Pd-catalyzed coupling of arcyriarubin A (**131**) was used to generate arcyriaflavin A (**132**), followed by two-step reduction to give the staurosporine aglycone (**1**) via compound **135** (Fig. 25a) [7–9]. In Hill's synthesis [7], treatment of dibromomaleimide (**133**) with indoylmagnesium bromide **134** directly delivered 29% of the coupling precursor **131**. Compound **131** was cyclized using $Pd(OAc)_2$ as the catalyst to give **132**, which, upon two-step reduction, furnished **1** in 47% combined yield. Alternate routes to the **131** precursor have also been reported [8, 9]. While neither arcyriarubin A (**131**) nor arcyriaflavin A (**132**) are biosynthetic precursors to staurosporine (**121**) [153], both molecules share clear structural features with **128**. Furthermore, the palladium-catalyzed aryl–aryl bond formation is analogous to the reaction catalyzed by StaP on chromopyrrolic acid (**128**). However, this synthetic strategy also involves *over*-oxidizing a precursor and then reducing the structure back to the staurosporine aglycone (**1**); by contrast, the biosynthetic strategy is a sequential set of oxidations, without any reductions, to give rise to **1**.

Another example of a partially biomimetic strategy is shown in Fig. 25b. Here the analogy is to the starting materials: two indole-containing molecules, **136** and **137**, which were dimerized via formation of the same bond as seen in the reaction carried out by StaD on two molecules of indole-3-pyruvic acid imine (**125**). This bond – between the β-carbons of the L-tryptophan precursor (**123**) – once formed biosynthetically leads to **127**, which spontaneously forms chromopyrrolic acid (**128**) in oxygen (Fig. 24). In the biosynthetic route, this reaction is between two unactivated carbons, which would seem to render it challenging, although low levels of **128** formation are seen between indole-3-acetic acid, L-tryptophan, and

Table 1 Synthetic routes to the staurosporine aglycone (**1**)

Citation(s)	Synthetic intermediate(s)	Key reaction(s) to give K252c	Yield from starting materials (%)
[2]	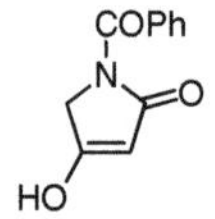 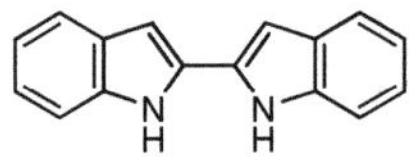	Condensation/ photocyclization	46
[3]	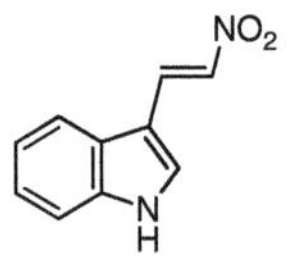 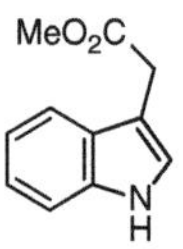	Intermolecular Michael addition	5
[4]	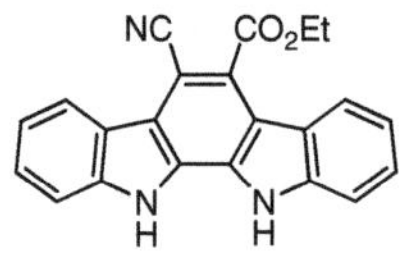	Reductive cyclization	37
[5]	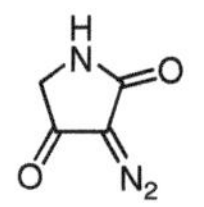 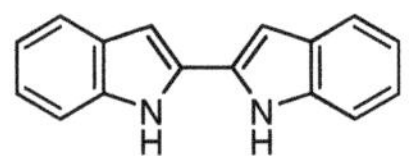	Cycloaromatization	25
[6]	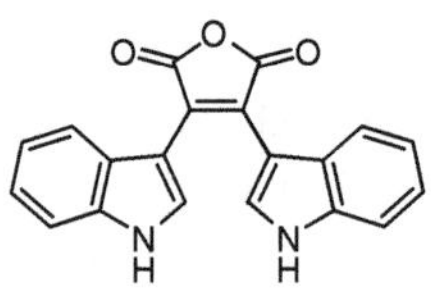	Photocyclization	25
[7]	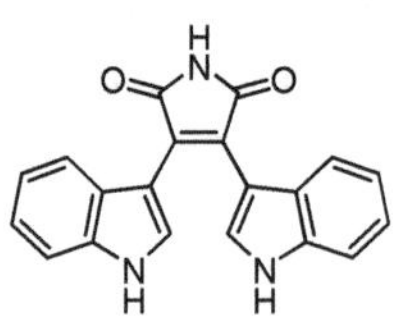	Biaryl coupling	14
[10]	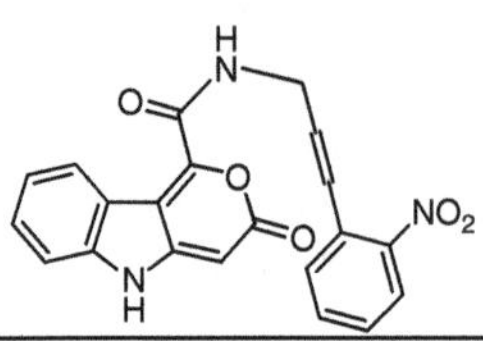	Diels–Alder reaction/nitrene-mediated cyclization	8

(*continued*)

Table 1 (continued)

Citation(s)	Synthetic intermediate(s)	Key reaction(s) to give K252c	Yield from starting materials (%)
[11]	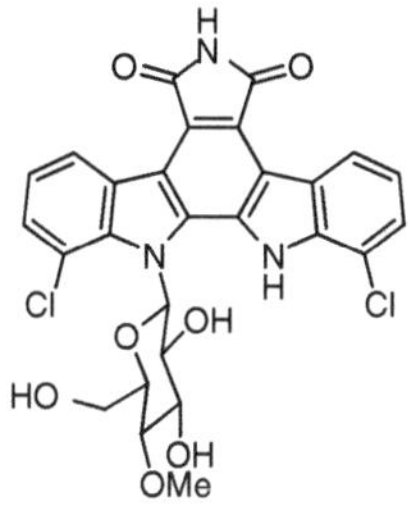	*N*-Deglycosylation/ Clemmensen reduction	80% to rebeccamycin aglycone; to stauroporinone not reported
[12]	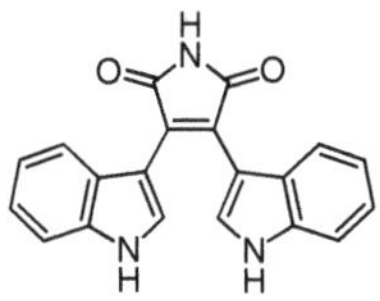	Zinc amalgam mediated cyclization	13
[13]	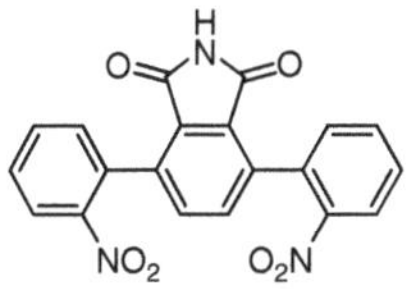	Deoxygenation of nitro groups	12
[14]	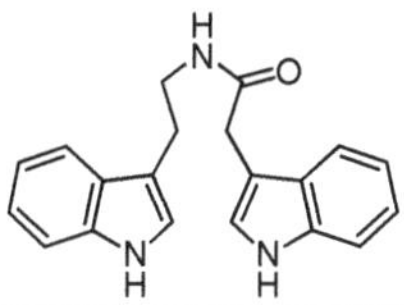	Oxidation/ photocyclization	10

NH_4^+, without enzyme [158]. In the synthetic case, the reaction goes forward as a Michael addition of **137** to **136** to give **138** in 10% yield. Because of the wrong oxidation state of the nitrogen, direct cyclization of **138** – as seen in the spontaneous cyclization of **127** in the respective biosynthetic step – is not possible. Instead, reduction of the nitro function with Raney-Ni/H_2 produced the free amine, which was cyclized using NaH to give **139**. Final oxidative cyclization with DDQ gave the product **1** in 52% yield from **138** (Fig. 25b) [3]. Again, this synthesis has features that are biosynthetically reminiscent – first, the use of two indole containing precursors that will each provide either the "left" or the "right" side of the final product, as seen biosynthetically, and second, the closure of the upper (eventually pyrrole) ring prior to aryl–aryl coupling to give the central ring of the structure.

Finally, a "reverse" biomimetic strategy can be seen from using rebeccamycin (**124**), a natural product produced at approximately 700 mg L^{-1} of culture [11], as a starting material. As described above, the biosynthetic routes to both **124** and **121** are closely related, particularly between the aglycone portions of the molecules. The synthesis of K252c (**1**) from rebeccamycin (**124**) was accomplished by initial cleavage of the sugar from the core structure to give the rebeccamycin aglycone,

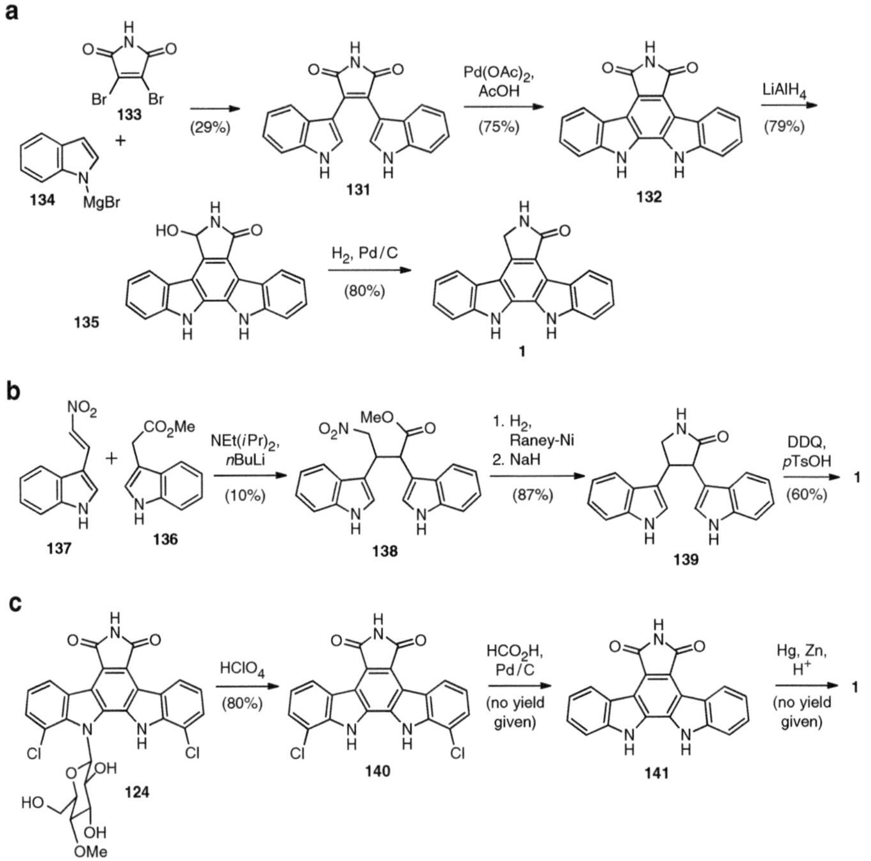

Fig. 25 A sample of synthetic routes **(a–c)** to the staurosporine aglycone (**1**) with biomimetic "features"

dichloro arcyriaflavin A (**140**), in 80% yield. Pd/C-catalyzed dechlorination gave rise to rebeccamycin aglycone (**141**), which after Clemmensen reduction yielded **1** (Fig. 25c) [11].

Other synthetic strategies have nothing in common with the biosynthetic route. A particularly creative synthesis was presented by the research group of Moody, who aimed to use no protecting groups on the nitrogens [10]. All atoms of the staurosporine aglycone scaffold were incorporated from the beginning by condensation of **142**, in turn derived from aldehyde **143** by formal reductive amination (58% yield), with **144** and oxalyl chloride to give **145** in 76% yield. Saponification of the ester **145** gave **146**, which was transformed into **147** by lactone formation. Intramolecular Diels–Alder reaction of the benzylic double bond with the pyrone substructure in **147** accompanied by decarboxylation gave **148** in 26% yield from **145**. Final cyclization to give the product **1** was achieved by treatment of **148** with triethyl phosphite (54% yield) (Fig. 26a) [10]. The only

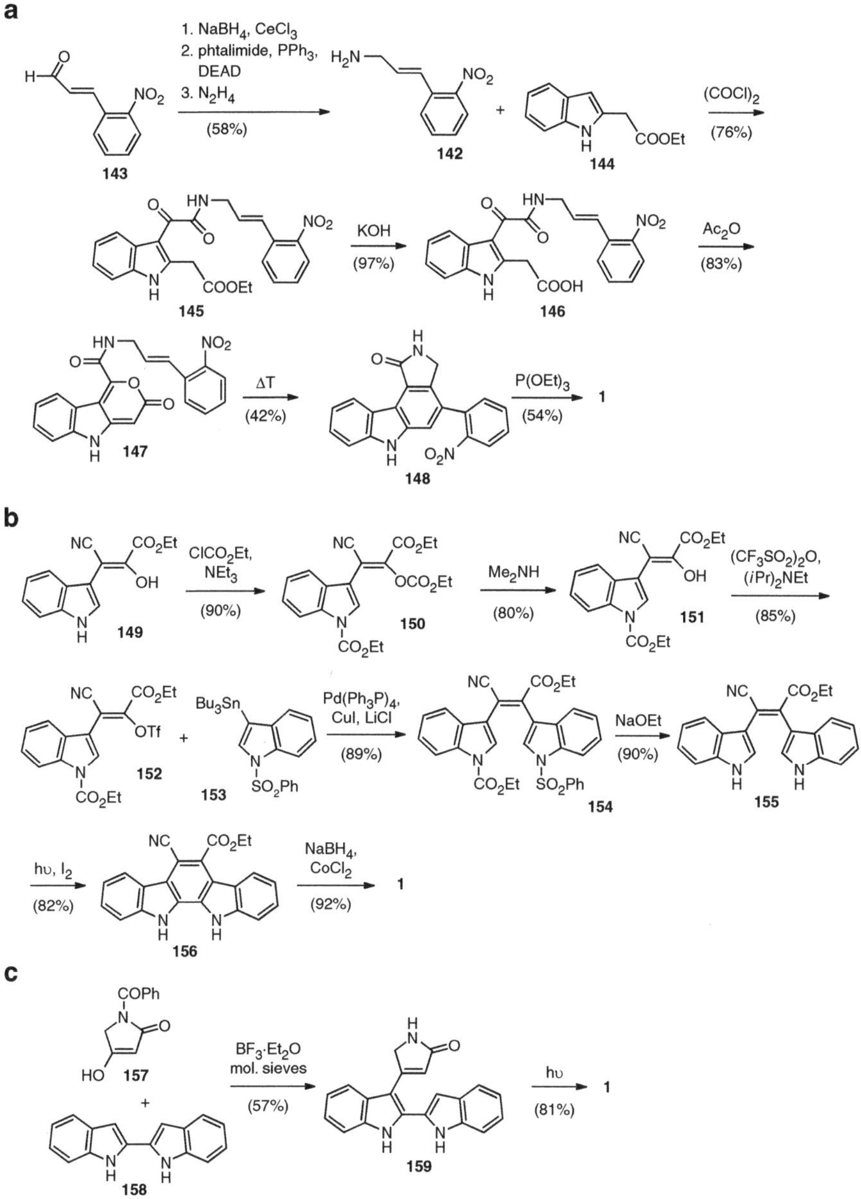

Fig. 26 A selection of synthetic routes (**a–c**) to the staurosporine aglycone (**1**) lacking biomimetic "features"

sense in which this strategy relates to a biosynthetic route is that all carbons and nitrogens present in the final structure are derived from an early coupling reaction.

A synthetic route presented by Beccalli [4] builds up the indolo[2,3-*a*]carbazole portion of the molecule first. Formation of the left molecular half of **1** proceeded from indole **149** by *N*-protection to give **150**, cleavage of the undesired carbonate group to furnish **151**, and final *O*-triflation yielding 61% of **152**. Pd-catalyzed coupling of **152** with tributyl stannate **153** delivered bisindole **154** in 89% yield. Ethoxide-mediated deprotection of both nitrogens to give **155** followed by photocyclization of the central ring led to **156**, which was – in a single step – transformed into the desired product **1** by reduction of the cyano group using $NaBH_4/CoCl_3$ and in situ formation of the final pyrrolo ring system (Fig. 26b) [4].

Finally, the most recent, and highest-yielding strategy, is derived from a route literally perpendicular to the natural strategy: instead of dimerization using L-tryptophans (**123**), the bisindole portion and the pyrrole portion were fused. Specifically, **157** and **158** were coupled through a Lewis-acid catalyzed electrophilic substitution reaction to give **159**. Photocyclization then resulted in generation of **1** in only two steps and 46% overall yield (Fig. 26c) [2].

The rich number of synthetic strategies to the staurosporine aglycone – which is rare among other natural products – demonstrates the fallacy of concluding strongly, for a given molecule, whether the best synthetic route is likely to be biomimetic. For the staurosporine aglycone (**1**), while some synthetic routes have features that are biochemically "reminiscent," others have nothing to do with the natural route. The probable impossibility of making a completely biomimetic route in this case, though, suggests that biosynthesis and total synthesis occupy two unique niches, which might intersect but are unlikely to replace one another.

6 Phenazines: Pyocyanin

Pyocyanin (**160**) is a blue-colored phenazine (**161**) molecule produced by *Pseudomonas aeruginosa* (Fig. 27). The pigment was first observed as a blue pus in the wounds of infected patients; by 1859, the pigment was isolated as a chloroform extraction of a wound dressing, and the correct structure was identified in 1938 [168]. Although pyocyanin (**160**) has antibiotic properties [169], its main importance to human health is as a critical virulence factor excreted by *P. aeruginosa* during lung infections [170] and, in particular, in lung disease in cystic fibrosis patients. Pyocyanin (**160**) is thought to exhibit its effects through generation of

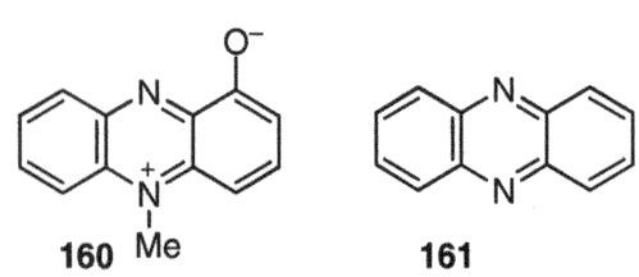

Fig. 27 Pyocyanin (**160**) is a phenazine (**161**) natural product

reactive oxygen species, which target, among others, the vacuolar ATPase [171], and lead to the induction of neutrophil apoptosis [172], inhibition of uptake of apoptotic cells by macrophages [173], and other effects [174]. Interestingly, the function of pyocyanin (**160**) in the producing bacterium is now under active investigation, and the postulated roles include intracellular redox balance in the absence of other electron acceptors [175], promoting mineral reduction [176], functioning as a terminal signaling molecule in the quorum sensing network [177], and controlling colony biofilm physiology [178].

The biosynthetic route to pyocyanin (**160**) has been studied for decades. As in the case of K252c (**1**), described above, no single, one-pot, total biosynthesis has been reported for **160**; however, each of the biosynthetic enzymes required for its production has been characterized *in vitro*. All known phenazine natural products which share a core phenazine structure are thought to derive from a largely conserved biosynthetic pathway, with the possible exception of the phenazine from *Methanosarcina mazei* Gö1 [168, 179], whose genome lacks the "usual" phenazine biosynthetic gene cluster. Early isotopic labeling studies established that pyocyanin (**160**) is derived from the shikimic acid pathway [180], and more recent work has resulted in the sequencing of the biosynthetic gene clusters for phenazines in a variety of bacterial species [181–186].

Pyocyanin (**160**) is derived from the shikimate pathway, and one protein, PhzC, is equivalent to enzymes that catalyze the first step in this pathway, converting erythrose 4-phosphate (**162**) and phosphoenolpyruvic acid (**163**) to 3-deoxy-D-arabinoheptulosonate 7-phosphate (**164**) (Fig. 28). The equivalent enzyme in the shikimate pathway is thought to be feedback regulated, and PhzC is likely to shunt intermediates toward the shikimate pathway in preparation for pyocyanin (**160**)

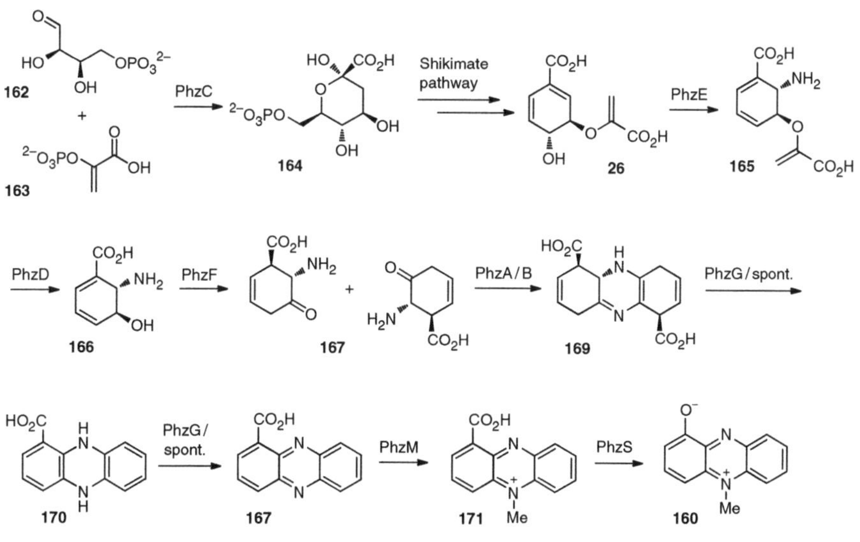

Fig. 28 Overall biosynthetic route to pyocyanin (**160**)

biosynthesis when the shikimate enzyme is inactive [181]. Subsequent enzymes in the shikimate pathway, which construct chorismic acid (**26**), are thought to be constitutively active [187], and there are no "replacement" enzymes for them in the pyocyanin biosynthetic route.

Chorismic acid (**26**), thus, represents the first "divergence" point of pyocyanin from other biosynthetic pathways. The first authentic pyocyanin biosynthetic enzyme is PhzE, which has sequence similarity to anthranilate synthases, which generate anthranilate from chorismate. PhzE is thought to catalyze the conversion of chorismic acid (**26**) to amine **165**. Compound **165** is in turn a substrate for PhzD, an isochorismatase that catalyzes the hydrolysis of the vinyl ether to **166** and pyruvate [188, 189].

Incubation of PhzF, PhzA, PhzB, and PhzG converts **166** to phenazine-1-carboxylic acid (**167**). The specific roles of each enzyme are still not fully elucidated; however, a series of impressive studies, which include crystal structures of each enzyme, have begun to reveal the chemistry involved in the conversion [190–193]. While the highest rates of conversion are achieved with all four enzymes, interestingly, PhzF alone is able to generate, at low levels, the desired product **167**. The mystery of why PhzF can generate **167** on its own was resolved when it was realized that PhzF catalyzed the formation of the ketone **168**, a molecule that can, in turn, undergo spontaneous dimerization, decarboxylation, and oxidation to give low levels of **167** [190, 192]. When PhzA/B are present, however, the reactions are dramatically accelerated. PhzA/B, which are highly related enzymes (80% identity), form a heterodimer and accelerate the head-to-tail dimerization of two molecules of **168**, giving rise to a tricyclic intermediate thought to be dicarboxylic acid **169**. This molecule is also likely to be oxygen-sensitive, and can spontaneously decarboxylate to give **170** [193]. Finally, PhzG, which has structural similarity to other known flavin-dependent oxidases, is thought to catalyze the oxidation of a tricyclic intermediate such as **170** to **167** through unknown mechanisms [191]. Regrettably, however, its activity with any proposed intermediates still remains to be established, and the possibility that some of the chemistry is spontaneous still remains.

The final conversions of **167** to **160** are catalyzed by PhzM and PhzS, two enzymes that are likely to interact, at least transiently, during catalysis. PhzM is related structurally to SAM-dependent methyltransferases, and it is thought to catalyze *N*-methylation of **167** to give *N*-methyl-phenazine-1-carboxylic acid (**171**) [194]. PhzS, a structural relative of flavin-dependent hydroxylases, is thought to catalyze the decarboxylation and oxidation of **171** [195]. However, while activity for the two-enzyme reaction has been verified *in vitro*, experimental proof for either single reaction is not yet available. PhzM is inactive by itself with **167** and only reacts when PhzS is also present [194]. The reaction of PhzS alone has been similarly challenging to determine; the putative substrate **171** is unstable and therefore has not yet been tested [195]. The currently available evidence suggests that PhzM and PhzS form at least a transient complex during catalysis, and that the intermediate **171**, generated by PhzM from **167**, is shuttled to PhzS, where flavin-dependent decarboxylation and oxidation reactions occur to give rise to the final product **160**.

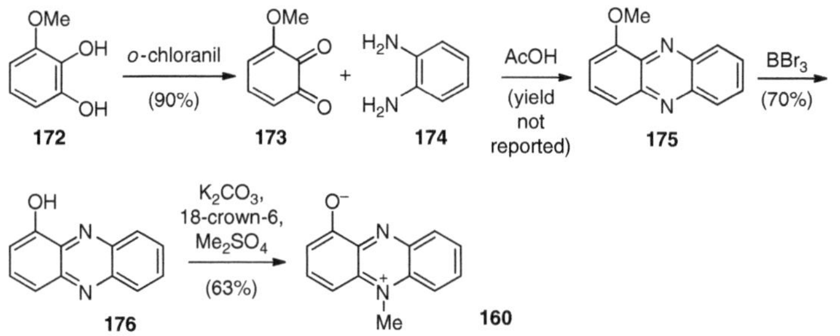

Fig. 29 Synthetic route to pyocyanin (**160**)

Overall, the biosynthesis of **160** is characterized by the dimerization of **168** to give the central structure of the molecule. This head-to-tail dimerization strategy is efficient, using the same substrate twice, and is a sensible route, given the existence of the shikimate pathway, which provides, in turn, a precursor to **168**. An analogous dimerization route can be seen for the biosynthesis of K252c (**1**), described in Sect. 5, where two molecules of indole-3-pyruvic acid imine (**125**), derived in turn from L-tryptophan (**123**), are dimerized to give an intermediate that leads to chromopyrrolic acid (**128**). In both cases, the monomer precursors, either **168** or **125**, serve as both nucleophiles and electrophiles, and are activated to react by the presence of the appropriate enzymes.

Interestingly, the synthetic route to **160** utilized a similar strategy of condensing the two outer rings of the desired product to give the central heterocycle. The synthesis commenced with diol **172**, which was oxidized to furnish 3-methoxy-1,2-benzoquinone (**173**) in 90% yield. Condensation of electrophile **173** with **174** delivered phenazine **175**, which was *O*-deprotected to give **176** and selectively *N*-methylated to furnish the natural product **160** (Fig. 29) [196, 197]. As in the biosynthetic route, this synthetic strategy utilizes a key coupling reaction of **173** and **174** to give **175**. Here, however, the nucleophilic nitrogens are all located in one molecule **174** whereas the electrophilic carbonyl oxygens are contained entirely in a separate compound **173** [194, 195]. In the case of pyocyanin (**160**), the synthesis forms the desired product from simple starting materials in a very concise way. The comparison of the biosynthesis and synthesis of **160** thus evidences a striking example for a natural product which can more effectively be prepared by chemists than by nature.

7 Enterocin- and Wailupemycin-Type Polyketides

Another chemically and biosynthetically intriguing suite of secondary metabolites is produced by "*Streptomyces maritimus*." The main product is the structurally complex enterocin (**177**) (Fig. 30) [198], which has also been isolated from terrestrial *Streptomyces* [199–201] and from an ascidian [202]. *S. maritimus* also

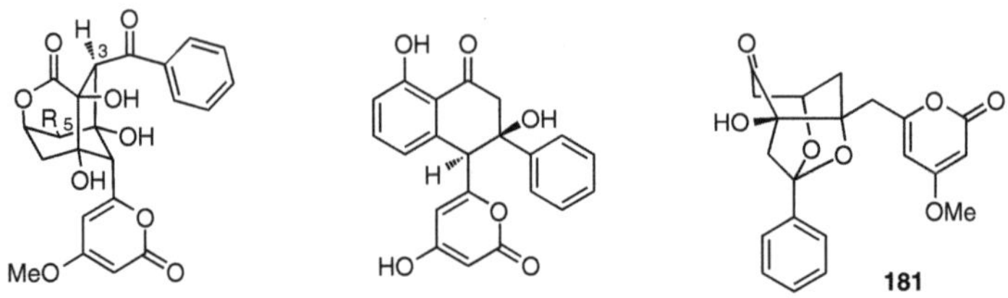

177 (R = OH)
178 (R = H)
179 (C3 epimer, R = H)

180

181

Fig. 30 Selection of enterocin- and wailupemycin-type natural products produced by *S. maritimus*

produces simple analogs of **177**, differing only in the oxygenation pattern as in 5-deoxyenterocin (**178**), or in the absolute configuration as in 3-*epi*-5-deoxyenterocin (**179**). In addition, the organism biosynthesizes the wailupemycins. The molecular structures of these compounds range from relatively simple α-pyrones like wailupemycin D (**180**) to the complex tricyclic acetal wailupemycin B (**181**). The unusual molecular architectures of enterocin (**177**) and wailupemycin B (**181**), and the bacteriostatic properties of these metabolites against *E. coli* and species within the genera *Corynebacterium*, *Proteus*, *Sarcina*, and *Staphylococcus* [198, 199], have triggered investigations of their biosynthesis at the genetic and biochemical level, as well as total synthetic efforts.

The α-pyrone moiety observed in **177** and in the wailupemycins constitutes a typical structural motif of compounds produced by genetically engineered iterative type II polyketide synthase (PKS) systems, such as in mutacin (**182**) (Fig. 31) [203]. These PKSs usually catalyze the formation of (poly)cyclic aromatic natural products such as doxorubicin (**183**) and actinorhodin (**184**) [204, 205]. The enzymes involved in the formation of aromatic PKS metabolites are related to type II fatty acid synthases. At least three elements are necessary to render such PKS systems functional: two β-ketosynthase subunits, KS_{α} and KS_{β}, and an acyl carrier protein (ACP). In typical type II PKS systems, these genes are accompanied by cyclases and aromatases which funnel the nascent highly reactive poly-acetate precursor into the desired aromatic PKS products. The latter can be further functionalized by tailoring enzymes which can, for example, catalyze oxidation, alkylation, and glycosylation reactions. The apparent structural similarity of some known iterative type II PKS shunt products with the observed α-pyrones of *S. maritimus* suggested using a type II PKS based genetic probe to identify the enterocin biosynthesis cluster. These investigations led to the discovery of the enterocin biosynthetic machinery *enc* consisting of 20 putative open reading frames (ORFs) in a contiguous 21.3-kb region of the cosmid [206, 207].

The *enc* cluster contains four genes (*encH*, *encI*, *encJ*, *encP*) involved in the biosynthesis of the unusual benzoyl-CoA (**185**) starter from phenylalanine (**186**) via a plant-like β-oxidation mechanism (Fig. 32) [208–210]. This pathway is initiated by the unique phenylalanine ammonia-lyase EncP [211], which catalyzes the generation of cinnamic acid (**187**) from **186**. The cinnamate-CoA ligase EncH

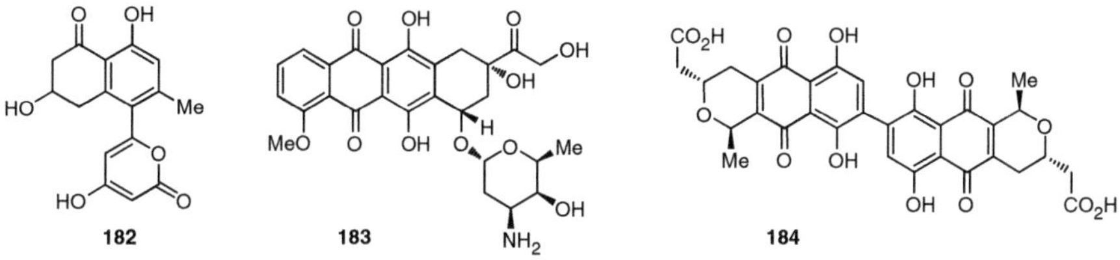

Fig. 31 Structures of mutacin (**182**), doxorubicin (**183**), and actinorhodin (**184**)

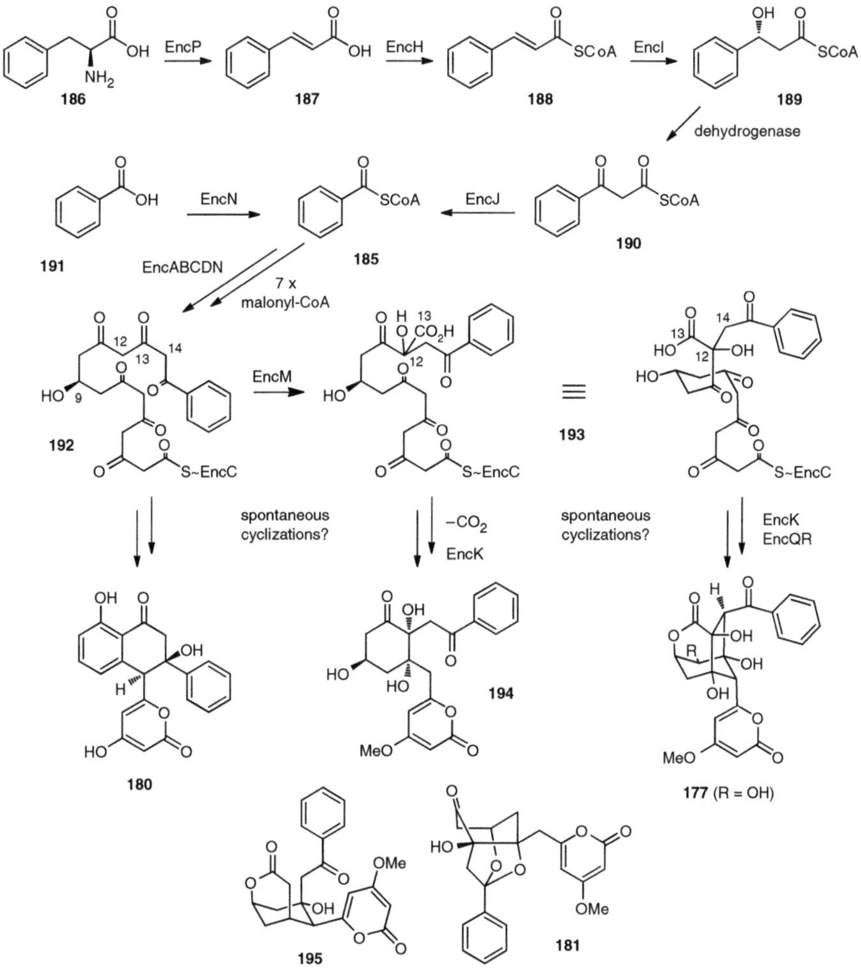

Fig. 32 Biosynthetic pathway leading to enterocins and wailupemycins

then forms **188**, which is oxidized by action of the cinnamoyl-CoA hydratase EncI to give 3-hydroxy-3-phenylpropionyl CoA (**189**). The required dehydrogenase for the following oxidation of **189** to diketone **190** is missing in the *enc* cluster and is thus thought to be entirely supplied from primary metabolism. Subsequently, the keto thiolase EncJ catalyzes the last step to the starter unit **185**. The benzoate-CoA ligase EncN additionally facilitates the use of exogenic benzoic acid (**191**) in the pathway. The *enc* minimal PKS, EncABC, which catalyzes the iterative assembly of the polyketide chain, comprises the typical KS_α (EncA) and KS_β (EncB) ketosynthase subunits and an ACP (EncC). The latter is primed with **185** in a type II NRPS-like fashion by EncN [212]. Loading of the malonyl-CoA extender units is presumably catalyzed by the malonyl-CoA:ACP transacylase, EncL, but can be complemented by enzymes from primary metabolism. In addition, the production of wailupemycins or enterocins requires the presence of the ketoreductase (KR) EncD, which catalyzes regioselective reduction (possibly of the still growing polyketide chain) [213] to furnish **192**, the central biosynthetic precursor of the *enc* pathway (Fig. 32). Spontaneous cyclization of **192** gives rise to wailupemycin D (**180**). In contrast to all other type II PKS gene sets, the *enc* cluster does not harbor any cyclases or aromatases, but a flavin-dependent oxygenase EncM. This enzyme putatively catalyzes α-oxidation at C12 of **192** and Favorskii-type rearrangement to give **193** [214]. Cyclization of **193**, followed by selective *O*-methylation of the α-pyrone system, by the methyltransferase EncK, opens the pathway to wailupemycins A (**194**) and, with additional loss of CO_2, to **181** and wailupemycin C (**195**). On the other hand, spontaneous cyclization and ester formation, followed by EncK mediated *O*-methylation and oxygenation at C5 by EncQR furnishes **177**.

The exact order of biosynthetic transformations was firmly established by the enzymatic total biosynthesis of enterocin (**177**) [215]. Incubation of an EncA–EncB heterodimer, holo-EncC, and a malonyl-CoA:ACP transacylase from *Streptomyces glaucescens* (SgFabD) [216] with malonyl-CoA yielded several highly unstable nonaketides of unknown structure (Fig. 33). The formation of wailupemycin-type compounds was achieved by addition of the ketoreductase EncD together with its cofactor NADPH. With this set of enzymes the natural products wailupemycins F (**196**) and G (**197**) [217] were produced as the main products. Addition of the "favorskiiase" EncM to the enzymatic mixture allowed for the production of desmethyl-5-deoxyenterocin (**198**), along with **196** and **197**. EncM, thus, not only catalyzes the oxidative rearrangement, but also initiates two subsequent aldol reactions and heterocycle formations [214], which leads to the generation of a remarkable six chiral centers with perfect regio- and stereocontrol starting from a highly reactive, linear and achiral precursor. Production of 5-deoxyenterocin (**178**) was achieved by adding the methyltransferase EncK and SAM to the above enzyme mixture. Final oxidation at C5 succeeded using the P450 hydroxylase EncR, ferredoxin, ferredoxin-$NADP^+$ reductase, and catalase. High turnover rates for this hydroxylation to **177**, however, were only achieved after an extractive work-up of **178** and renewed reaction setup. This can be attributed to the inhibitory effect of SAM on cytochrome P450 enzymes [218]. Adding CoA and the *Rhizobium*

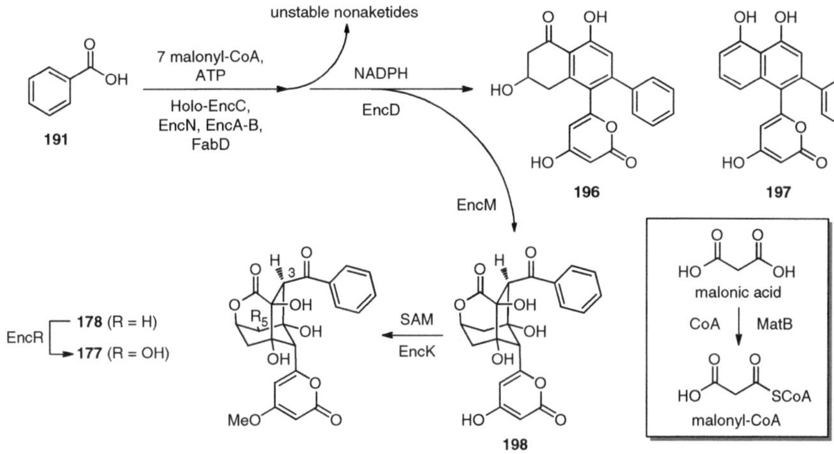

Fig. 33 One-pot total *in vitro* biosynthesis of enterocin (**180**) and wailupemycin F (**199**) and G (**200**)

leguminosarum malonyl-CoA synthase MatB [219] directly to the above enzyme mixture allowed for the substitution of up to 90% malonyl-CoA by malonic acid without lowering chemical yields (Fig. 33).

Using this chemo-biosynthetic toolbox consisting of EncA/B, EncC, EncD, EncK, EncM, EncN, EncR, SgFabD, and MatB in combination with the respective cofactors, as well as benzoic and malonic acid as carbon sources, the *in vitro* total synthesis of enterocin (**177**) was accomplished in a remarkable 25% overall yield. The construction of ten C–C bonds, five C–O bonds, and seven chiral centers to give the highly complex tricyclic structural framework of **177** was achieved using seven equivalents malonyl-CoA and two equivalents NADPH per **191**, SAM, and ATP [215]. The utility of this total biosynthesis was recently further expanded by using a series of halogen and hydroxyl-substituted benzoic acid derivatives, such as **199** and **200**, as well as heterocyclic aromatic precursors like **201**, as unnatural starter units to produce novel wailupemycin and enterocin derivatives, such as **202–204** and **205–207**, respectively (Fig. 34) [220]. This purely enzymatic approach to generate molecular diversity eliminates unpredictable factors that are usually associated with more classical *in vivo* mutasynthetic experiments, such as precursor uptake, toxicity, transport, and metabolism. Consequently, this *in vitro* biosynthesis method resulted in the successful preparation of many more structural variants of wailupemycins and enterocins than previous related *in vivo* work [221].

The first and so far only total synthesis of a member of the wailupemycin family, wailupemycin B (**181**), was reported by the Bach group [222, 223]. In their retrosynthetic considerations, they aimed for an initial stereoselective generation of the central cyclohexane ring system which would allow for the flexible attachment of the peripheral substituents at a late stage. This strategy would facilitate the preparation of structural analogs of wailupemycin B for use in investigations of structure–activity relationships (SARs). A key precursor of this synthetic strategy was

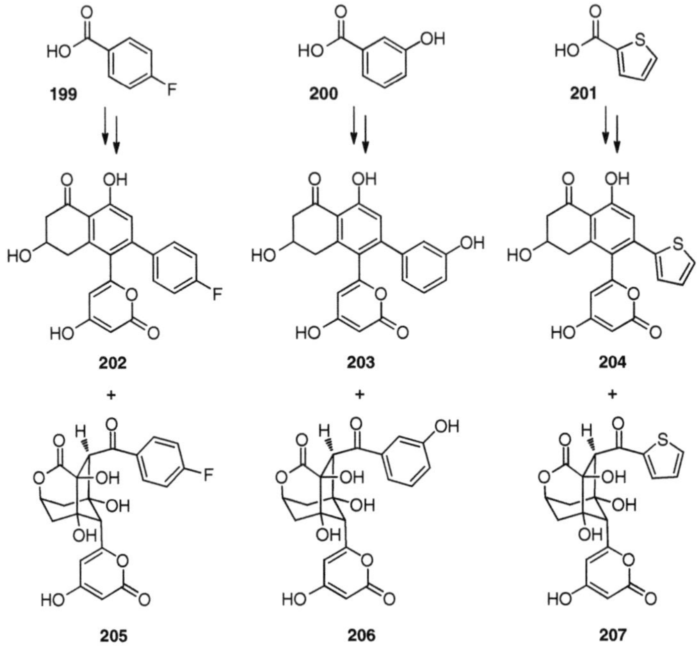

Fig. 34 Selection of novel wailupemycin **202–204** and enterocin analogs **205–207** generated by *in vitro* biosynthesis using unnatural starter units **199–201**

cyclohexane **208**, which is accessible from ketone **209** (Fig. 35). It is interesting to note that in order for the envisioned nucleophilic attack at C12 to proceed from the desired side when installing the C12 side chain, the configuration at C9 of **209** had to be inverted as compared to the natural product. The reason is that bulky groups at C9 and C11 adopt equatorial positions, thus locking **209** into a chair conformation, which is predominantly attacked from the face delivering the correct diastereomer.

Ketone **209** was prepared starting from (*S*)-(+)-carvone via alkene **210** [224], with subsequent ozonolytic cleavage of the double bond in 42% overall yield (Fig. 36). Addition of allylmagnesium bromide and MEM protection of the tertiary alcohol furnished **211** in 80% yield and good diastereoselectivity (*de* > 90%). Selective cleavage of the TMS ether and 2-iodoxybenzoic acid (IBX)-mediated oxidation gave ketone **212** (93% yield). Subsequent addition of the dianion of 4-hydroxy-6-methylpyran-2-one (**213**) yielded **214** in 82% (*de* > 90%). After *O*-methylation of the pyrone moiety using dimethylsulfate and selective removal of the TBDPS protective group, the released secondary alcohol was oxidized and stereoselectively reduced with L-selectride to install the correct absolute configuration at C9. Protection of the *syn* hydroxy groups at C7 and C9 as isopropylidene acetal delivered alkene **215** (five steps, 72% overall yield), which possessed all the stereogenic elements of the final product **181**. Oxidative cleavage of the terminal double bond in **215** resulted in an aldehyde at C14, which was subsequently functionalized by nucleophilic attack with PhMgBr and reoxidized using Dess–

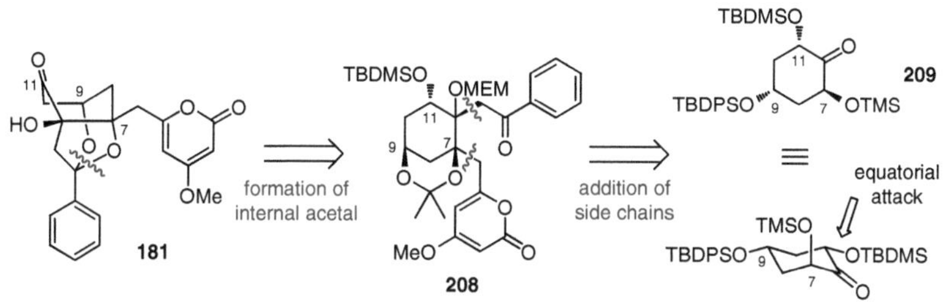

Fig. 35 Retrosynthetic analysis of wailupemycin B (**181**) by the Bach group

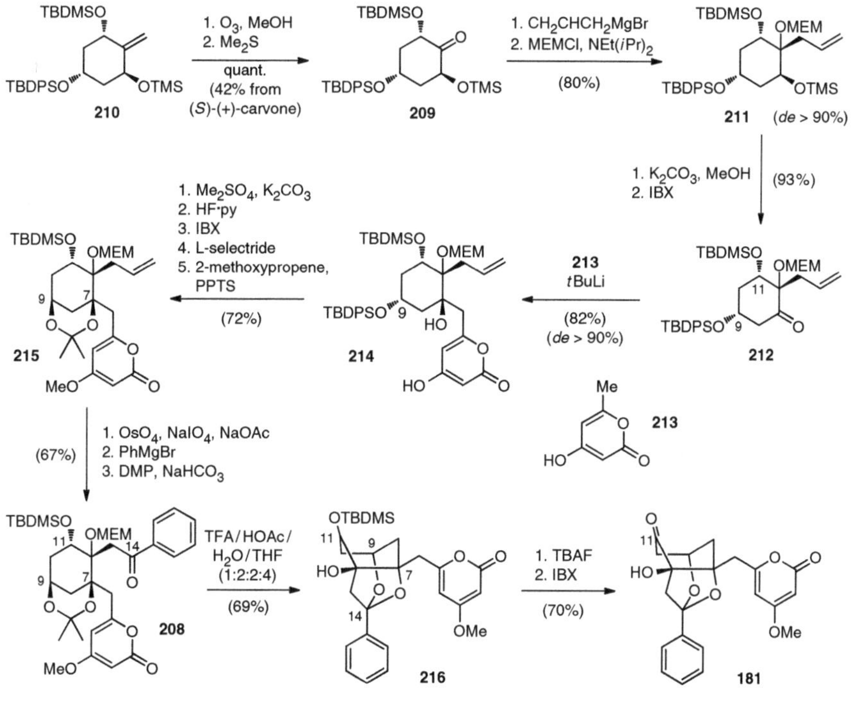

Fig. 36 Total synthesis of wailupemycin B (**181**)

Martin periodinane to deliver the desired key intermediate, ketone **208**. The total synthesis of **181** was finalized in two more steps (49% overall yield) by removal of the acetal and the MEM protective group with simultaneous formation of the desired intramolecular ketal **216**, followed by cleavage of the C11 TBDMS ether and in situ oxidation with IBX. The highly complex secondary metabolite **181** was thus accessed in 16 linear steps starting from **210** with an impressive 14% overall yield (Fig. 36).

In contrast to all other examples mentioned in this review, the structurally diverse compounds derived from the *enc* cluster have either been accessed by *in vitro* biosynthesis *or* by total organic synthesis, and not by both methods. The total biosynthesis of enterocin (**177**) shows the power of biosynthetic enzymes to prepare complex natural products, even when used in a test tube. The preparation of this secondary metabolite can be achieved with a relatively small number of pathway-specific enzymes and their respective cofactors. It is interesting to note, however, that the *in vitro* reconstitution of the *enc* pathway has so far not been proven to be suitable to access the whole set of different metabolites formed in the natural producer, *S. maritimus*. This raises questions about the true biosynthetic routes to these metabolites, which had so far been assumed to be initiated by spontaneous cyclization and decarboxylation reactions of intermediate linear poly-keto precursors. But this should finally lead to the formation of identical products in the case of the total biosynthetic approach. A reason for this discrepancy might be, e.g., an additional contribution of other enzymes in the *enc* pathway (with no assigned function) actively funneling a fraction of the precursor molecules into the formation of wailupemycins like **181**.

The total synthesis of **181** by Bach et al. highlights the capability of modern synthetic methods to generate challenging molecular frameworks. The pathway chosen to prepare **181** is completely independent from any biosynthetic considerations, but still utilizes a natural product, (*S*)-(+)-carvone, as the structural and stereochemical basis. While the ability to target a particular natural product in such a creative and directed way is extremely valuable, the resulting very specialized routes also carry a typical disadvantage of complex synthetic approaches, namely the necessity to develop distinctively different chemical accesses even to biosynthetically highly related metabolites. As an example, the Bach group attempted to convert intermediate **217** of their wailupemycin B synthesis into the natural product wailupemycin A (**194**) (Fig. 37). The only apparent differences in the core structures of **217** and the desired product **194** were the absolute configurations at C7 and C9, the latter being easily adjustable by an oxidation-stereoselective reduction sequence as described above for the synthesis of **181**. Inversion of the stereocenter at C7 was planned to be performed by oxidative cleavage of the C7–C12 bond in **217** using $Pb(OAc)_4$ leading to **218** and subsequent recyclization to give the required *cis* arrangement of the C7 and C12 hydroxy functions in **219**. Unfortunately this ring closure was not possible under any conditions tested, thereby demonstrating the often observed lack of flexibility of highly specific and complex synthetic routes.

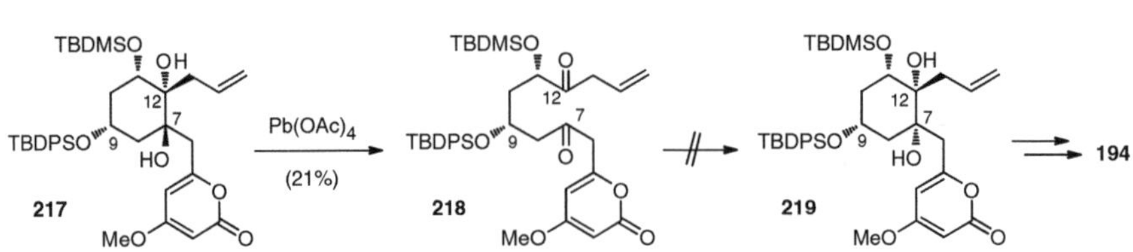

Fig. 37 Attempted total synthesis of wailupemycin A (**194**) from **217**, a late intermediate of the wailupemycin B (**181**) synthesis

8 Conclusions

The recent development of innovative techniques in modern natural product research, such as molecular cloning, bioinformatics, and genomics, have facilitated the functional characterization of a huge variety of biosynthetic pathways. The most striking feature of the majority of these routes to secondary metabolites developed by nature is the strict sequential assembly of the core molecular structures using simple building blocks. This holds true for linear modular biosynthetic machineries, like the NRPS systems involved in yersiniabactin (**24**) biosynthesis (see Sect. 3), as well as for iterative systems, e.g., the enterocin PKS II cluster (see Sect. 7). In the NRPS pathway leading to **24**, for example, the seemingly trivial sequential assembly is facilitated by tethering the nascent molecules on PCP domains for selective activation and further functionalization (see Figs. 6 and 7). Using this simple biosynthetic logic, the head groups of the growing molecule are rendered highly chemically reactive thioesters and, additionally, kept close to the subsequently incorporated biosynthetic building blocks. The molecular portions of the biosynthetic intermediates that are not to be involved in any chemistry are embedded into passive sections of the NRPS proteins and are thus chemically inert. Such an intrinsic protection of the growing pathway intermediate by the biosynthetic enzyme is most impressive in the case of the enterocin type II PKS system, where a highly reactive polyketo chain is repeatedly reacted at the ACP bound head group only. The resulting regioselectivity in chain elongation is paired with the prevention of undesired cyclization reactions of the polyketide in the enzyme pocket (see Fig. 32). The enzymes involved in these processes thus perform as biocatalysts for highly regio- and stereoselective reactions and simultaneously act as global protective groups for the remainder of the biosynthetic precursor. These desirable characteristics of NRPS and PKS enzymes can impressively be utilized in total biosynthetic experiments. However, due to the observed, usually strict, selectivity of these enzymes, such approaches can only be used to generate the target molecule of the respective pathway or close structural analogs thereof. Application of NRPS and PKS genes for the total de novo design of new molecules is currently not possible.

In organic synthesis such a remarkable selectivity is so far unreached. Classic synthetic routes, instead, utilize chemical protective groups to prevent undesired side reactions. In addition, more complex molecules are usually assembled in a convergent manner. By constructing several parts of the molecule separately, as seen for yersiniabactin (**24**) (see Figs. 8–10), the number of reactive functional groups – and the possible need of the same number of (orthogonal) protective groups – can thus be significantly decreased. In addition, the molecular parts can be synthesized in parallel, leading to a more efficient approach. From the perspective of step and atom economy, however, the use of protective groups is highly undesirable. In the case of the synthesis of wailupemycin B (**181**), for example, comparison of the atomic masses of the three protective groups to the remaining atoms of the early precursor **210** leads to a ratio of 3:1 in favor of the seemingly unnecessary, and subsequently lost, chemical material in addition to the multiple steps required to install and remove these groups (see Fig. 36).

The adaptation of biosynthetic concepts to total synthesis may help eliminate such total synthetic drawbacks. For example, the development of strategies involving highly chemoselective transformations may avoid the excessive use of protective groups and will reduce the number of steps to a particular goal [225, 226]. This can be achieved by the development of very specific routes to a single target molecule, like in the recent protective-group-free synthesis of intricarene (**220**) (Fig. 38a) [227, 228]. In biosynthesis, final tailoring steps are often used to increase the structural complexity and diversity of a molecule after the production of the central scaffold. Examples for late biosynthetic modifications presented in this review are the oxidative alterations of (+)-5-epi-aristolochene (**2**) to give capsidiol (**12**) in terpene biosyntheseis (see Fig. 2), or of 5-deoxyenterocin (**178**) to furnish enterocin (**177**) (see Fig. 33) [215]. The development of similar strategies in total synthesis – preparation of the core structure of a certain molecule and late stage selective functionalization, for example of C–H bonds – can lead to major advances in organic chemistry. Such an approach has recently been realized in the total synthesis of structurally diverse eudesmane terpenes, like eudesmantetraol (**221**), by applying site-selective C–H oxidations starting from dihydrojunenol (**222**) as the sole precursor molecule (Fig. 38b) [229]. Another fascinating example for late-stage C–H functionalization is the oxidation of artemisinin (**4**) to give the 10-hydroxy analog **223** [230]. This reaction can be carried out using an iron based small molecule catalyst **224** together with H_2O_2 and is superior in reaction speed and yields of the analogous microbiological transformation using living cells of *Cunninghamella echinulata* (Fig. 38b) [231]. The implementation of such biosynthetic logic into organic chemical concepts to reduce molecular complexity throughout the pathway will continue to help improve total synthetic efforts.

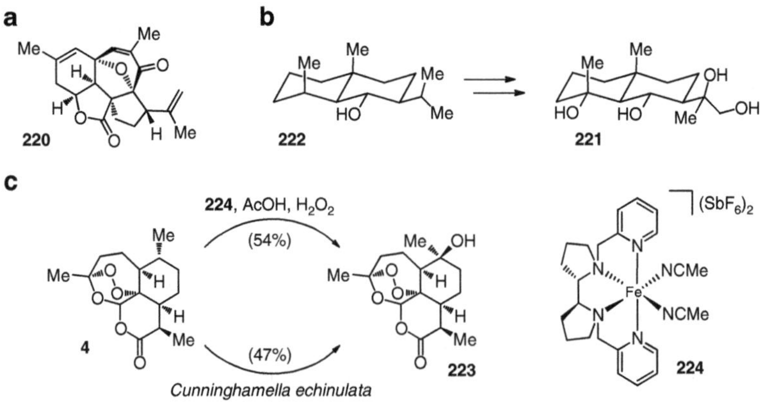

Fig. 38 Examples of molecules prepared following new concepts in total synthesis: (**a**) intricarene (**220**) generated by protective group-free synthesis and (**b**) late-stage site-selective C–H oxidations to generate eudesmane-type terpenes like **222** or (**c**) to prepare the hydroxylated artemisinin derivative **223**

Taking advantage of readily available precursors from the natural chiral pool, for example using amino acids as in the siderophore syntheses (see Sects. 3 and 4) or employing (*S*)-(+)-carvone as a precursor in the synthesis of wailupemycin B (**181**) (see Fig. 36), has a long tradition in total synthesis. In this way the chemical route can take advantage of the chiral information provided by nature and use it to install further functional groups stereoselectively. The modeling of biosynthetic transformations to further generate molecular complexity on such scaffolds constitutes a worthwhile additional synthetic alternative. In the case of the synthesis of sesquiterpenes, for example, this has been successfully utilized to form the spirocenter of (–)-premnaspirodiene (**3**) using a biomimetic rearrangement reaction (see Fig. 5). Similar biomimetic transformations can be extended to most complex molecular frameworks. Of particular interest are biomimetic cascade reactions [232–234]. In Johnson's synthesis of progesterone (**225**), for example, trienynol **226** was cyclized by generation of carbocation **227**, which undergoes formation of the three ring systems in analogy to the respective biosynthetic polycyclization reaction to give progesterone precursor **228** [235] (Fig. 39). Other impressive examples can be found in the preparation of large natural polyethers [236] from their corresponding poly-epoxide precursors [237]. Using such a strategy, the Jamison lab recently used polyepoxide **229** in a bromine-induced polycyclization cascade via **230** to obtain polyether **231**, an advanced building block in their synthesis of *ent*-dioxepandehydrothyrsiferol (**232**) [238]. Further research in these fields will undoubtedly result in highly effective routes to other diverse natural products.

While the construction of countless different secondary metabolites from simple building blocks via a limited set of reactions, as realized by nature, can be very

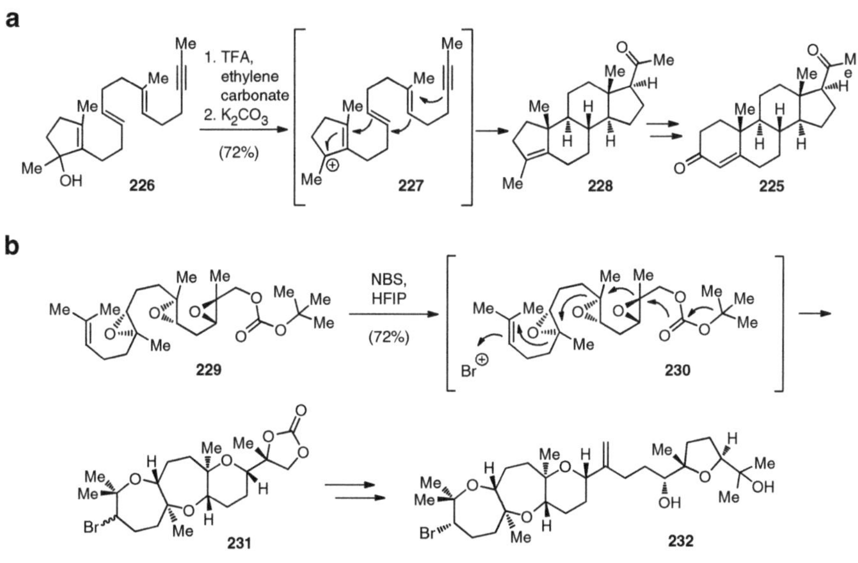

Fig. 39 Examples for biomimetic cascade reactions in total synthesis

effective, it also restricts the number of possible pathways to a certain compound. Thus, in terms of creativity, synthetic organic chemistry can clearly outclass total biosynthesis. The diverse routes to staurosporinone (**1**), even though it is a structurally simple molecule, testify to the huge variety of synthetic methods to construct a molecule of interest. The simplicity of the currently shortest staurosporinone synthesis (see Fig. 26c), taking advantage of readily available precursors, even outcompetes the efficiency of the respective natural pathway. The fact that being restricted to certain starting materials can lead to quite inefficient biosynthetic pathways can best be seen in the case of pyocyanin (**160**), where elaborate chiral precursors are transformed into a nonchiral phenazine following a complicated multistep enzymatic sequence (see Fig. 28). The synthetic route in this case shows a much more straightforward conversion of cheap and simple starting material into the desired phenazine (see Fig. 29).

This discussion of some of the key differences of total biosynthesis vs chemical synthesis shows that both approaches have clear advantages and disadvantages. While using biosynthetic enzymes for the *in vitro* reconstitution of whole pathways allows for the construction of remarkable molecular complexity using only simple starting materials, such work requires in-depth knowledge on the respective pathways and the production of sufficient amounts of all corresponding proteins in a soluble and functional form. These drawbacks lead to restrictions concerning the quantities of products that can be made by *in vitro* biosyntheses. The main application of total biosyntheses is thus still to answer mechanistic questions rather than the large-scale generation of metabolites or derivatives. Total synthesis, on the other hand, provides an excellent means to prepare larger amounts of a particular metabolite of interest with a high degree of creativity. One of the most challenging problems remaining in total synthesis is to achieve chemo- and stereoselective transformations in order to limit the need for seemingly superfluous protective groups, oxidation/reduction sequences, or other functional group alterations. This apparent lack of selectivity in chemical synthesis might be addressed using the power of biocatalysts. A combination of classic organic synthesis with highly selective enzymatic transformations in key synthetic steps, which could not effectively be realized by purely chemical means, should thus be considered as a powerful alternative in modern natural product chemistry. The remarkable chemo- and enantioselectivity of pathway-specific enzymes, and their ability to generate complex molecules from relatively simple (bio)synthetic intermediates, clearly has the potential to greatly impact the field of total synthesis. While applying enzymes for the preparation of important organic molecules has a long tradition, including in industrial chemistry [239], most applications to date still focus on the preparation of relatively small and structurally simple building blocks. Such methods include (dynamic) kinetic resolutions and deracemizations, reduction and oxidation reactions and, more recently, C–C bond formations, e.g., by using aldolases [240, 241]. Most total synthetic works, however, still avoid taking advantage of biocatalysts. In particular, pathway-specific enzymes have only rarely been used. An impressive recent example demonstrating the applicability of such an approach is the chemo-enzymatic total synthesis of the antiproliferative polyketide

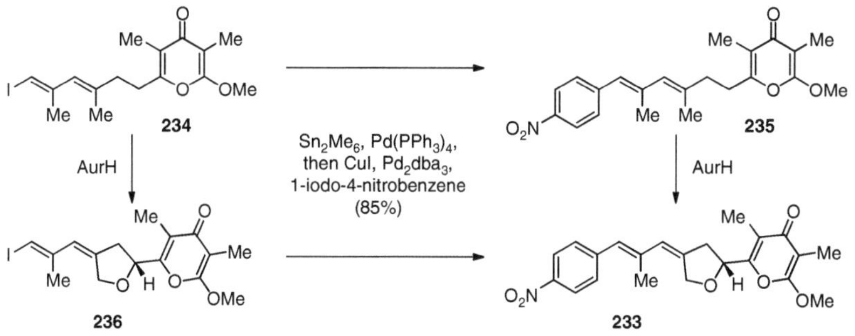

Fig. 40 Key steps in the chemo-enzymatic total synthesis of (+)-(*R*)-aureothin (**233**)

(+)-(*R*)-aureothin (**233**) (Fig. 40) [242, 243]. The synthetic precursor **234** was either first coupled with 1-iodo-4-nitrobenzene to yield **235** and subsequently oxidized using the pathway-specific cytochrome P450 monooxygenase (AurH) to give the natural product **233**, or first enzymatically oxidized to furnish **236** with subsequent coupling to deliver **233** (Fig. 40).

These examples clearly show that the combination of modern methods and concepts of natural product biosynthesis and enzymology with effective techniques of chemical synthesis can help to solve challenging problems in the preparation of complex molecules. Rapidly evolving methods for the optimization of certain desirable characteristics of a particular enzyme (e.g., directed alteration of enzyme properties like substrate specificity, selectivity, turn-over rates, and stability) will continue to facilitate the design of desired enzyme reactivity [241]. Increasing interdisciplinary cooperation between chemists, biochemists, and biologists will thus be vital for the successful continuation and development of modern natural product research.

Acknowledgments This work was supported by National Institutes of Health grants GM85770 and AI47818 to B.S.M., a postdoctoral fellowship from the German Academic Exchange Service (DAAD) to T.A.M.G. and a Howard Hughes Medical Institute sponsored Life Sciences Research Foundation postdoctoral fellowship to K.S.R.

References

1. Nicolaou KC, Vourloumis D, Winssinger N, Baran PS (2000) Angew Chem Int Ed 39:44–122
2. Gaudêncio SP, Santos MMM, Lobo AM, Prabhakar S (2003) Tetrahedron Lett 44: 2577–2578
3. Mahboobi S, Eibler E, Koller M, Kumar S, Popp A, Schollmeyer D (1999) J Org Chem 64:4697–4704
4. Beccalli EM, Gelmi ML, Marchesini A (1998) Tetrahedron 54:6909–6918
5. Wood JL, Stoltz BM, Dietrich HJ (1995) J Am Chem Soc 117:10413–10414
6. Xie J, Lown JW (1994) Tetrahedron Lett 35:5555–5558

7. Harris W, Hill CH, Keech E, Malsher P (1993) Tetrahedron Lett 34:8361–8364
8. Faul MM, Sullivan KA, Winneroski LL (1995) Synthesis 1511–1516
9. Faul MM, Winneroski LL, Krumrich CA (1998) J Org Chem 63:6053–6058
10. Moody CJ, Rahimtoola KF, Porter B, Ross BC (1992) J Org Chem 57:2105–2114
11. Fabre S, Prudhomme M, Rapp M (1992) Bioorg Med Chem Lett 2:449–452
12. Toullec D, Pianetti P, Coste H, Bellevergue P, Grand-Perret T, Ajakane M, Baudet V, Boissin P, Boursier E, Loriolle F et al (1991) J Biol Chem 266:15771–15781
13. Hughes I, Nolan WP, Raphael RA (1990) J Chem Soc Perkin Trans I 2475–2480
14. Sarstedt B, Winterfeldt E (1983) Heterocycles 20:469–476
15. Dewick PM (2001) Medicinal natural products – a biosynthetic approach, 3rd edn. Wiley, West Sussex
16. Davis EM, Croteau R (2000) Top Curr Chem 209:53–95
17. Sattely ES, Fischbach MA, Walsh CT (2008) Nat Prod Rep 25:757–793
18. Kirby J, Keasling JD (2009) Annu Rev Plant Biol 60:335–355
19. Whitehead IM, Threlfall DR, Ewing DF (1989) Phytochemistry 28:775–779
20. Takahashi S, Zhao Y, O'Maille PE, Greenhagen BT, Noel JP, Coates RM, Chappell J (2005) J Biol Chem 280:3686–3696
21. O'Maille PE; Chappell J; Noel JP (2004) Anal Biochem 335:210–217
22. Faraldos JA, Zhao Y, O'Maille PE, Noel JP, Coates RM (2007) Chembiochem 8:1826–1833
23. Back K, Chappell J (1995) J Biol Chem 270:7375–7381
24. Starks CM, Back K, Chappell J, Noel JP (1997) Science 277:1815–1820
25. Greenhagen BT, O'Maille PE, Noel JP, Chappell J (2006) Proc Natl Acad Sci USA 103:9826–9831
26. O'Maille PE, Malone A, Dellas N, Andes Hess B Jr, Smentek L, Sheehan I, Greenhagen BT, Chappell J, Manning G, Noel JP (2008) Nat Chem Biol 4:617–623
27. Whitehead IM, Ewing DF, Threlfall DR, Cane DE, Prabhakaran PC (1990) Phytochemistry 29:479–482
28. Whitehead IM, Threlfall DR, Ewing DF (1987) Phytochemistry 26:1367–1369
29. Yoshizawa Y, Yamaura T, Kawaii S, Hoshino T, Mizutani J (1994) Biosci Biotechnol Biochem 58:305–308
30. Blay G, Cardona L, Collado AM, García B, Morcillo V, Pedro JR (2004) J Org Chem 69:7294–7302
31. Birladeanu L (2003) Angew Chem Int Ed 42:1202–1208
32. Hwu JR, Wetzel JM (1992) J Org Chem 57:922–928
33. Neilands JB (1981) Annu Rev Nutr 1:27–46
34. Neilands JB (1981) Annu Rev Biochem 50:715–731
35. Butler A (1998) Science 281:207–209
36. Wake A, Misawa M, Matsui A (1975) Infect Immun 12:1211–1213
37. Haag H, Hantke K, Drechsel H, Stojiljkovic I, Jung G, Zahner H (1993) J Gen Microbiol 139:2159–2165
38. Drechsel H, Stephan H, Lotz R, Haag H, Zähner H, Hantke K, Jung G (1995) Liebigs Ann 1995:1727–1733
39. Miller MC, Parkin S, Fetherston JD, Perry RD, DeMoll E (2006) J Inorg Biochem 100:1495–1500
40. Perry RD, Fetherston JD (1997) Clin Microbiol Rev 10:35–66
41. Buchrieser C, Brosch R, Bach S, Guiyoule A, Carniel E (1998) Mol Microbiol 30:965–978
42. Chambers CE, McIntyre DD, Mouck M, Sokol PA (1996) Biometals 9:157–167
43. Valdebenito M, Bister B, Reissbrodt R, Hantke K, Winkelmann G (2005) Int J Med Microbiol 295:99–107
44. Henderson JP, Crowley JR, Pinkner JS, Walker JN, Tsukayama P, Stamm WE, Hooton TM, Hultgren SJ (2009) PLoS Pathog 5:e1000305
45. Bultreys A, Gheysen I, de Hoffmann E (2006) Appl Environ Microbiol 72:3814–3825

46. Perry RD, Balbo PB, Jones HA, Fetherston JD, DeMoll E (1999) Microbiology 145: 1181–1190
47. Heesemann J (1987) FEMS Microbiol Lett 48:229–233
48. Bearden SW, Perry RD (1999) Mol Microbiol 32:403–414
49. Rakin A, Saken E, Harmsen D, Heesemann J (1994) Mol Microbiol 13:253–263
50. Ferreras JA, Ryu J-S, Di Lello F, Tan DS, Quadri LEN (2005) Nat Chem Biol 1:29–32
51. Stirrett KL, Ferreras JA, Jayaprakash V, Sinha BN, Ren T, Quadri LEN (2008) Bioorg Med Chem Lett 18:2662–2668
52. Bearden S, Fetherston J, Perry R (1997) Infect Immun 65:1659–1668
53. Pelludat C, Rakin A, Jacobi CA, Schubert S, Heesemann J (1998) J Bacteriol 180: 538–546
54. Gehring AM, DeMoll E, Fetherston JD, Mori I, Mayhew GF, Blattner FR, Walsh CT, Perry RD (1998) Chem Biol 5:573–586
55. Fetherston JD, Bearden SW, Perry RD (1996) Mol Microbiol 22:315–325
56. Anisimov R, Brem D, Heesemann J, Rakin A (2005) Int J Med Microbiol 295:19–28
57. Brem D, Pelludat C, Rakin A, Jacobi CA, Heesemann J (2001) Microbiology 147:1115–1127
58. Fetherston JD, Bertolino VJ, Perry RD (1999) Mol Microbiol 32:289–299
59. Schubert S, Rakin A, Heesemen J (2004) Int J Med Microbiol 294:83–94
60. Buchrieser C, Rusniok C, Frangeul L, Couve E, Billault A, Kunst F, Carniel E, Glaser P (1999) Infect Immun 67:4851–4861
61. Antonenka U, Nölting C, Heesemann J, Rakin A (2005) Mol Microbiol 57:727–734
62. Miller DA, Luo L, Hillson N, Keating TA, Walsh CT (2002) Chem Biol 9:333–344
63. Pelludat C, Brem D, Heesemann J (2003) J Bacteriol 185:5648–5653
64. Kerbarh O, Ciulli A, Howard NI, Abell C (2005) J Bacteriol 187:5061–5066
65. Patel HM, Tao J, Walsh CT (2003) Biochemistry 42:10514–10527
66. Bobrov AG, Geoffroy VA, Perry RD (2002) Infect Immun 70:4204–4214
67. Geoffroy VA, Fetherston JD, Perry RD (2000) Infect Immun 68:4452–4461
68. Patel HM, Walsh CT (2001) Biochemistry 40:9023–9031
69. Darling P, Chan M, Cox AD, Sokol PA (1998) Infect Immun 66:874–877
70. Cox CD (1980) J Bacteriol 142:581–587
71. Liu PV, Shokrani F (1978) Infect Immun 22:878–890
72. Cox CD, Rinehart KLJ, Moore ML, Cook JC Jr (1981) Proc Natl Acad Sci USA 78: 4256–4260
73. Cobessi D, Celia H, Pattus F (2005) J Mol Biol 352:893–904
74. Tseng C-F, Burger A, Mislin G, Schalk I, Yu S, Chan S, Abdallah M (2006) J Biol Inorg Chem 11:419–432
75. Reimmann C, Patel HM, Serino L, Barone M, Walsh CT, Haas D (2001) J Bacteriol 183:813–820
76. Ino A, Murabayashi A (2001) Tetrahedron 57:1897–1902
77. Ino A, Hasegawa Y, Murabayashi A (1998) Tetrahedron Lett 39:3509–3512
78. Kemp DS, Carey RI (2002) J Org Chem 54:3640–3646
79. Ankenbauer RG, Toyokuni T, Staley A, Rinehart KL Jr, Cox CD (1988) J Bacteriol 170:5344–5351
80. Rinehart KL, Staley AL, Wilson SR, Ankenbauer RG, Cox CD (1995) J Org Chem 60: 2786–2791
81. Rivault F, Schons V, Liébert C, Burger A, Sakr E, Abdallah MA, Schalk IJ, Mislin GLA (2006) Tetrahedron 62:2247–2254
82. Zamri A, Abdallah MA (2000) Tetrahedron 56:249–256
83. Pollack JR, Neilands JB (1970) Biochem Biophys Res Commun 38:989–992
84. O'Brien I, Gibson F (1970) Biochim Biophys Acta 215:393–402
85. Shanzer A, Libman J, Lifson S, Felder CE (1986) J Am Chem Soc 108:7609–7619
86. Carrano CJ, Raymond KN (1979) J Am Chem Soc 101:5401–5404
87. Harris WR, Carrano CJ, Cooper SR, Sofen SR, Avdeef AE, McArdle JV, Raymond KN (1979) J Am Chem Soc 101:6097–6104
88. Konopka K, Neilands JB (1984) Biochemistry 23:2122–2127

89. Goetz DH, Holmes MA, Borregaard N, Bluhm ME, Raymond KN, Strong RK (2002) Mol Cell 10:1033–1043
90. Holmes MA, Paulsene W, Jide X, Ratledge C, Strong RK (2005) Stucture 13:29–41
91. Bister B, Bischoff D, Nicholson GJ, Valdebenito M, Schneider K, Winkelmann G, Hantke K, Süssmuth RD (2004) Biometals 17:471–481
92. Hantke K, Nicholson G, Rabsch W, Winkelmann G (2003) Proc Natl Acad Sci USA 100:3677–3682
93. Fischbach MA, Lin H, Liu DR, Walsh CT (2006) Nat Chem Biol 2:132–138
94. Luo M, Lin H, Fischbach MA, Liu DR, Walsh CT, Groves JT (2006) ACS Chem Biol 1:29–32
95. Fischbach MA, Lin H, Zhou L, Yu Y, Abergel RJ, Liu DR, Raymond KN, Wanner BL, Strong RK, Walsh CT, Aderem A, Smith KD (2006) Proc Natl Acad Sci USA 103: 16502–16507
96. de Lorenzo V (1984) Arch Microbiol 139:72–75
97. Lagos R, Wilkens M, Vergara C, Cecchi X, Monasterio O (1993) FEBS Lett 321:145–148
98. Thomas X, Destoumieux-Garzón D, Peduzzi J, Afonso C, Blond A, Birlirakis N, Goulard C, Dubost L, Thai R, Tabet J-C, Rebuffat S (2004) J Biol Chem 279:28233–28242
99. Nolan EM, Fischbach MA, Koglin A, Walsh CT (2007) J Am Chem Soc 129:14336–14347
100. Vassiliadis G, Peduzzi J, Zirah S, Thomas X, Rebuffat S, Destoumieux-Garzon D (2007) Antimicrob Agents Chemother 51:3546–3553
101. Kaiser A, Leistner E (1990) Arch Biochem Biophys 276:331–335
102. Liu J, Quinn N, Berchtold GA, Walsh CT (1990) Biochemistry 29:1417–1425
103. Tummuru MK, Brickman TJ, McIntosh MA (1989) J Biol Chem 264:20547–20551
104. Gehring AM, Bradley KA, Walsh CT (1997) Biochemistry 36:8495–8503
105. Rusnak F, Liu J, Quinn N, Berchtold GA, Walsh CT (1990) Biochemistry 29:1425–1435
106. Liu J, Duncan K, Walsh CT (1989) J Bacteriol 171:791–798
107. Gehring AM, Mori I, Walsh CT (1998) Biochemistry 37:2648–2659
108. Shaw-Reid CA, Kelleher NL, Losey HC, Gehring AM, Berg C, Walsh CT (1999) Chem Biol 6:385–400
109. Fischbach MA, Lin H, Liu DR, Walsh CT (2005) Proc Natl Acad Sci USA 102:571–576
110. Zhu M, Valdebenito M, Winkelmann G, Hantke K (2005) Microbiology 151:2363–2372
111. Lin H, Fischbach MA, Liu DR, Walsh CT (2005) J Am Chem Soc 127:11075–11084
112. Lagos R, Villanueva JE, Monasterio O (1999) J Bacteriol 181:212–217
113. Corey EJ, Bhattacharyya S (1977) Tetrahedron Lett 45:3919–3922
114. Rastetter WH, Erickson TJ, Venuti MC (1980) J Org Chem 45:5011–5012
115. Rogers HJ (1995) J Chem Soc Perkin Trans I 3073–3075
116. Shanzer A, Libman J (1983) J Chem Soc Chem Commun 846–847
117. Marinez ER, Salmassian EK, Lau TT, Gutierrez CG (1996) J Org Chem 61:3548–3550
118. Meyer M, Telford JR, Cohen SM, White DJ, Xu J, Raymond KN (1997) J Am Chem Soc 119:10093–10103
119. Ramirez RJA, Karamanukyan L, Ortiz S, Gutierrez CG (1997) Tetrahedron Lett 38:749–752
120. Rastetter WH, Erickson TJ, Venuti MC (1981) J Org Chem 46:3579–3590
121. Gong H, Gagné MR (2008) J Am Chem Soc 130:12177–12183
122. Silakowski B, Kunze B, Nordsiek G, Blöcker H, Höfle G, Müller R (2000) Eur. J Biochem 267:6476–6485
123. Miyanaga S, Obata T, Onaka H, Fujita T, Saito N, Sakurai H, Saiki I, Furumai T, Igarashi Y (2006) J Antibiot (Tokyo) 59:698–703
124. Keating TA, Marshall CG, Walsh CT (2000) Biochemistry 39:15522–15530
125. Gaitatzis N, Kunze B, Müller R (2001) Proc Natl Acad Sci USA 98:11136–11141
126. Silakowski B, Schairer HU, Ehret H, Kunze B, Weinig S, Nordsiek G, Brandt P, Blöcker H, Höfle G, Beyer S, Müller R (1999) J Biol Chem 274:37391–37399
127. Griffiths GL, Sigel SP, Payne SM, Neilands JB (1984) J Biol Chem 259:383–385
128. Keating TA, Marshall CG, Walsh CT (2000) Biochemistry 39:15513–15521

129. Butterton JR, Choi MH, Watnick PI, Carroll PA, Calderwood SB (2000) J Bacteriol 182:1731–1738
130. Zhou Z, Lai JR, Walsh CT (2007) Proc Natl Acad Sci USA 104:11621–11626
131. Bergeron RJ, Garlich JR, McManis JS (1985) Tetrahedron 41:507–510
132. Sakakura A, Umemura S, Ishihara K (2008) Chem Commun 3561–3563
133. Omura S, Iwai Y, Hirano A, Nakagawa A, Awaya J, Tsuchiya H, Takahashi Y, Masuma R (1977) J Antibiot (Tokyo) 30:275–282
134. Xie Q, Wang Y, Huang Y, Wu Y, Ba F, Liu Z (2002) Int J Syst Evol Microbiol 52: 1815–1820
135. Tamaoki T, Nomoto H, Takahashi I, Kato Y, Morimoto M, Tomita F (1986) Biochem Biophys Res Commun 135:397–402
136. Hoehn P, Ghisalba O, Moerker T, Peter HH (1995) J Antibiot (Tokyo) 48:300–305
137. Takahashi I, Kobayashi E, Asano K, Yoshida M, Nakano H (1987) J Antibiot (Tokyo) 40:1782–1784
138. Lapenna S, Giordano A (2009) Nat Rev Drug Discov 8:547–566
139. Akinaga S, Sugiyama K, Akiyama T (2000) Anticancer Drug Des 15:43–52
140. Link JT, Raghavan S, Danishefsky SJ (1995) J Am Chem Soc 117:552–553
141. Link JT, Raghavan S, Gallant M, Danishefsky SJ, Chou TC, Ballas LM (1996) J Am Chem Soc 118:2825–2842
142. Wood JL, Stoltz BM, Goodman SN (1996) J Am Chem Soc 118:10656–10657
143. Wood JL, Stoltz BM, Goodman SN, Onwneme K (1997) J Am Chem Soc 119:9652–9661
144. Onaka H, Taniguchi S, Igarashi Y, Furumai T (2002) J Antibiot (Tokyo) 55:1063–1071
145. Kim SY, Park JS, Chae CS, Hyun CG, Choi BW, Shin J, Oh KB (2007) Appl Microbiol Biotechnol 75:1119–1126
146. Gao Q, Zhang C, Blanchard S, Thorson JS (2006) Chem Biol 13:733–743
147. Sánchez C, Butovich IA, Braña AF, Rohr J, Méndez C, Salas JA (2002) Chem Biol 9: 519–531
148. Onaka H, Taniguchi S, Igarashi Y, Furumai T (2003) Biosci Biotechnol Biochem 67:127–138
149. Balibar CJ, Walsh CT (2006) Biochemistry 45:15444–15457
150. Sánchez C, Zhu L, Braña AF, Salas AP, Rohr J, Méndez C, Salas JA (2005) Proc Natl Acad Sci USA 102:461–466
151. Nishizawa T, Aldrich CC, Sherman DH (2005) J Bacteriol 187:2084–2092
152. Howard-Jones AR, Walsh CT (2005) Biochemistry 44:15652–15663
153. Howard-Jones AR, Walsh CT (2006) J Am Chem Soc 128:12289–12298
154. Ryan KS, Howard-Jones AR, Hamill MJ, Elliott SJ, Walsh CT, Drennan CL (2007) Proc Natl Acad Sci USA 104:15311–15316
155. Nishizawa T, Gruschow S, Jayamaha DH, Nishizawa-Harada C, Sherman DH (2006) J Am Chem Soc 128:724–725
156. Wang Y, Chen H, Makino M, Shiro Y, Nagano S, Asamizu S, Onaka H, Shaik S (2009) J Am Chem Soc 131:6748–6762
157. Makino M, Sugimoto H, Shiro Y, Asamizu S, Onaka H, Nagano S (2007) Proc Natl Acad Sci USA 104:11591–11596
158. Asamizu S, Kato Y, Igarashi Y, Furumai T, Onaka H (2006) Tetrahedron Lett 47:473–475
159. Sánchez C, Braña AF, Méndez C, Salas JA (2006) Chembiochem 7:1231–1240
160. Salas AP, Zhu L, Sánchez C, Braña AF, Rohr J, Méndez C, Salas JA (2005) Mol Microbiol 58:17–27
161. Sánchez C, Salas AP, Braña AF, Palomino M, Pineda-Lucena A, Carbajo RJ, Méndez C, Moris F, Salas JA (2009) Chem Commun 4118–4120
162. Nakanishi S, Matsuda Y, Iwahashi K, Kase H (1986) J Antibiot (Tokyo) 39:1066–1071
163. Ryan KS, Drennan CL (2009) Chem Biol 16:351–364
164. Sánchez C, Méndez C, Salas JA (2006) Nat Prod Rep 23:1007–1045
165. Chae CS, Park JS, Chung SC, Kim TI, Lee SH, Yoon KM, Shin J, Oh KB (2009) Bioorg Med Chem Lett 19:1581–1583

166. Penn K, Jenkins C, Nett M, Udwary DW, Gontang EA, McGlinchey RP, Foster B, Lapidus A, Podell S, Allen EE, Moore BS, Jensen PR (2009) ISME J 3:1193–1203
167. Howard-Jones AR, Walsh CT (2007) J Am Chem Soc 129:11016–11017
168. Mentel M, Ahuja EG, Mavrodi DV, Breinbauer R, Thomashow LS, Blankenfeldt W (2009) Chembiochem 10:2295–2304
169. Hassan HM, Fridovich I (1980) J Bacteriol 141:156–163
170. Lau GW, Ran H, Kong F, Hassett DJ, Mavrodi D (2004) Infect Immun 72:4275–4278
171. Ran H, Hassett DJ, Lau GW (2003) Proc Natl Acad Sci USA 100:14315–14320
172. Allen L, Dockrell DH, Pattery T, Lee DG, Cornelis P, Hellewell PG, Whyte MK (2005) J Immunol 174:3643–3649
173. Bianchi SM, Prince LR, McPhillips K, Allen L, Marriott HM, Taylor GW, Hellewell PG, Sabroe I, Dockrell DH, Henson PW, Whyte MK (2008) Am J Respir Crit Care Med 177:35–43
174. Look DC, Stoll LL, Romig SA, Humlicek A, Britigan BE, Denning GM (2005) J Immunol 175:4017–4023
175. Price-Whelan A, Dietrich LE, Newman DK (2007) J Bacteriol 189:6372–6381
176. Hernandez ME, Kappler A, Newman DK (2004) Appl Environ Microbiol 70:921–928
177. Dietrich LE, Price-Whelan A, Petersen A, Whiteley M, Newman DK (2006) Mol Microbiol 61:1308–1321
178. Dietrich LE, Teal TK, Price-Whelan A, Newman DK (2008) Science 321:1203–1206
179. Deppenmeier U, Johann A, Hartsch T, Merkl R, Schmitz RA, Martinez-Arias R, Henne A, Wiezer A, Baumer S, Jacobi C, Bruggemann H, Lienard T, Christmann A, Bomeke M, Steckel S, Bhattacharyya A, Lykidis A, Overbeek R, Klenk HP, Gunsalus RP, Fritz HJ, Gottschalk G (2002) J Mol Microbiol Biotechnol 4:453–461
180. Millican RC (1962) Biochem Biophys Acta 57:407–409
181. Mavrodi DV, Ksenzenko VN, Bonsall RF, Cook RJ, Boronin AM, Thomashow LS (1998) J Bacteriol 180:2541–2548
182. Pierson LS III, Gaffney T, Lam S, Gong F (1995) FEMS Microbiol Lett 134:299–307
183. Mavrodi DV, Bonsall RF, Delaney SM, Soule MJ, Phillips G, Thomashow LS (2001) J Bacteriol 183:6454–6465
184. Giddens SR, Feng Y, Mahanty HK (2002) Mol Microbiol 45:769–783
185. Haagen Y, Gluck K, Fay K, Kammerer B, Gust B, Heide L (2006) Chembiochem 7:2016–2027
186. Saleh O, Gust B, Boll B, Fiedler HP, Heide L (2009) J Biol Chem 284:14439–14447
187. Olekhovich IN, Kotik EA, Formichev YK (1994) Genetika 30:285–286
188. McDonald M, Mavrodi DV, Thomashow LS, Floss HG (2001) J Am Chem Soc 123:9459–9460
189. Parsons JF, Calabrese K, Eisenstein E, Ladner JE (2003) Biochemistry 42:5684–5693
190. Blankenfeldt W, Kuzin AP, Skarina T, Korniyenko Y, Tong L, Bayer P, Janning P, Thomashow LS, Mavrodi DV (2004) Proc Natl Acad Sci USA 101:16431–16436
191. Parsons JF, Calabrese K, Eisenstein E, Ladner JE (2004) Acta Crystallogr D Biol Crystallogr 60:2110–2113
192. Parsons JF, Song F, Parsons L, Calabrese K, Eisenstein E, Ladner JE (2004) Biochemistry 43:12427–12435
193. Ahuja EG, Janning P, Mentel M, Graebsch A, Breinbauer R, Hiller W, Costisella B, Thomashow LS, Mavrodi DV, Blankenfeldt W (2008) J Am Chem Soc 130:17053–17061
194. Parsons JF, Greenhagen BT, Shi K, Calabrese K, Robinson H, Ladner JE (2007) Biochemistry 46:1821–1828
195. Greenhagen BT, Shi K, Robinson H, Gamage S, Bera AK, Ladner JE, Parsons JF (2008) Biochemistry 47:5281–5289
196. Müller PK, Krohn K, Mühlradt PF (1989) Infect Immun 57:2591–2596
197. Surrey AR (1967) Pyocyanine. In: Horning EC (ed) Organic syntheses collection, vol 3. Wiley, New York, pp 753–756

198. Sitachitta N, Gadepalli M, Davidson BS (1996) Tetrahedron 52:8073–8080
199. Miyairi N, Sakai H-I, Konomi T, Imanaka H (1976) J Antibiot (Tokyo) 29:227–235
200. Tokuma Y, Miyairi N, Morimoto Y (1976) J Antibiot (Tokyo) 29:1114–1116
201. Seto H, Sato T, Urano S, Uzawa J, Yonehara H (1976) Tetrahedron Lett 17:4367–4370
202. Kang H, Jensen PR, Fenical W (1996) J Org Chem 61:1543–1546
203. Zhang H-I, He X-g, Adefarati A, Gallucci J, Cole SP, Beale JM, Keller PJ, Chang C-j, Floss HG (1990) J Org Chem 55:1682–1684
204. Hertweck C, Luzhetskyy A, Rebets Y, Bechthold A (2007) Nat Prod Rep 24:162–190
205. Hertweck C (2009) Angew Chem Int Ed 48:4688–4716
206. Piel J, Hoang K, Moore BS (2000) J Am Chem Soc 122:5415–5416
207. Piel J, Hertweck C, Shipley PR, Hunt DM, Newman MS, Moore BS (2000) Chem Biol 7:943–955
208. Hertweck C, Moore BS (2000) Tetrahedron 56:9115–9120
209. Xiang L, Moore BS (2003) J Bacteriol 185:399–404
210. Hertweck C, Jarvis AP, Xiang L, Moore BS, Oldham NJ (2001) Chembiochem 10:784–786
211. Xiang L, Moore BS (2002) J Biol Chem 277:32505–32509
212. Izumikawa M, Cheng Q, Moore BS (2006) J Am Chem Soc 128:1428–1429
213. Hertweck C, Xiang L, Kalaitzis JA, Cheng Q, Palzer M, Moore BS (2004) Chem Biol 11:461–468
214. Xiang L, Kalaitzis JA, Moore BS (2004) Proc Natl Acad Sci USA 101:15609–15614
215. Cheng Q, Xiang L, Izumikawa M, Meluzzi D, Moore BS (2007) Nat Chem Biol 3:557–558
216. Florova G, Kazanina G, Reynolds KA (2002) Biochemistry 41:10462–10471
217. Xiang L, Kalaitzis JA, Nilsen G, Chen L, Moore BS (2002) Org Lett 4:957–960
218. Caro AA, Cederbaum AI (2005) Biochem Pharmacol 69:1081–1093
219. Pohl NL, Hans M, Lee HY, Kim YS, Cane DE, Khosla C (2001) J Am Chem Soc 123:5822–5823
220. Kalaitzis JA, Cheng Q, Thomas PM, Kelleher NL, Moore BS (2009) J Nat Prod 72:469–472
221. Kalaitzis JA, Izumikawa M, Xiang L, Hertweck C, Moore BS (2003) J Am Chem Soc 125:9290–9291
222. Kirsch S, Bach T (2003) Angew Chem Int Ed 42:4685–4687
223. Kirsch SF, Bach T (2005) Chem Eur J 11:7007–7023
224. Hatakeyama S, Numata H, Osanai K, Takano S (1989) J Org Chem 54:3515–3517
225. Young IS, Baran PS (2009) Nat Chem 1:193–205
226. Baran PS, Maimone TJ, Richter JM (2007) Nature 446:404–408
227. Roethle PA, Hernandez PT, Trauner D (2006) Org Lett 8:5901–5904
228. Roethle PA, Trauner D (2006) Org Lett 8:345–347
229. Chen K, Baran PS (2009) Nature 459:824–828
230. Chen MS, White MC (2007) Science 318:783–787
231. Zhan J, Guo H, Dai J, Zhang Y, Guo D (2002) Tetrahedron Lett 43:4519–4521
232. Nicolaou KC, Montagnon T, Snyder SA (2003) Chem Commun 551–564
233. Nicolaou KC, Edmonds DJ, Bulger PG (2006) Angew Chem Int Ed 45:7134–7186
234. Nicolaou KC, Chen JS (2009) Chem Soc Rev 38:2993–3009
235. Johnson WS, Gravenstock MB, McCarry BE (1971) J Am Chem Soc 93:4332–4334
236. Nicolaou KC, Frederick MO, Aversa RJ (2008) Angew Chem Int Ed 47:7182–7225
237. Vilotijevic I, Jamison TF (2009) Angew Chem Int Ed 48:5250–5281
238. Tanuwidjaja J, Ng S-S, Jamison TF (2009) J Am Chem Soc 13:12084–12085
239. Breuer M, Ditrich K, Habicher T, Hauer B, Kesseler M, Stuermer R, Zelinski T (2004) Angew Chem Int Ed 43:788–824
240. Drauz K, Waldmann H (1995) Enzyme catalysis in organic synthesis. Wiley, Weinheim
241. Gotor V (2008) Asymmetric organic synthesis with enzymes. Wiley, Weinheim
242. Werneburg M, Hertweck C (2008) Chembiochem 9:2064–2066
243. Richter MEA, Traitcheva N, Knuepfer U, Hertweck C (2008) Angew Chem Int Ed 47: 8872–8875

Index

MIX
Papier aus verantwortungsvollen Quellen
Paper from responsible sources
FSC® C105338

If you have any concerns about our products,
you can contact us on
ProductSafety@springernature.com

In case Publisher is established outside the EU,
the EU authorized representative is:
Springer Nature Customer Service Center GmbH
Europaplatz 3, 69115 Heidelberg, Germany

Printed by Libri Plureos GmbH
in Hamburg, Germany